International Conference on Informatics, Technology and Engineering (InCITE 2019)

Enhancing Engineering Innovation Towards a Greener Future

IOP Conference Series: Materials Science and Engineering
Volume 703

Bali, Indonesia
22 - 23 August 2019

Editors:

Namuel Daniel Pah
Markus Hartono
Oki Muraza

ISBN: 978-1-7138-1033-9
ISSN: 1757-8981

Printed from e-media with permission by:

Curran Associates, Inc.
57 Morehouse Lane
Red Hook, NY 12571

Some format issues inherent in the e-media version may also appear in this print version.

Institute of Physics
Dirac House, Temple Back
Bristol BS1 6BE UK

Phone: 44 1 17 929 7481
Fax: 44 1 17 920 0979

techtracking@iop.org

Additional copies of this publication are available from:

Curran Associates, Inc.
57 Morehouse Lane
Red Hook, NY 12571 USA
Phone: 845-758-0400
Fax: 845-758-2633
Email: curran@proceedings.com
Web: www.proceedings.com

TABLE OF CONTENTS

GREEN MANUFACTURING AND GREEN PROCESSES

POWER SYSTEM AND GREEN ENERGY MANAGEMENT

THE ROLE OF IT IN INNOVATION ENHANCEMENT

IOP Conf. Series: Materials Science and Engineering **703** (2019) 011001 doi:10.1088/1757-899X/703/1/011001

A collaborative activity jointly organised by:

FACULTY OF ENGINEERING
UNIVERSITAS SURABAYA

THE UNIVERSITY of ADELAIDE

InCITE 2019

INTERNATIONAL CONFERENCE ON INFORMATICS, TECHNOLOGY, AND ENGINEERING

22-23 AUGUST 2019

Enhancing Engineering Innovation Towards A Greener Future

EDITOR:
Asst. Prof. Namuel Daniel Pah, Ph.D.
Assoc. Prof. Markus Hartono, Ph.D.
Assoc. Prof. Oki Muraza, Ph.D.

PUBLISHER:
UNIVERSITAS SURABAYA

UBAYA
UNIVERSITAS SURABAYA

IOP Conf. Series: Materials Science and Engineering **703** (2019) 011001 doi:10.1088/1757-899X/703/1/011001

Preface

Welcome Remarks, Chair of the Steering Committee

It is a great pleasure to welcome all of you to Bali and to the International Conference on Informatics, Technology, and Engineering 2019 (InCITE 2019) held by the Faculty of Engineering, University of Surabaya (UBAYA) in collaboration with The University of Adelaide, Australia and Sirindhorn International Institute of Technology (Thammasat University), Thailand. The first InCITE has been successfully held in Bali, Indonesia in 2017. We are very delighted to host the second InCITE here in Bali, Indonesia again.

There are more than 75 presentations in this conference. We welcome leading experts not only from Indonesia, but also from different parts of the world. The experts will share the knowledge and experiences in the fields of informatics, technology, science, and engineering. The main theme of this conference is **Enhancing Engineering Innovation Towards A Greener Future** in response to several world challenges including sustainable development, global convergence of information and communications technologies, climate change and global warming as well as the depletion of unrenewable natural resources. We hope this conference will provide you a good opportunity to get to know each other better and consolidate bonds of friendship and mutual trust.

We would like to express our sincere gratitude to the Keynote and Plenary speakers, International Scientific Committee, Steering Committee, and Organising Committee for their huge efforts to make this conference successful.

Thank you all for your support and attendance at InCITE 2019. Please enjoy the conference and Bali !

Asst. Prof. Djuwari, Ph.D.

IOP Conf. Series: Materials Science and Engineering **703** (2019) 011001 doi:10.1088/1757-899X/703/1/011001

Preface

Welcome Remarks,
Chair of The Organizing Committee

Welcome to Bali, Indonesia to all delegates and presenters. It is my pleasure and privilege to welcome all of you to the 2nd (second) International Conference on Informatics, Technology, and Engineering 2019 (InCITE 2019) held by the Faculty of Engineering, University of Surabaya (UBAYA) in collaboration with The University of Adelaide, Australia and Sirindhorn International Institute of Technology (Thammasat University), Thailand.

InCITE 2019 has received more than 75 papers to be presented in this conference. All papers represent four following parallel clusters: Green Design and Innovation, Green Manufacturing and Green Processes, Power System and Green Energy Management, and The Role of IT in Innovation Enhancement. Each cluster supports the main theme of the conference, which is **Enhancing Engineering Innovation Towards A Greener Future.** The engineering innovation is the key to increase our awareness in maintaining the sustainable growth and development in the world.

The Organising Committee of InCITE 2019 would like to express our sincere gratitude for the tremendous supports and contributions from many parties. The supports from The Faculty of Engineering of UBAYA, keynote and plenary speakers, our International Scientific Committee, the Steering and Organising Committees are really acknowledged.

The last but not the least, thank you for your supports, enjoy the conference and we hope through this meeting all of you can extend your networks and collaborations.

Asst. Prof. Putu Doddy Sutrisna, Ph.D.

IOP Conf. Series: Materials Science and Engineering **703** (2019) 011001 doi:10.1088/1757-899X/703/1/011001

Conference Organizers

SCIENTIFIC COMMITTEE

- Prof. Willy Susilo, Ph.D. (University of Wollongong, AUSTRALIA)
- Prof. Dr. Anton Satria Prabuwono (King Abdulaziz University, SAUDI ARABIA)
- Assoc. Prof. Oki Muraza, Ph.D. (King Fahd University of Petroleum & Minerals, KINGDOM OF SAUDI ARABIA)
- Prof. Ravindra S. Goonetilleke, Ph.D. (Hong Kong University of Science & Technology, PRC)
- Assoc. Prof. Tan Kay Chuan, Ph.D. (National University of Singapore, SINGAPORE)
- Asst. Prof. Aldy Gunawan, Ph.D. (Singapore Management University, SINGAPORE)
- Asst. Prof. Hendry Raharjo, Ph.D. (Chalmers University of Technology, SWEDEN)
- Assoc. Prof. Dr. A. F. M. Saifuddin Saif (American International University, BANGLADESH)
- Asst. Prof. Itthisek Nilkhamhang, Ph.D. (Sirindhorn International Institute of Technology, THAILAND)
- Assoc. Prof. Akawut Siriruk, Ph.D. (Suranaree University of Technology, THAILAND)
- Assoc. Prof. Avirut Chinkulkijniwat, Ph.D. (Suranaree University of Technology, THAILAND)
- Assoc. Prof. Peerapong Uthansakul, Ph.D. (Suranaree University of Technology, THAILAND)
- Assoc. Prof. Dr. Andi Cakravastia Arisaputra Raja (Institut Teknologi Bandung, INDONESIA)
- Assoc. Prof. Dr. Anas Maruf (Institut Teknologi Bandung, INDONESIA)
- Assoc. Prof. Yassierli, Ph.D. (Institut Teknologi Bandung, INDONESIA)
- Prof. Dr. Ali Altway (Institut Teknologi Sepuluh Nopember, INDONESIA)
- Prof. Dr-Ing. I Made Londen Batan (Institut Teknologi Sepuluh Nopember, INDONESIA)
- Assoc. Prof. Setiyo Gunawan, Ph.D. (Institut Teknologi Sepuluh Nopember, INDONESIA)
- Prof. Renanto Handogo, Ph.D. (Institut Teknologi Sepuluh Nopember, INDONESIA)
- Prof. Mauridhi Hery Purnomo, Ph.D. (Institut Teknologi Sepuluh Nopember, INDONESIA)
- Prof. Nur Iriawan, Ph.D. (Institut Teknologi Sepuluh Nopember, INDONESIA)
- Prof. I Nyoman Pujawan, Ph.D. (Institut Teknologi Sepuluh Nopember, INDONESIA)
- Asst. Prof. Budi Hartono, Ph.D. (Universitas Gadjah Mada, INDONESIA)
- Prof. Sarjiya, Ph.D. (Universitas Gadjah Mada, INDONESIA)
- Asst. Prof. Nemuel Daniel Pah, Ph.D. (Universitas Surabaya, INDONESIA)
- Assoc. Prof. Markus Hartono, Ph.D (Universitas Surabaya, INDONESIA)
- Prof. Joniarto Parung, Ph.D. (Universitas Surabaya, INDONESIA)
- Prof. Lieke Riadi, Ph.D. (Universitas Surabaya, INDONESIA)

IOP Conf. Series: Materials Science and Engineering **703** (2019) 011001 doi:10.1088/1757-899X/703/1/011001

Conference Organizers

STEERING COMMITTEE

Chair:

Asst. Prof. Djuwari, Ph.D.

Honorary Members:

Prof. David Lewis, Ph.D.
Prof. Joniarto Parung, Ph.D.
Prof. Lieke Riadi, Ph.D.
Asst. Prof. Dr. Steve Kardinal Jusuf
Asst. Prof. Dr. Itthisek Nilkhamhang

Members:

Assoc. Prof. Amelia Santoso, Ph.D.
Mr. Agung Prayitno
Assoc. Prof. Emma Savitri, Ph.D.
Assoc. Prof. Markus Hartono, Ph.D., CHFP.
Assoc. Prof. Budi Hartanto, Ph.D.
Mr. Sunardi Tjandra
Assoc. Prof. Eric Wibisono, Ph.D.
Asst. Prof. Nemuel Daniel Pah, Ph.D.
Assoc. Prof. Elieser Tarigan, Ph.D.
Assoc. Prof. Jaya Suteja, Ph.D.
Prof. Joniarto Parung, Ph.D.
Assoc. Prof. Hudiyo Firmanto, Ph.D.
Assoc. Prof. Restu Kartiko Widi, Ph.D.

ORGANIZING COMMITTEE

Chair	:Asst. Prof. Putu Doddy Sutrisna, Ph.D.
Vice Chair	:Dr. Delta Ardy Prima
Secretary	:Ms. Aprilia Karina
Treasurers	:Ms. Dhiani Tresna Absari
Secretariat	:Maria Agatha E.Gunawan, Ph.D.
	Asst. Prof. Lanny Sapei, Ph.D.
	Mr. Rahman Dwi Wahyudi
	Ms. Yenny Sari
	Ms. Yuana Elly Agustin
	Ms. Susana Limanto
	Ms. Monica Widiasri

Conference Organizers

Program	:Mr. Yunus Fransiscus Ms. Melissa Angga Mr. I Made Ronyastra Mr. Henry Hermawan Mr. Felix Handani Ms. Indri Hapsari Mr. Mochammad Arbi Hidayat
Website	:Mr. Daniel Soesanto Mr. Marcellinus Ferdinand Suciadi
Design	:Ms. Tyrza Adelia
Sponsorship	:Assoc. Prof. Susila Candra, Ph.D.
Logistic:	Mr. Arief Rachman Hakim Mr. Muhamad Yulham Effendy

REVIEWER

- Prof. David Lewis, Ph.D. (University of Adelaide, AUSTRALIA)
- Prof. Willy Susilo, Ph.D. (University of Wollongong, AUSTRALIA)
- Dr. Jingwei Hou (University of Queensland, AUSTRALIA)
- Asst. Prof. Hendry Raharjo, Ph.D. (Chalmers University of Technology, SWEDEN)
- Prof. Dr. Anton Satria Prabuwono (King Abdulaziz University, SAUDI ARABIA)
- Assoc. Prof. Oki Muraza, Ph.D. (King Fahd University of Petroleum & Minerals, KINGDOM OF SAUDI ARABIA)
- Prof. Dr. Winarto Kurniawan (Tokyo Institute of Technology, JAPAN)
- Dr. Wahyudiono (Nagoya University, JAPAN)
- Prof. Ravindra S. Goonetilleke, Ph.D. (Hong Kong University of Science & Technology, PRC)
- Asst. Prof. Dr. Steve Kardinal Jusuf (Singapore Institute of Technology, SINGAPORE)
- Assoc. Prof. Tan Kay Chuan, Ph.D. (National University of Singapore, SINGAPORE)
- Asst. Prof. Aldy Gunawan, Ph.D. (Singapore Management University, SINGAPORE)
- Assoc. Prof. Dr. A. F. M. Saifuddin Saif (American International University, BANGLADESH)
- Asst. Prof. Itthisek Nilkhamhang, Ph.D. (Sirindhorn International Institute of Technology, THAILAND)
- Assoc. Prof. Akawut Siriruk, Ph.D. (Suranaree University of Technology, THAILAND)
- Assoc. Prof. Avirut Chinkulkijniwat, Ph.D. (Suranaree University of Technology, THAILAND)
- Assoc. Prof. Peerapong Uthansakul, Ph.D. (Suranaree University of Technology, THAILAND)
- Asst. Prof. Dr. Phuong Lan Tran Nguyen (Can Tho University, VIETNAM)
- Assoc. Prof. Dr. Anas Maruf (Institut Teknologi Bandung, INDONESIA)
- Dr. Khoiruddin (Institut Teknologi Bandung, INDONESIA)
- Assoc. Prof. Dr. Veinardi Suendo (Institut Teknologi Bandung, INDONESIA)

Conference Organizers

REVIEWER

- Assoc. Prof. Dr. Andi Cakravastia Arisaputra Raja (Institut Teknologi Bandung, INDONESIA)
- Assoc. Prof. Yassierli, Ph.D. (Institut Teknologi Bandung, INDONESIA)
- Prof. Dr. Judy Retti B. Witono (Universitas Parahyangan, INDONESIA)
- Asst. Prof. Budi Hartono, Ph.D. (Universitas Gadjah Mada, INDONESIA)
- Prof. Sarjiya, Ph.D. (Universitas Gadjah Mada, INDONESIA)
- Asst. Prof. Dr. Hendri Himawan Triharminto (Akademi Angkatan Udara Yogyakarta, INDONESIA)
- Assoc. Prof. Dr. Djoko Budiyanto Setyohadi (Universitas Atmajaya Yogyakarta, INDONESIA)
- Prof. Dr. Ali Altway (Institut Teknologi Sepuluh Nopember, INDONESIA)
- Prof. Renanto Handogo, Ph.D. (Institut Teknologi Sepuluh Nopember, INDONESIA)
- Prof. Mauridhi Hery Purnomo, Ph.D. (Institut Teknologi Sepuluh Nopember, INDONESIA)
- Prof. Nur Iriawan, Ph.D. (Institut Teknologi Sepuluh Nopember, INDONESIA)
- Asst. Prof. Astria Nur Irfansyah, Ph.D.(Institut Teknologi Sepuluh Nopember, INDONESIA)
- Prof. Dr-Ing. I Made Londen Batan (Institut Teknologi Sepuluh Nopember, INDONESIA)
- Assoc. Prof. Setiyo Gunawan, Ph.D. (Institut Teknologi Sepuluh Nopember, INDONESIA)
- Prof. I Nyoman Pujawan, Ph.D. (Institut Teknologi Sepuluh Nopember, INDONESIA)
- Asst. Prof. Rr. Poppy Puspitasari, S, Ph.D (Universitas Negeri Malang, INDONESIA)
- Asst. Prof. Ratna Surya Alwi, S.T., M.Si., Ph.D (Universitas Fajar Makassar, INDONESIA)
- Prof. Joniarto Parung, Ph.D. (Universitas Surabaya, INDONESIA)
- Prof. Lieke Riadi, Ph.D. (Universitas Surabaya, INDONESIA)
- Asst. Prof. Nemuel Daniel Pah, Ph.D. (Universitas Surabaya, INDONESIA)
- Assoc. Prof. Markus Hartono, Ph.D (Universitas Surabaya, INDONESIA)
- Asst. Prof. Dr. Hazrul Iswadi (Universitas Surabaya, INDONESIA)
- Asst. Prof. Gunawan, Ph.D (Universitas Surabaya, INDONESIA)
- Assoc. Prof. Dr. Evy Herowati (Universitas Surabaya, INDONESIA)
- Assoc. Prof. Dr. Amelia Santoso (Universitas Surabaya, INDONESIA)
- Assoc. Prof. Eric Wibisono, Ph.D. (Universitas Surabaya, INDONESIA)
- Asst. Prof. Dr. Joko Siswantoro (Universitas Surabaya, INDONESIA)
- Assoc. Prof. Dr. Budi Hartanto (Universitas Surabaya, INDONESIA)
- Asst. Prof. Dr. Delta Ardy Prima (Universitas Surabaya, INDONESIA)
- Asst. Prof. Jimmy (Universitas Surabaya, INDONESIA)
- Assoc. Prof. Lisana (Universitas Surabaya, INDONESIA)
- Assoc. Prof. Dr. Emma Savitri (Universitas Surabaya, INDONESIA)
- Assoc. Prof. Restu Kartiko Widi, Ph.D. (Universitas Surabaya, INDONESIA)
- Assoc. Prof. Akbarningrum Fatmawati (Universitas Surabaya, INDONESIA)

IOP Conf. Series: Materials Science and Engineering **703** (2019) 011001 doi:10.1088/1757-899X/703/1/011001

Conference Organizers

REVIEWER

- Assoc. Prof. Akbarningrum Fatmawati (Universitas Surabaya, INDONESIA)
- Asst. Prof. Putu Doddy Sutrisna, Ph.D. (Universitas Surabaya, INDONESIA)
- Asst. Prof. Djuwari, Ph.D. (Universitas Surabaya, INDONESIA)
- Asst. Prof. Elieser Tarigan, Ph.D. (Universitas Surabaya, INDONESIA)
- Assoc. Prof. The Jaya Suteja, Ph.D (Universitas Surabaya, INDONESIA)
- Assoc. Prof. Dr. Susila Candra (Universitas Surabaya, INDONESIA)
- Asst. Prof. Sunardi Tjandra (Universitas Surabaya, INDONESIA)
- Asst. Prof. Yuwono Budi Pratiknyo (Universitas Surabaya, INDONESIA)

CONFERENCE ORGANIZING COMMITTEE:
FACULTY OF ENGINEERING, UNIVERSITAS SURABAYA
DEAN BUILDING TB 2, RAYA KALIRUNGKUT
SURABAYA, 60293, INDONESIA
PHONE: +62-31-2981150, FAX: +62-31-2981151
E-MAIL: incite@unit.ubaya.ac.id
WEBSITE: https://incite.ubaya.ac.id; http://teknik.ubaya.ac.id

IOP Conf. Series: Materials Science and Engineering **703** (2019) 011001 doi:10.1088/1757-899X/703/1/011001

Conference Photographs

IOP Conf. Series: Materials Science and Engineering **703** (2019) 011001 doi:10.1088/1757-899X/703/1/011001

IOP Conf. Series: Materials Science and Engineering **703** (2019) 011002 doi:10.1088/1757-899X/703/1/011002

Peer review statement

All papers published in this volume of *IOP Conference Series: Materials Science and Engineering* have been peer reviewed through processes administered by the proceedings Editors. Reviews were conducted by expert referees to the professional and scientific standards expected of a proceedings journal published by IOP Publishing.

IOP Conf. Series: Materials Science and Engineering **703** (2019) 012001 doi:10.1088/1757-899X/703/1/012001

The use of blockchain to support sustainable supply chain strategy

J Parung

Department of Industrial Engineering, University of Surabaya, Raya Kalirungkut, Surabaya, 60293, Indonesia

E-mail: jparung@staff.ubaya.ac.id

Abstract. This paper aims to identify, and analyze how the use of blockchains technology can support sustainable supply chain management strategies. This support is seen from environmental, economic, and social aspects. In addition, it explains the various advantages of using Blokchain technology to reduce the use of transportation and other resources that have the potential to damage the environment. From an economic aspect, this technology has the potential to increase the efficiency of cost and time. Blockchain technology also has the potential to provide social benefits for the companies involved because of the increasingly good reputation of the organization. Three examples of implementation of the blockchain in three different industries are also provided in this paper. This paper also points out the potential disadvantages of using this technology for supply chains that can threaten supply chain sustainability.

1. Introduction

In the 21st century, awareness of environmental care in organizations is a trend. Environmental awareness of organizations grows towards sustainability which is not only related to care for the environment but also economic and social sectors. The concept of sustainability has been introduced to many fields as a strategy to improve the operations of the organization. This strategy has also been implemented in the supply chain to support efforts to improve organizational performance in the supply chain. This strategy is called Sustainable Supply Chain Management (SSCM).

The process to achieve sustainability in the supply chain often does not run smoothly due to various factors including inaccurate data or lack of smooth flow of information between members in the supply chain, data asynchronous, and inefficiency in fulfilling supply and demand. These constraints occur mainly because the availability of data on each supply chain member is centralized and often kept secret for the benefit of the member concerned so that it often cannot be anticipated by other members. Related to this situation, another approach is needed to reduce the constraints in SSCM. The approach that has been tested in several industries is the use of blockchain technology.

Blockchain is a new technology for storing and moving data. All data are held in one place. The data is then atomized and spreadover thousands of points throughout the network [1]. Distribution of data on the blockchain is based on the principle of decentralization, so intermediaries between members in the supply chain can be eliminated. Thus, the need for banks or other financial intermediaries can be eliminated. Transactions can be directly carried out between buyers and sellers. Payments can be done quickly and smart contracts are executed after all the underlying conditions or obligations are fulfilled [2].

2. Literature Review

This blockchain technology is believed to be able to overcome the lack of smooth flow of data and inefficiency in a sustainable supply chain.

2.1. Supply Chain Management and Sustainable Supply Chain Management

The term Supply Chain Management (SCM), according to the Council of Supply Chain Management (CCSCM), refers to the planning and management of all activities involved in procurement, conversion, and all logistics management activities. Importantly, this also includes coordination and collaboration with member's partners, which can be suppliers, intermediaries, third-party service providers, and customers. In essence, supply chain management integrates supply and demand management within and throughout the company. SCM's main objective is to meet consumer demand with more efficient use of resources including labor, inventory, stock and distribution capacity. In this context, SCM aims to integrate key business functions and business processes within and throughout the company into cohesive and high-performance business models that can quickly react to dynamic market demand and rapidly changing features.

In each supply chain many members are involved in performing different tasks. As a result, the whole process becomes broad, long, slow, and inefficient. In this long process, the supply chain does not have an audit or centralized mechanism to ensure that contracts between supply chain actors are enforced according to the rules. Consumers do not have a way to know whether goods such as shoes that they have bought are fake, the drugs he bought were original, or whether the food he consumed was contaminated with bacteria or not. In this case, conventional supply chains have serious transparency and traceability issues.

Sustainability is known as an integral component of supply chain management, but no single definition of Sustainable Supply Chain (SSC) is universally accepted [3]. On the other hand, SSC is often explained as a green supply chain or triple bottom line supply chain. SSC is a growing movement to show the concern of SC members for the environment without forgetting economic or social factors as covered in the triple bottom line. SSC also define as a strategic step, with transparent integration, to achieve organizational, social, environmental, and economic goals in systemic coordination of business processes among key organizations to improve the long-term economic performance of each company and its supply chain [4].

Furthermore, the SSC is defined as "the management of material, information, and capital flows as well as cooperation between companies throughout the supply chain while taking the objectives of the three dimensions of sustainable development, namely, economic, environmental, and social, into accounts originating from needs of customers and stakeholders [5]. In a sustainable supply chain, environmental and social criteria must be met by members to remain in the supply chain, while it is expected that competitiveness will be maintained by meeting customer needs and related economic criteria.

In all the SSC definitions discussed above, it can be seen that the integration and objectives expected by SSCM are always emphasized so that the achievement of SSC objectives must be carried out with a strategy that integrates and considers the internal and external conditions of the organizations involved in SSC.

2.2. Blockchain and Cryptocurrency

Blockchain is a digital information recording method that can record data using the logbook approach and with certain characteristics. Operationally, Blockchain is known as a large decentralized and distributed ledger that keeps records of digital transactions in such a way that makes them accessible and visible to many members in a gated network [6].

Blockchain consists of cryptographically connected data blockchains. Blocks are chained in sequence using cryptographic hashes. A hash is a fixed-length number that originates from the message or document provided. Cryptography is the study of mathematical techniques related to aspects of information security, such as data confidentiality, data validity, data integrity, and data authentication. Cryptocurrency is a digital currency created using the concept of cryptography. Cryptography was developed for online information and financial exchange activities. Thus, cryptocurrency becomes a currency that cannot be falsified because the security code is quite sophisticated. One type of cryptocurrency that is most widely known and in demand right now is Bitcoin. This currency was developed by Satoshi Nakamoto [7].

3. How does Blockchain technology support SSCM strategy?

To answer the question above, let's take a case example. If we buy drugs at the pharmacy, then, of course, we want to get a guarantee from the pharmacist, that the drug is safe, has not expired, is accurate, and is genuine. But the problem is that pharmacists cannot convince us that the drug is exactly what we expect it to be, so to ensure that there is a need for technology that can help us. Suppose we tag the microchip as an internet implementation for a drug package. Then, we trace the journey of the drug through each stage of the supply chain sequence.

The process to track the origin of raw materials, production locations, product carriers, storage and retailers to buy products can be done easily using the Internet of things (IoT) combined with blockchain technology. The buyer only needs to scan the package using his cellphone to retrace the supply chain route that the product goes through before it reaches us. The conditions described above indicate that consumers can now reject a product if the source or ingredients are suspicious or if the trip from the supplier to the retail store changes the quality of the product.

We know that the conventional supply chain management system is a centralized SCM system. This system cannot present data in real-time. The data sent will take time to process and present it, so the use of blockchain technology allows all members in any SC to access the network safely and quickly. Because blockchain is a decentralized database, no one controls or owns it and after the data is uploaded to the blockchain, the data cannot be changed so that the data cannot be corrupted or faked. Blockchain technology provides an opportunity to improve supply chain operations in a transformable manner.

The great potential of blockchain technology has moved various industries to conduct trials and even implement this technology in their companies. But a common question that often arises for researchers is how is blockchain step used in SSCM?

Blockchain is a simple, clever and quick way to convey transaction information between two parties automatically. If one party buys books from one supplier, then the supplier starts the process by making a book block and this block is verified by a group of computers connected to the Internet. Next the verified book block is added to the chain, which is distributed in a special book network (blockchain). On the verified block, a unique note will be written in the form of a secret code (cryptography) and can only be opened by buyers who pay using cryptocurrency (for example bitcoin), see figure 1 for the steps.

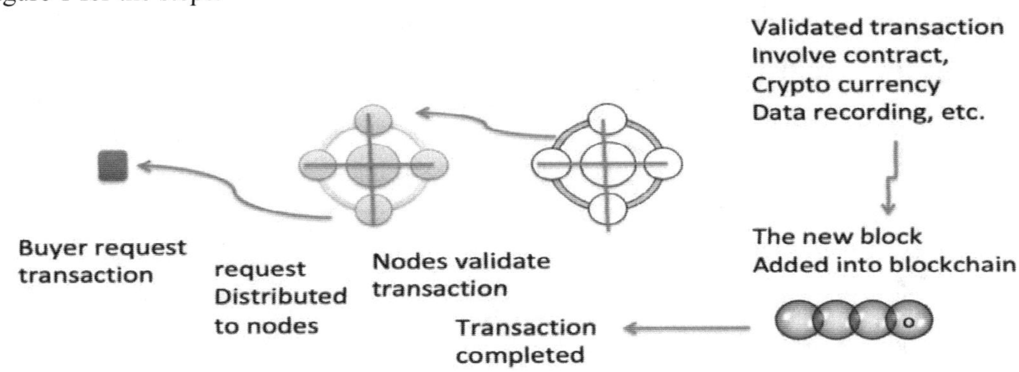

Figure 1. Blockchain steps in a suppllier-buyer transaction (adapted from World Economic Forum)

From various literatures, it is found that several large international organizations have tested and implemented it for various purposes. Blockchain has been applied for tracking and tracing the origin of the product as was done in the Abu Dhabi National Oil Company (ADNOC) in collaboration with IBM. The idea was to track oil delivery from well to customers, while simultaneously automating transactions along the way. ADNOC, which produces oil about 3 million barrels per day, will benefit from tracking all the oil produced and reducing the time and costs associated with shipping if blockchain technology is fully implemented [8].

A research conducted by Walmart in collaboration with JD.com, IBM, and Tsinghua University found *Escherichia coli* bacteria that contaminate the food sold. It took a long time to find where the food was contaminated so the research was trying to increase food transparency and shipment efficiency with blockchain technology. The efforts were divided into two sections: Walmart and JD.com handled production and shipment of produce, while IBM and Tsinghua University handled the research and maintaining the blockchain. This project was completed in 2017 and Walmart announced that it would require suppliers of its domain and other leafy greens to upload their data to the blockchain by September 2019 [9].

As food safety is the main concern of every individual, and then the blockchain technology is expected to be used by every supply chain food corporation that guarantees the delivery of quality products to the public. Guaranteeing healthy and safe food and high quality directly or indirectly enhances the company's reputation in the eyes of consumers. Unilever, Kroger, Nestle, and Tyson Foods all plan to collaborate as the project advances, with more food corporations to join along the way. Furthermore, a software company Provenance is collaborating with designer Martine Jarlgaard to support the efforts of more transparency in the fashion industry [10]. Their first-ever garment tracked with blockchain was presented at the Danish Fashion Show in 2017. The goal is to track every aspect of a garment's life through all the development phases. With more transparency, consumers will always know the clothing they purchase is legitimate and that it is produced in factories providing acceptable working conditions. This example shows that blockchain applications directly support sustainability of supply chain management in terms of social aspects.

In order to understand how blockchain technology is used in the supply chain, please see figure 2, which is the adoption of the Applicature consultant below (applicature.com).

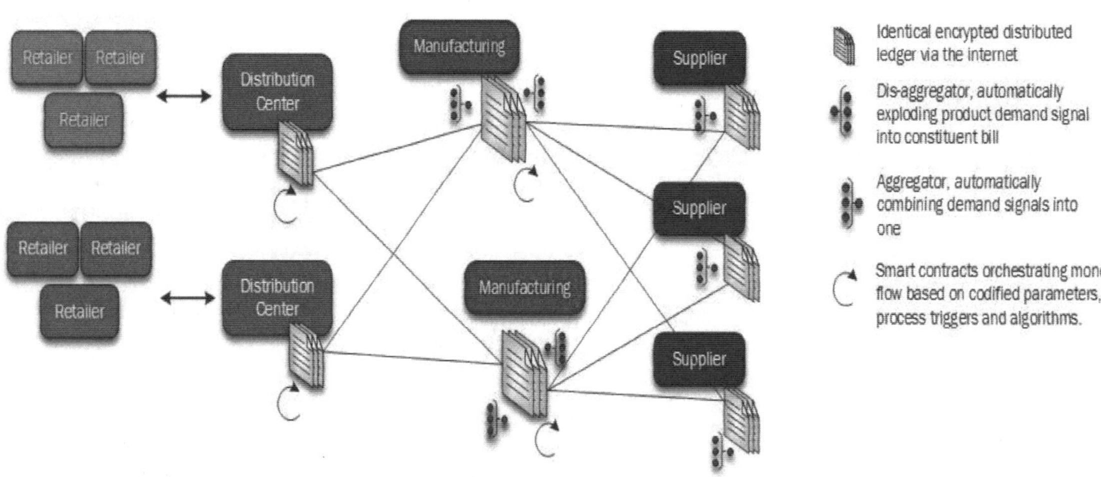

Figure 2. Blockchain in Supply Chain (adapted from Applicature consultant)

In addition to the various benefits of using blockchain in SCM, there are potential losses that can also be identified. Various disadvantages that SC members may face, that is, SC is difficult to achieve its full potential because it is blocked from areas that might be more beneficial individually for one SC member [11]. Another obstacle, namely too much SC transparency, can endanger data privacy issues.

Another thing is the companies can no longer maintain a kind of information asymmetry to maintain an organization's competitive advantage and to further reduce the risk of information leakage.

4. Conclusion

The use of blockchain can support the SSCM strategy to reduce environmental impacts while still providing economic benefits and providing social benefits for all stakeholders in the supply chain. Environmental benefits include pollution reduction due to the use of transportation in more appropriate shipping. It provides economic benefits for members because it can reduce costs by adopting more efficient production and transportation methods. On other hand, this kind of businesses can increase revenue by reducing delays. While, the social impact is the customers feel safe, comfortable in using the product because they can transparently trace the product flow while increasing reputation for all members in the supply chain.

References

[1] Marsal-Llacuna ML 2018 Future living framework: is blockchain the next enabling network? *Technol. Forecasting and Social Change* **128** pp 226-34
[2] Al-Saqaf W and Seidler N 2017 Blockchain technology for social impact: opportunities and challenges ahead *J. Cyber Policy* **2** pp 338-54
[3] Chkanikova O 2012 *Sustainable Supply Chain Management: Theoretical Literature Overview* (Sweden: International Institute for Industrial Environmental Economics, Lund University)
[4] Carter, C.R. and Rogers, D.S. 2008 A framework of sustainable supply chain management: moving toward new theory, Int. J. Phys. Distribution & Logistics Manage. **38** pp 360-38
[5] Seuring S and Müller M 2008 From a literature review to a conceptual framework for sustainable supply chain management *J. Cleaner Prod.* **16** pp 1699-710
[6] Treiblmaier H 2018 The impact of the blockchain on the supply chain: a theory-based research framework and a call for action *Supply Chain Manage.: An Int. J.* **23** pp 545-59
[7] Nakamoto S 2008 *Bitcoin: A peer-to-peer electronic cash system*
[8] (https://www.worldoil.com/news/2018/12/10/) accessed at 29st July 2019, 8.00 pm
[9] (https://www-03.ibm.com/press/us/en/pressrelease/53487.wss). Accessed at 30th July 2019, 7.00 am
[10] (https://www.provenance.org/case-studies/martine-jarlgaard). Access at 30th July 2019, 9.00 pm
[11] Hald KS and Kinra A, 2019 How the blockchain enables and constrains supply chain performance *Int. J. Phys. Distrib. Logist. Manage.* **49** pp 376-97

Green chemical engineering: challenges in chemical industrial processes for a better life

L Riadi[1,2]

[1]Department of Chemical Engineering,University of Surabaya, Raya Kalirungkut, Surabaya, 60293, Indonesia.
[2]Center for Environmental Studies, University of Surabaya, Raya Kalirungkut, Surabaya 60293, Indonesia.

E-mail: lieke@staff.ubaya.ac.id

Abstract. Almost thirty years away from sustainable development Rio summit, we still need to put a big effort in promoting and implementing industrial sustainable process. Green chemical engineering has been involved in industrial processes and chemical engineers are working to develop solutions for green and economic production of product using clean technologies, solving energy and raw materials shortage, sustainable water supply and wastewater treatment, carbon capture and utilization. This paper aims to highlight developments, challenges and best practices in green chemical engineering for both green product engineering and green process engineering. Incorporating the green engineering concept in the curriculum at university level to create a critical mass of engineers and scientists in green engineering is also included.

1.Introduction

One of the Sustainable Development Goals as Indonesia commitment by 2030 is "sustainable consumption and production'. It is related to green industrial process and product. Chemical products have contributed in supplying the needs and demands of our daily life. The products have been produced by chemical processes which consume huge amounts of natural resources and also generates unwanted by-products. It caused environmental pollution and the issue of climate change has been raised due to the rapid consumption of fossil-based energy. Nowadays, green chemical engineering which is a combination between green chemistry and green engineering has been involved in industrial processes to optimize the design of chemical processes and efficient use of materials and catalysts. As one of the most efficient means to achieve sustainable development in chemical industrial processes, green chemical engineering has been developed to address the needs of the green chemical industries and also for further studies on new technologies and processes. Green chemical engineering can be classified in two categories: green product engineering and green process engineering (Figure 1) [1]. Green product engineering includes the developments and use of new catalysts and renewable resources such as renewable energy and biomass. Green process engineering includes process intensification, new reaction media, energy conservation, CO_2 emission reduction, novel reactors and new separation techniques. Green product engineering has an objective to change or modify product designs by using non-toxic compounds or biodegradable materials during the production process. Green product engineering involves product lifecycle, from manufacturing process to distribution, from use to disposal or reuse and recycling. Green process engineering aims to reduce energy, water consumption and also undesired by products during the production process [2]. It includes reducing gas, energy or water emissions, improving resource and energy efficiency which involves in both novel separation technology and reactors design. Green process engineering will lead to sustainability of various competitive advantages and obtain both cost efficiency and profitability [2].

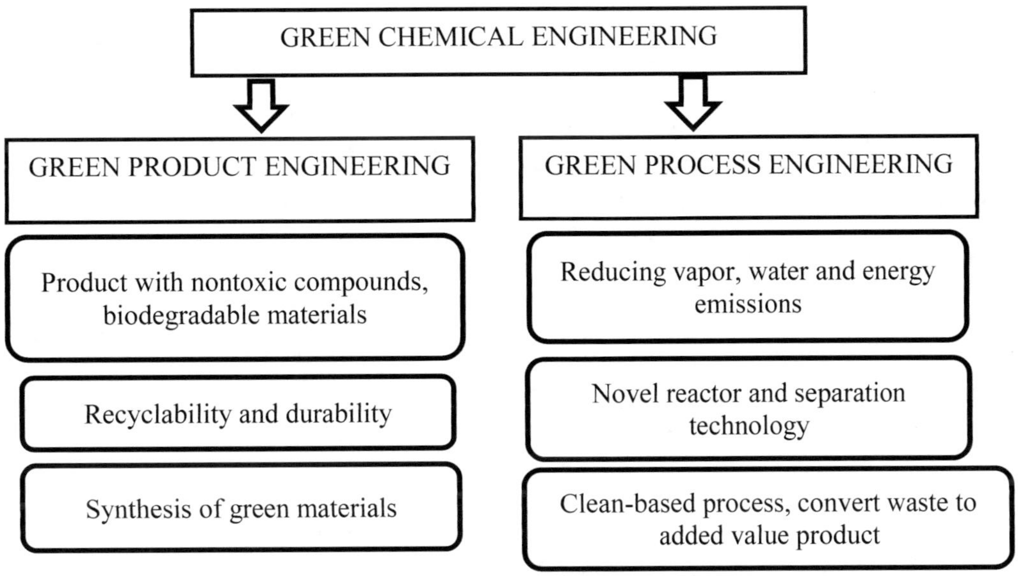

Figure 1. Green Chemical Engineering Concept

2.Green product engineering, challenges and best practice

There are three main focus areas of Green product engineering: 1) how to use alternative synthesis pathways which is more "green", 2) the design of safer chemicals that are less toxic or safer with regards to hazard and exposure and 3) the use of alternative reaction conditions [3]. Figure 2 shows the ideal concept of green material synthesis which creates a lot of challenges in implementation, as 100% yield is unlikely to achieve. Identifying synthesis pathways requires creative thinking in chemistry and process design, since the pathways are big and large scopes. The applications in chemical industrial processes are so complex and the quantitative and qualitative design tools are not available. So, virtual reality powered by supercomputing and big data analysis is needed. Hence, a lot of research has to be conducted which make the tools emerge. It is a big challenge. An example of safer chemicals used is in the synthesis of adipic acid (Figure 3) which uses glucose as alternative of benzene [4]. The best practice of choosing alternative synthesis pathways is the synthesis of Ibuprofen, a pain reliever from 2-methypropylbenzene [4]. The synthesis involves Friedel-Crafts chemistry. The long steps reaction with acid catalyst ($AlCl_3$) lead to several by products and also the uses of many solvents (Figure 4). The new approach using recoverable Hydrogen Fluoride (HF) leads to shorter pathway and more "green". As alternative reaction conditions, temperature and initial concentration are highlighted in this paper. Working on higher or lower temperature depends on the type,order of reaction and also activation energy in each reaction. The illustration of a parallel reaction for the first order reaction with reaction constant of kp and kw as seen in equations(1) and (2) shows that higher temperature is a better operation condition with Ep >Ew, the opposite is true for Ep <Ew. It can be explained by equation (3) which results in bigger ratio of kp/kw. The condition is also valid for first order series reaction as presented in equation (4). The notations in the equations are described as: A is frequency factor, Ep and Eware activation energy of P and W, R is a gas constant and T is absolute temperature. The notation of M, P and W represents raw material, product and waste (undesired product).

$$M \overset{kp}{\rightarrow} P \qquad (1)$$

$$M \overset{kw}{\longrightarrow} W \quad (2)$$

$$\frac{kp}{kw} = \frac{A_p e^{-(\frac{Ep}{RT})}}{A_w e^{-(\frac{Ew}{RT})}} \quad (3)$$

$$M \overset{kp}{\rightarrow} P \overset{kw}{\longrightarrow} W \quad (4)$$

In the series and parallel reactions, the effect of raw material initial concentration can be sensitive. The rates of product and waste formation are expressed in equations (5) and (6) where [M] is the raw material concentration, n_p and n_w are the order of reactions. The ratio of these rates shows the selectivity to product formation.If$n_p > n_w$, the increase of [M] will increase product, but if $n_p < n_w$ the increase of [M] will reduce product and lead to the formation of waste.

$$rate\ of\ product\ formation\ = kp\ [M]^{n_p} \quad (5)$$

$$rateof\ waste\ formation = kw[M]^{n_w} \quad (6)$$

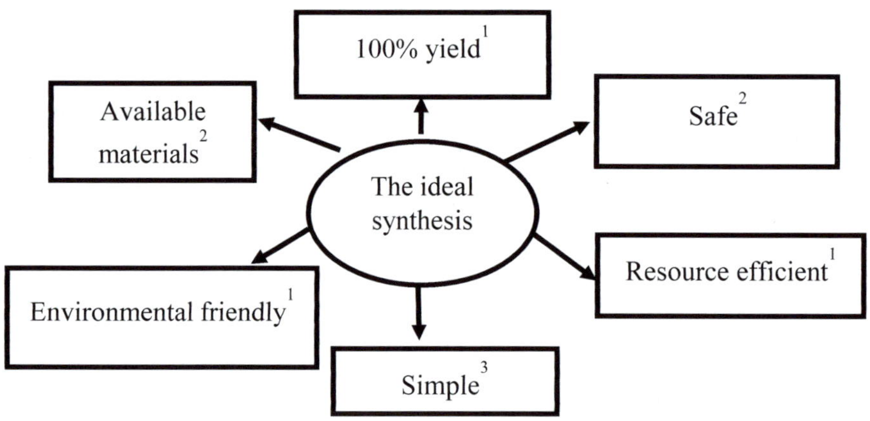

Figure 2. The ideal green product synthesis, where 1,2 and 3 indicate the aforementioned three main focus areas.

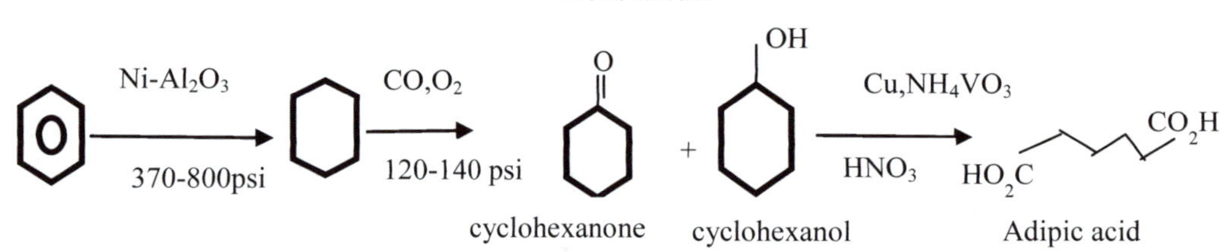

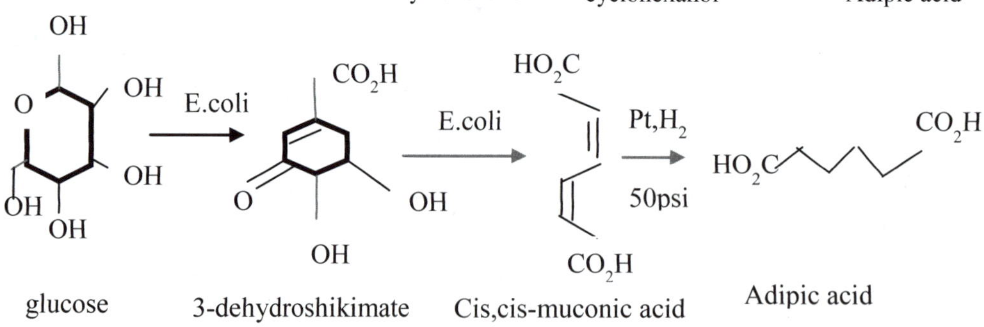

Figure 3. Previous (above) and alternative synthetic (below) pathways for adipic acid synthesis

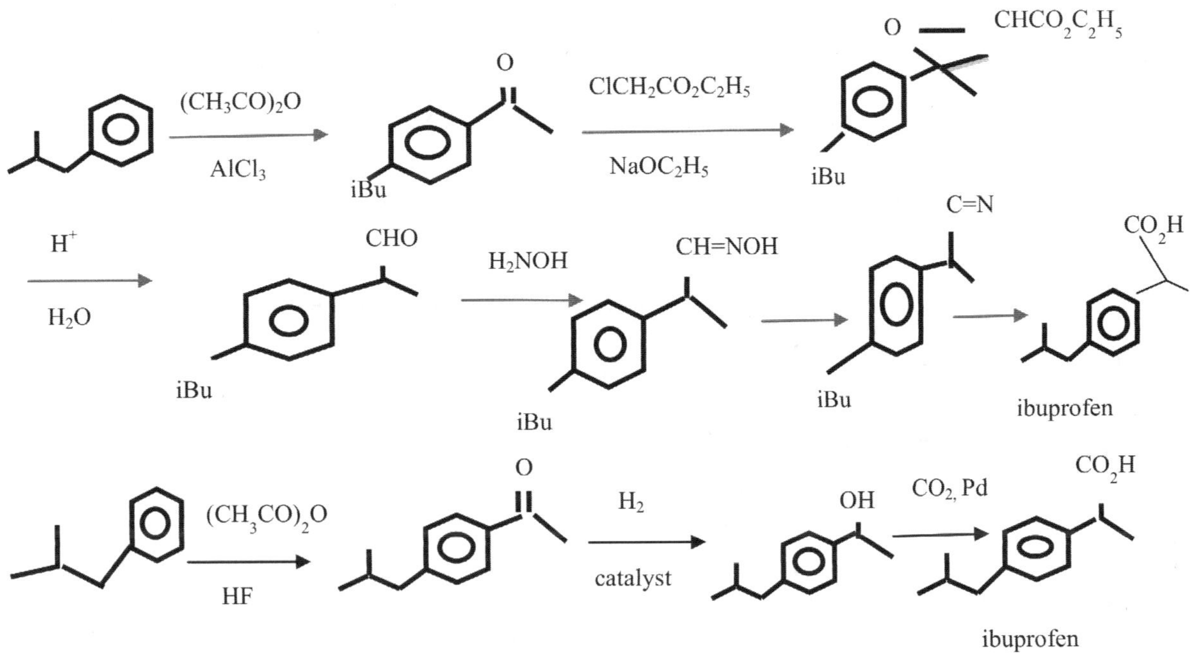

Figure 4. Previous (above) and alternative synthesis (below) pathways for Ibuprofen synthesis.

One example of green products is biodegradable plastic which is bio-based and biodegradable. The product is made from starch, vegetable oil and other materials from nature without polyethylene and polypropylene .It dissolves in hot water and is degradable after 2 months by composting system.

3. Green process engineering, challenges and best practice

Green Process Engineering involves the design, commercialization and use of processes to limit environmental burden by minimizing pollution at the source and risk to human health and the environment. It also contributes to the technological modernization of the economy. The criteria for Green Process Engineering can be categorized into two: 1) it must be based on new technological knowledge, 2) it must be able to be implemented. The significant opportunities are in the development of advanced materials, growth renewable energy, whereas the barriers are the difficulty in quantifying the true costs of waste and the potential savings offered by Green Process Engineering leading to the availability of comprehensive proof of good environmental and economic performance of proposed green chemical processes. Many research and manufacturing firms have started to tackle this problem. The best practice in Green Process Engineering discussed in this paper are :1) VOCs (volatile organic compounds) recovery from printing process and 2) water reduction in a process of recycle polypropylene (PP) woven sack to make polypropylene pellet for PP rope. There are two VOC recovery processes in printing industry that will be discussed in the paper [5]. The first process is considering to treat the VOC, while the second process is a new development to recover the VOC. The new technology of VOC recovery in printing process is a parallel system of advanced materials as adsorbents. Ethanol is used extensively in the printing industry. The major component of volatile effluents in printing industries are ethanol (30-50 wt%) and ethyl acetate (10-15 wt %). The flow sheet in Figure 5 shows the VOC which was treated using incineration leading to energy and CO_2 emissions. The green process engineering tried to solve the incineration process with other green VOC recovery technology (Figure 6), the dotted rectangle is the challenge for green process engineering to develop new technology in a green separation process. The new development green process engineering for such issue is presented in Figure 7 by using advanced materials which have been developed in emerging catalysis and separation applications.

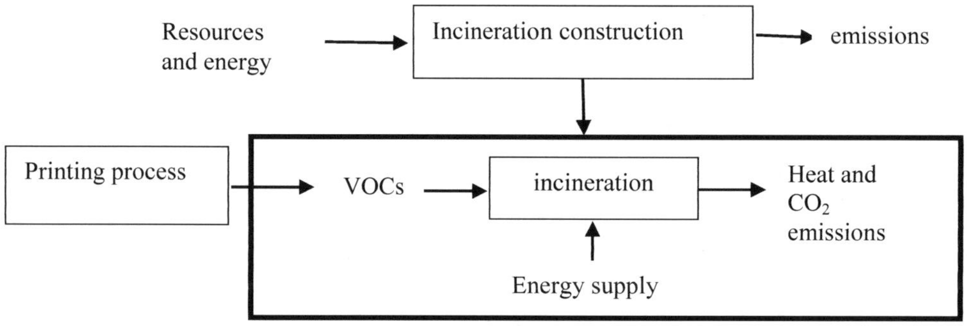

Figure 5. Diagram of incineration for VOC treatment in printing process, the box with solid line is the system

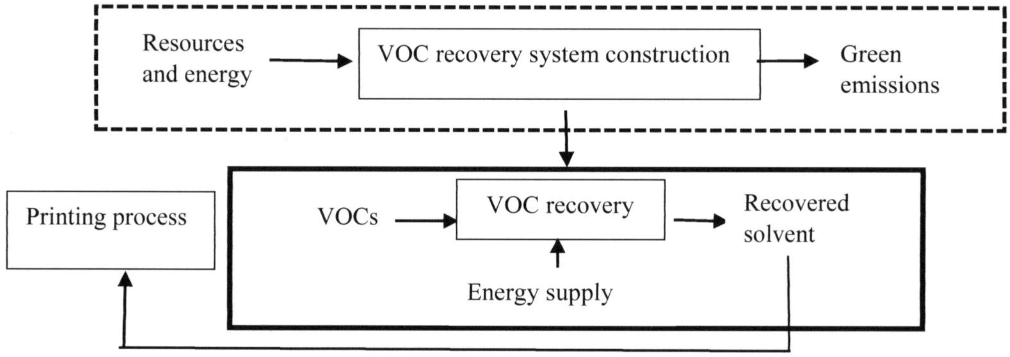

Figure 6. The concept of green process engineering in VOC recovery by designing green separation process.

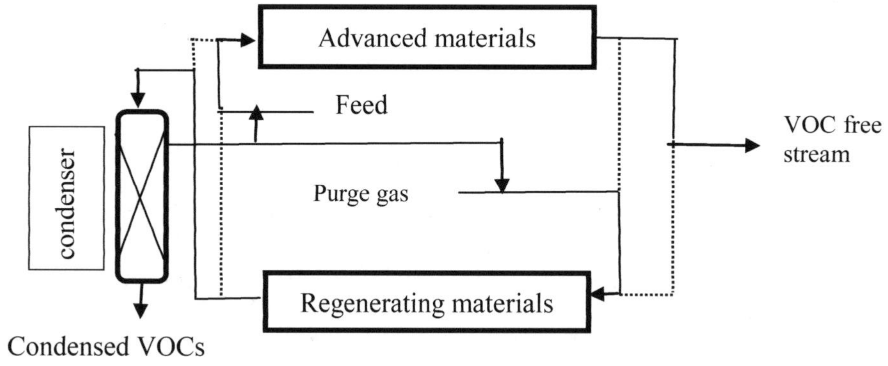

Figure 7. Green separation process for VOC recovery in printing industry.

The other best practice for green processing engineering is the water reduction in a recycle process of polypropylene (PP) woven sack to make polypropylene pellet for PP rope. The project has been conducted in East Java, Indonesia. The process diagram is presented in Figure 8. We changed the angle of the conveyor to reduce water loss in rinsing and transporting processes. Consequently, there was 54.17 % reduction of water losses from the initial water outlet flowrate of 0.51 m^3/minutes. Hence wastewater treatment in the process is more economically efficient.

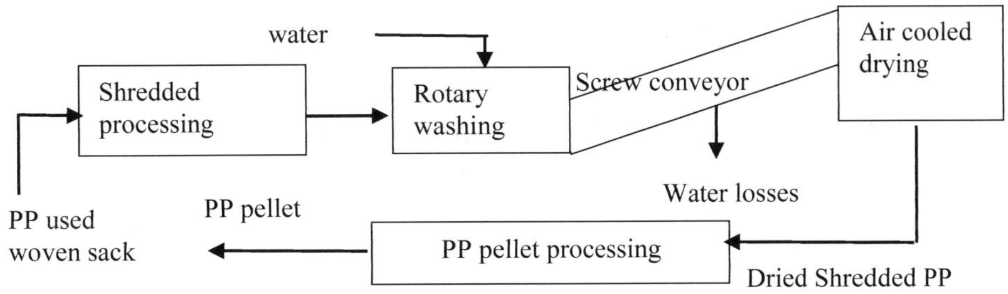

Figure 8. Schematic diagram of recycle PP woven sack processing

4. Incorporating the green engineering in the curriculum

Since green engineering principles should be used by all engineers, incorporating the concept to undergraduate curriculum is important. Integrating green engineering concepts into various chemical engineering courses is challenging. We propose to incorporate the concept in material and energy balances, thermodynamics especially in single phase and multiphase systems, reaction engineering, separation process, and process design courses. The material and energy balance course is an initial place to put the green engineering concept in chemical engineering. Table 1 shows some conceptual green engineering topics proposed in material and energy (M&E) balance course.

Table 1. Conceptual green engineering topics in M&E balancescourse [6]

Green engineering topic	M&E balances topic
Balances on recycle operation in green engineering process; green chemistry in stoichiometry; combustion process and environmental impact	Fundamental of material balances
Various form of energy in a green engineering process	Energy and energy balances
Use of heat capacity and phase change calculation; recovery of energy in process-energy integration; mixing and solution issues in green engineering	Non reactive process balance
Energy use in green chemistry reaction, combustion process; overall mass and energy balances in green engineering on an overall plant design basis	Reactive process balance

5.Conclusions

Engineering solutions to promote green engineering are required to adapt consumer demands and minimize environmental impact. We can put together the concept of green chemistry and green engineering at the earliest stages to make an effective strategy for minimizing waste, maximizing efficiency and increasing profitability. The need to incorporate green engineering concept in the curriculum at university level to create a critical mass of engineers and scientists for future development and challenges is inevitable since the multidisciplinary approach in solving a complex problems is needed.

References

[1] Chen J F 2017 *Engineering* **3** 283-284.
[2] Xie X, Huo J and Zuo H 2019 *Journal of Business Research* https://doi.org/10.1016/j.jbusres.2019.01.010
[3] Serna J G, P′erez-Barrig′on L and Cocero M J 2007 *Chem. Eng. J.* **133** 7-30.
[4] Allen D T and Shonnard D A 2002 *Green engineering: environmentally conscious design of chemical processes* (New Jersey, Prentice Hall) 180-9
[5] Lapkin A, Joyce L, and Crittenden B 2004 *Environ. Sci.Technol.* **38** 5815-23.
[6] Slater C S and Hesketh R P 2004 *CEE* **38** 48-53.

Xylanase production from combined *Reutealis trisperma* with potato dextrose broth by *Tricodherma reesei*: the effect of pretreatment

Y E Agustin[1], L Riadi[1,2] and T P Utami[3]

[1] Chemical Engineering Department, University of Surabaya, Raya Kalirungkut, Surabaya, Indonesia
[2] Center for Environmental Studies, University of Surabaya, Raya Kalirungkut, Surabaya, Indonesia
[3] CV. Energi Baru Sentosa, Raya Bambe, Driyorejo, Gresik, Indonesia

E-mail: yuana@staff.ubaya.ac.id

Abstract. Xylanase is an important enzyme in pulp and paper, food, feed, and textile industries. It is produced by fermentation process with *Tricodherma reesei* as fungi, which grows on several carbon sources such as cellulose and xylan. *Reutealis trisperma* (kemiri sunan) have been used for biodiesel production after being mechanically pressed. The solid waste, which is called *Reutealis trisperma* cake (RTC) contains carbon and nitrogen sources which are valuable to be used for xylanase production by *T. reesei*. Shake flask systems were used to grow *T. reesei* in medium consists of RTC and Potato Dextrose Broth (PDB). The systems were maintained at 28°C, 200 rpm at initial pH of 6.0 for 84 hours. Two forms RTC used in this study were natural and alkaline pretreated. The RTC fermentations were performed for different concentration of PDB (15g/L and 20 g/L) and also for different concentration of RTC (13.33 g/L; 16.67 g/L and 20 g/L). Pretreated systems gave better yield of xylanase production compared to natural systems. Higher concentration of PDB for pretreated system was found to yield lower xylanase production. The highest xylanase activity (198.1 U/mL) was obtained in media containing 15 g/L PDB and 20 g/L pretreated RTC.

1. Introduction

Reutealis trisperma known as "Kemiri Sunan" is a shady tree which is often found in East and West Java. The seeds have been used for biodiesel production after being mechanically pressed to get non-edible oil [1]. The cake called *Reutealis trisperma* cake (RTC) consists of protein, carbohydrate (sugar), fat, moisture, ash, and others (Table 1). Since its protein and sugar content were quite high, we tried to use RTC as nitrogen and carbon sources in a media for *Trichoderma reesei* to produce xylanase. Xylan is a hemicellulose which is normally found in rice plant, wheat, and many other plants. Xylanase is an enzyme produced by fermentation process with liquid culture and has been used as pre-bleaching of kraft pulp to minimize use of harsh chemicals in the subsequent treatment stages of kraft pulp, to improve the ability for feed meal to be easily digested, modified agent for cereal based food, to covert lignocellulose and agricultural waste to fermented products [2][3][4]. Recently, the conversion of lignocellulosic agricultural waste to useful products has been very attractive. Agricultural wastes such as banana peels, rice straw, sugarcane bagasse, wheat bran and soybean hulls have been used as the substrate for xylanase enzyme production [5]. There is no study of using *Reutealis trisperma* pressed cake in xylanase production, hence our study gives new insight of using the cake in xylanase production. Since there is no xylanase yet produced in the country, xylanase is imported for domestic demand. The use of xylanase in pulp bleaching process reduces up to 30 % of chlorine used in the pulp bleaching, leading to 15-20% reduction of chlorine release in the effluent [6]. *Trichoderma reesei* is a mesophilic filamentous fungus which is used at industrial scale for the production of enzymes hydrolyzing biomass such as cellulase and xylanase since it easily grows and well adapted at several substrates in fermentor cultivation condition [7]. *Trichoderma reesei* has two specific gens (*xyn1* and *xyn2*) that specific for xylan degradation and produce endo-β-1,4-xylanases from simple carbon source (xylan) [8][9]. Other fungi, *Aspergillus niger* is also known to produce some enzymes, such as carbohydrases and protease using biomass [10]. Fungal culture for enzyme production is affected by the differences in the nutrient composition. Fungal cultivation on pre-treated corn stover has been reported to perform better in enzyme production [11]. Thus, we used RTC in natural and pretreated forms. Both natural and pretreated RTC were used as additional carbon and nitrogen sources for Potato Dextrose Broth (PDB) in a growth media for *T. reesei* to produce xylanase. This research aims to study xylanase production by *T. reesei* using RTC and PDB and to compare xylanase activity in media contains several concentrations of PDB and pretreated RTC.

2. Materials and methods

2.1. Cake preparation

Reutealis trisperma cake (RTC) used in this research was obtained from Energi Baru Santosa Company, Gresik. Table 1 shows reported compositions of this material [12].

Table 1. Characteristic of RTC

	Content (%w/w)
Protein	34.03
Moisture	6.32
Carbohydrate	18.56
Fat	29.5
Ash	7.76
Others	3.83

Ground RTC (40 mesh) was heated in oven at 50°C. The dried ground RTC is a natural RTC (NRTC), the pretreated RTC was subjected to alkaline solution pretreatment (Alk-RTC). For alkaline pretreatment, RTC was mixed with 0.6% (w/v) aqueous sodium hydroxide (NaOH) with 1: 30 (w/v) solid-liquids ratio. The mixture was heated at 50°C and agitated at 160 rpm for 25 minutes. Finally, The Alk-RTC was washed and dried at 50°C overnight until it reached constant weight. All chemicals used were analytical grade and purchased from Merck.

2.2. Media

Two different media were used in the experiment, Natural RTC (NRTC) with PDB and alkaline pretreated RTC (Alk-RTC) with PDB. Generally, a 24 g/L PBD was used to cultivate *T. reesei.* The sugar content in RTC (Table1) was used as basic calculation to calculate the combined use of RTC and PDB to be equivalent with 24 g/L PDB. A 20 g/L RTC which consists of 3.7 g/L sugar and 20 g/L PDB will give the sugar content equivalent to 24 g/L PDB. We started with 20 g/L RTC and 15 g/L PDB used in the initial experiment, the composition of PDB and RTC both Alk-treated and natural (untreated) were varied by scaling down 17-20% the amount of pressed cake to optimize media used for the systems. Both NRTC and Alk-RTC without PDB were used as control since we would like to investigate the effect of pressed cake for enzyme induction potential in shake flask systems. PDB was purchased from Sigma Aldrich. All media were adjusted to pH 6.0 prior to autoclave (TOMY,ES-315, Japan) sterilization at 121°C for 15 minutes.

2.3. Culture conditions

Tricoderma reesei was obtained from Gadjah Mada University and stored on Potato Dextrose Agar (PDA) at 4°C and subcultured monthly. *T. reesei* were inoculated into 20 mL culture bottle that contain Potato Dextrose Broth (PDB) for 3 days at 28°C as pre-culture. After 3 days, pre-culture was added into shake flask systems which consisted of media that contained RTC and PDB at various compositions. Culture was incubated for 84 hours in an orbital shaker operating at 200 rpm for 28°C with initial pH of 6.0. Daily samples were removed aseptically in a laminar flow hood for measurement of pH and enzyme activity. Sample was separated between the solid state and supernatant using centrifuge at 10,000 rpm for 10 minutes and clear supernatant was collected and used for enzyme assay. The analyses were done in duplicate and the mean value presented.

2.4. Enzyme assay

Xylanase activity was determined by a modified method from Bailey [13]. The method was best applied to properly diluted samples with xylanase activities in the range of 0.5-2 U/mL. A 1 wt% beechwood xylan (Sigma Aldrich, St. Louis, MO) was prepared in 200 mL 0.05 M sodium citrate buffer (pH 5.3). A 100 ∫L enzyme broth was added to 25 mL test tube and then followed by 900 ∫L of suspended xylan in buffer. For blanks, only suspended xylan was added to the test tube. The samples and blanks were incubated at 50°C for 5 minutes in a water bath. After 5 minutes, a 3 mL DNS reagent was added to the test tubes to terminate the enzyme reaction. A 100 µL enzyme-containing sample was added to the corresponding blank. The samples and blanks were boiled for 10 minutes to develop color and added with distilled water up to 25 mL. The absorbance values were measured using UV-Vis Spectrophotometer (Hewlett Packard 8453) at 560 nm. Xylanase activity was measured based on xylose release, which can be represented by following equation:

$$\boldsymbol{Xylanase\left(\frac{U}{mL}\right) = \frac{xylose\ released(mg)}{(5\ mins)(0.1\ mL\ enzyme\ sample)} \times \frac{1\ mmol}{150.13\ mg} \times \frac{1000\ \mu mol}{1\ mmol}}$$

$$\boldsymbol{= 13.32 \times xylose\ released\ (mg)} \quad (1)$$

3. Results and discussion

In this research, we used RTC as additional substrate to grow *Trichoderma reesei* for xylanase production. The variables used in this research both for NRTC and Alk-RTC were 13.33 g/L, 16.67 g/L and 20 g/L. The PDB concentrations were 15 g/L and 20 g/L.

3.1. Effect of natural and pretreated RTC on xylanase production

There were two systems used in the experiment, both NRTC with PDB and Alk-RTC with PDB. The PDB used was 15 g/L. The initial pH used for this study was 6.0. The xylanase activity profiles for all systems can be seen in Figure 1. The Alk-RTC gave higher production of xylanase compare to NRTC. The better performance of Alk-RTC in xylanase production may be due to better access to protein in the cake since the alkaline treatment opens the cellulose structure for cell metabolism [14]. Higher concentration of Alk- RTC gave better xylanase activity during the fermentation period since the protein and carbon sources were available enough for cell to induce xylanase production. The NRTC system shows that higher concentration of NRTC resulted in lower xylanase activity which implied that NRTC prevented access to the protein for cell metabolism. With the exception of 20 g/L of pretreated RTC, most of the optimum production of xylanse occurred on the second day. Though, it was a delay of optimum xylanase production on 20 g/L pretreated RTC, the xylanase produced in the system was higher than other pretreated RTC concentration which is due to more nitrogen and carbon sources in the pressed cake. After achieving the optimum production, the xylanase activity decreased which may be due to its interaction with the other component in media which is known as fermentation product [15]. The decrease of xylanase activity can be indicated by the significant decrease of pH as shown in Figure 2.

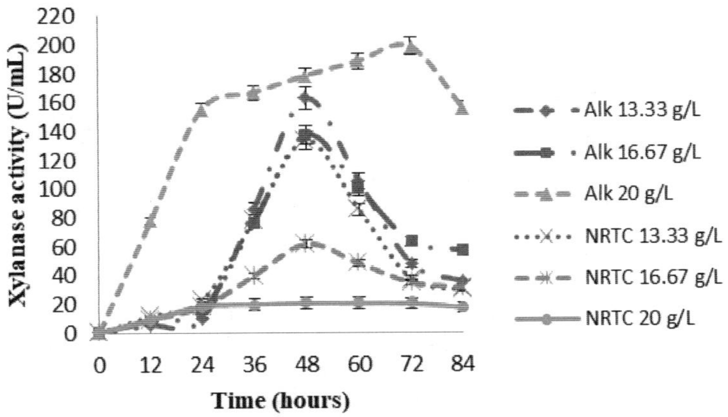

Figure 1. Profiles on xylanase activity for RTC in different concentration of RTC and 15 g/L PDB.

Table 2. Maximum xylanase activity at different concentration of RTC with 15 g/L PDB

RTC (g/L)	Xylanase activity (U/mL)	
	NRTC	Alk-RTC
13.33	133.378	162.28
16.67	61.262	137.12
20	20.534	198.13

Table 2 shows the maximum xylanase activity at different concentration of natural and pretreated RTC. It is clear that pretreated RTC gave a better xylanse activity compare to natural RTC. Natural RTC consists of lignin and cellulose which may hinder *T. reesei* to use the carbon and nitrogen sources in RTC. Having said that, the lower concentration of NRTC used in the production of xylanase may less inhibit xylanase induction, hence higher xylanase activity was found in the lower concentration of NRTC. However, in Alk-RTC (16.67 g/L and 20 g/L) the production of xylanase is 2 fold and approximately 10 fold compared to the same composition used of NRTC, which suggests adequate accessibility of the RTC's proteins and carbon sources to fungus. Other studies reported the effect of natural and pretreated soybean hull in a fermentation process to produce xylanase enzyme showed that the enzyme production was largely unaffected by the presence of pretreated biomass or natural biomass [16].

3.2. Effect of alkaline-pretreated RTC and PDB concentrations on the xylanase activity

Based on data in Table 2, we used 2 concentrations of Alk-RTC (13.33 g/L and 20 g/L) to compare the xylanse activity at different concentration of PDB (15 g/L and 20 g/L).

Table 3. Summary of evaluated fermentation system at different concentration alkaline pretreated (Alk) RTC and PDB

system	Carbon and nitrogen sources	Initial pH	Maximum xylanase activity (U/mL)
1	Alk 13.33 g/L , PDB 15 g/L	6.0	162.28 ± 0.97
2	Alk 20g/L , PDB 15 g/L	6.0	198.13 ± 0.03
3	Alk 13.33 g/L , PDB 20 g/L	6.0	151.19 ± 0.72
4	Alk 20g/L , PDB 20 g/L	6.0	157.93 ± 0.17
5	Alk 20 g/L as control	6.0	13.5 ± 5.1
6	NRTC 20 g/L as control	6.0	8.7 ± 0.64

Table 3 shows the best xylanase activity was achieved in a system with 20g/L Alk-RTC and 15 g/L PDB. Natural RTC and alkaline pretreated RTC were used as control. As has been mentioned in the above explanation, higher concentration pretreated RTC in systems with 15 g/L PDB will give higher xylanase activity. It seems that the lignin content in the cake has been degraded by alkali treatment, hence it give access for the cell to induce xylan. On the other hand, higher PDB concentration for both 13.33 g/L and 20 g/L of Alk- RTC gave similar maximum xylanase activities which were noted in system 3 and 4. The excess of PDB may hinder the production of xylanase which may be due to substrate inhibition. The data shows that natural RTC and pretreated RTC were able to induce xylanase production though the activity was low. The pretreated RTC gave better xylanase activity than natural RTC. It shows that alkaline pretreatment for RTC can give access for cell metabolism since the structure of lignin has been broken by the pretreatment process. There was more than 10 fold higher of xylanase activity in media using both Alk and PDB compare to Alk as control. It shows that carbon source in PDB enhanced the production of xylanase compare to Alk itself. Previous studies on xylanase production using *Trichoderma reesei* were conduct at various substrates. Xiong et al, have studied xylanase production by *T. reesei* RUT-30 which were grown at different pH medium with lactose as the main carbon source. The result showed highest xylanase activity (approximately 94.7± 4.8IU/mL) was achieved at pH 6 after 5 days of fermentation. Our result showed that Alk- RTC (20 g/L) combine with PDB (15 g/L) produced higher xylanase activity (approximately 198.13± 0.03 U/mL) compared to Xiong et.al [17].

3.3. pH profiles during fermentation process

The measurement of pH during fermentation process was obtained to indicate the growth of *Trichoderma reesei* related to the induction of xylanase. The decrease of pH means that *T. reesei* takes the TCA cycle to produce organic acid such as citric, oxalic and acetic acid [18]. However, the increase of pH shows that *T. reesei* takes protein to support its growth which may not go to TCA cycle which will end up with a little production of xylanase. The pH profile during fermentation process for alkaline-pretreated (Alk) RTC at various concentrations of RTC and PDB are shown in Figure 2.

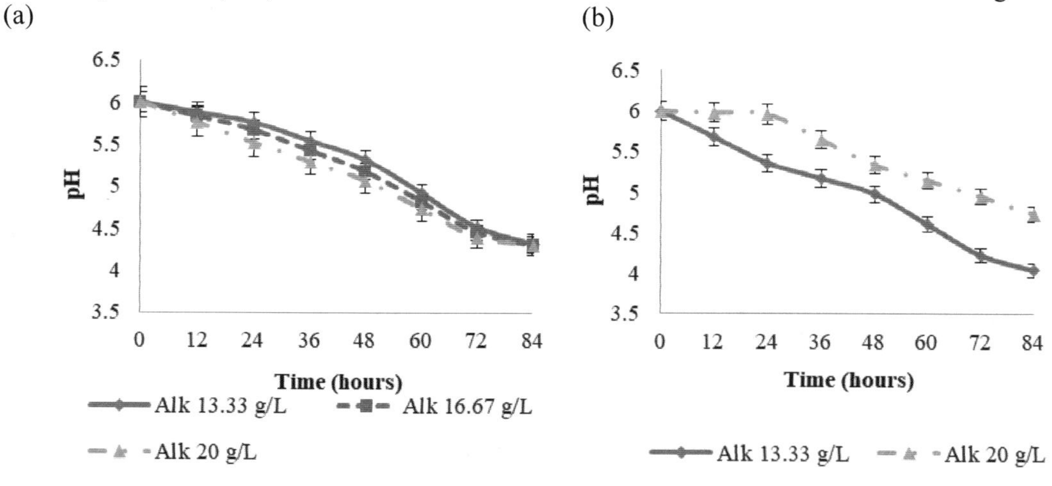

Figure 2. Profiles of pH during fermentation process for alkaline-pretreated (Alk) RTC at various concentration of RTC and PDB concentration (a) 15 g/L (b) 20 g/L

The initial pH used in this research was 6.This pH was taken regarding with previous research from Bailey et al. They found that the highest value of xylanase production was achieved when initial pH for fermentation in the pH range 6-7 [19]. From Figure 2, it can be seen that pH decreased during fermentation process. It means that *T. reesei* grows through carbohydrate metabolism pathway TCA cycle. It produced organic acid, especially citric acid as their final product that slowly decreased pH in the reactor along with the time of fermentation [20]. Most of maximum xylanase activities were always achieved on 48 hours of fermentation (pH 5- 5.5). The xylanase production decreased significantly at pH below 4.5 after 72 hours of fermentation as can be seen in Figure 1 and Figure 2.

4. Conclusions

Shake flasks systems indicated that *Reutealis trisperma* cake (RTC) can be used as additional carbon and nitrogen sources for xylanase production. Pretreated RTC has been proved to give higher xylanase activity compare to natural RTC both in PDB 15 g/L. The maximum xylanase activity was performed at the system with 20 g/L pretreated RTC and 15 g/L PDB which was 198.13 ± 0.03 U/mL. The fungi were able to use RTC to induce xylan for xylanase production. The pH profile during the fermentation decreased along with the time of fermentation which presumably indicated the production of organic acid during *Trichoderma reesei* growth.

Acknowledgment

The work was supported by a grant from Energi Baru Sentosa Company. The authors thank to Daniel Novendo and Evan Setiawan for their help during the research work.

References

[1] Holilah H, Didik P, Titie P U, Eka B S, Yusuf M Z, Hasliza B, Hamzah F, Ratna E and Juwari J 2015 *Biomass Convers. Biorefin.* **5** 347-353
[2] Beg Q K, Kapoor M, Mahajan L and Hoondal G S 2001 *Appl. Microbiol. Biotechnol.* **56** 326-338
[3] Kuhad R C and Ajay S 2008 *Crit. Rev. Biotech.* **13** 151-172
[4] Ratanakhanokchai K, Kyu K L and Morakot T 1999 *Appl. Environ. Microbiol.* **65** 694-697
[5] Fang T J, Bo C L and Shu C L 2010 *New Biotechnol.* **27** 25-32
[6] Motta F L, Andrade C C P and Santana M H A 2013 *InTechOpen*
[7] Seiboth B, Ivanova C and Seidl S V 2011 *InTechOpen.*
[8] Tenkanen M, Jurgen P and Kaisa P 1992 *MicrobTechnol.* **14** 566-574
[9] Zeilinger S, Robert L M, Martin S, Petra H and Christian P K 1996 *Journal of Biological Chemistry* **271** 25624-25629
[10] Li Q, Abdullah A L, Anthony M C & Lu K J 2017 *J. Biotechnol.* **248** 35-42
[11] Juhasz T, Szengyel Z, Reczey K, Siika AM and Viikari L 2005 *Process Biochem.* **40** 3519-3525
[12] Lestari Y E and Monika T 2016 *Research report, Chemical Engineering Department, University of Surabaya,* Surabaya.
[13] Bailey M J, Biely P and Poutanen K 1992 *J. Biotechnol.* **23** 257-270
[14] Gangadharan D, Sivaramakrishman S, Nampoothiri K M and Pandey A 2006 *Biotechnol.* **44** 269-274
[15] Septiningrum K and Chandra A P 2011 *Jurnal Riset Industri* **5** 87-97
[16] Coffman AM, Qian L and Lu K J 2014 *J. Am. Oil. Chem. Soc* **91** 1331-1338
[17] Xiong H V, Weymarn N, Leisola M and Turunen O 2004 *Process Biochem.* **39** 729-733.
[18] Betiku E, Emeko H A and Solomon B O 2016 *Heliyon* 2(2) e00082
[19] Bailey M J, Johanna B and Liisa V 1993 *Appl. Microbiol. Biotechnol.* **40** 224- 229
[20] Jouhten P, Esa, P, Tiina P, Markku S, Merja P and Hannu M 2009 *BMC Syst. Biol.* **3** 1-16

Regulatory performance of two different tuning methods for milk cooling control system

R Agustriyanto

Chemical Engineering Department, University of Surabaya, Jl.Raya Kalirungkut, Surabaya, Indonesia, 60293

E-mail: rudy.agustriyanto@staff.ubaya.ac.id

Abstract. Temperature is one of the essential factors for bacterial growth in milk. The higher the temperature, the faster the growth of bacteria. It is empirically proven that the bacteria will stop growing at the temperature of about 4°C. In this research, continuous milk cooling process was simulated and then controlled by using Proportional - Integral (PI) feedback control system. The regulatory performance of the two different tuning methods were then analyzed (i.e Tyreus - Luybean and Hagglund – Astorm). Their SSE (sum squared of errors) were compared. It was found that Tyreus – Luyben method gave better regulatory performance than Hagglund – Astorm.

1. Introduction

Milk is collected, transported and delivered to milk cooling centres in a number of ways, as shown in Figure 1.

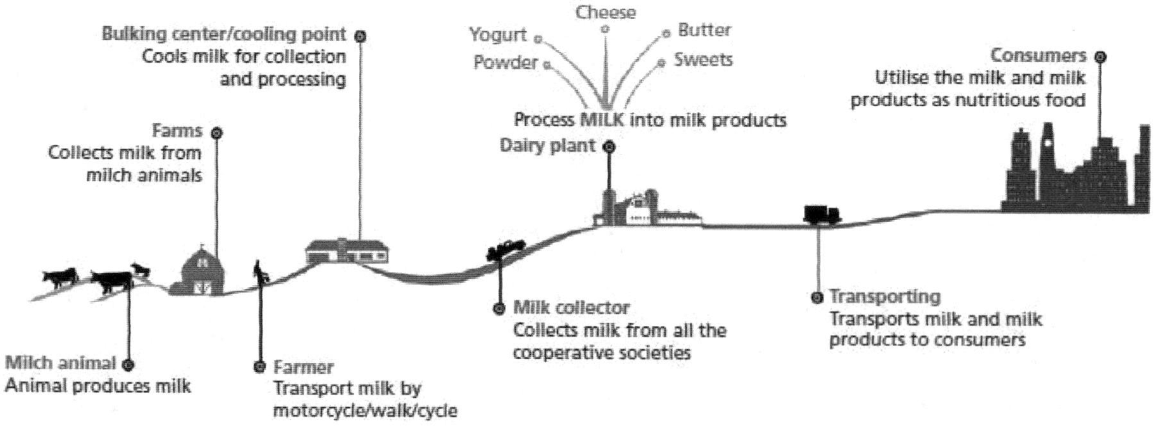

Figure1. Rural dairy collection [1]

Warm fresh milk should be cooled immediately after milking to preserve the quality and prevent spoilage. Cooling to 4°C within 3 to 4 hours is essential, but more rapid cooling is much preferred [1].

IOP Conf. Series: Materials Science and Engineering **703** (2019) 012004 doi:10.1088/1757-899X/703/1/012004

Normally, milk arrives at the milk cooling centres in the morning and late in the afternoon / early evening. Therefore, batch milk cooling is performed as such in cooperatives (i.e Koperasi Unit Desa SAE in Pujon, KUD Batu and elsewhere). However, in this study continuous milk cooling process was simulated as it was more applicable for larger capacity and it gave more rapid cooling.

Proportional Integral (PI) controllers were the most commonly used form of Proportional Integral Derivative (PID) controllers, accounting for over 90% of industrial PID applications. Tuning a PI controller involved setting the controller gain (k_c) and the reset time (τ_I) [7]. There was a strong evidence that PI and PID controllers remained poorly understood and, in particular, poorly tuned in many applications [9].

The aim of this research was to compare regulatory performance of two different tuning methods (i.e Tyreus - Luyben and Hagglund – Astorm) for controller parameter determination while servo performance for this application had been published elsewhere [2].

2. Simulation

The system being studied was the milk cooling system at Koperasi Unit Desa SAE Pujon [2] as shown Figure 2 and 3.

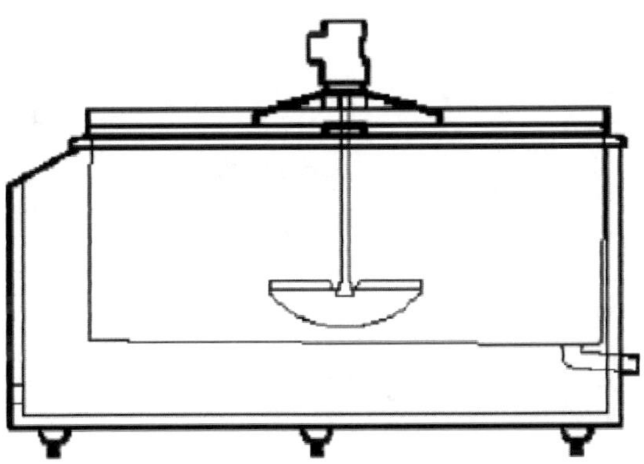

Figure 2. Milk cooling system seen from side

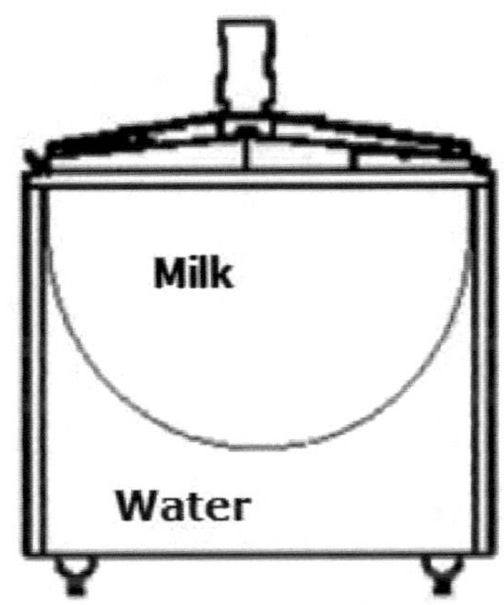

Figure 3. Milk cooling system seen from the front

For continuous system with Proportional – Integral (PI) controller installed as shown in Figure 4, its mathematical model can be derived based on energy balance. The final model in Laplace Transform is shown below [2]:

$$T_0(s)=\frac{T_i(s)}{\left(\frac{m}{w_s}s+1\right)}-\frac{\frac{Q(s)}{w_sC_p}}{\left(\frac{m}{w_s}s+1\right)}-\frac{\frac{(T_{0,s}-T_{i,s})W_s}{w_s}}{\left(\frac{m}{w_s}s+1\right)} \tag{1}$$

Where:

W_s = milk flowrate (kg/min)

$T_{0,s}$ = steady state temperature of inlet milk (°C)

$T_{i,s}$ = steady state temperature of outlet milk (°C)

m = mass of inlet milk(kg)

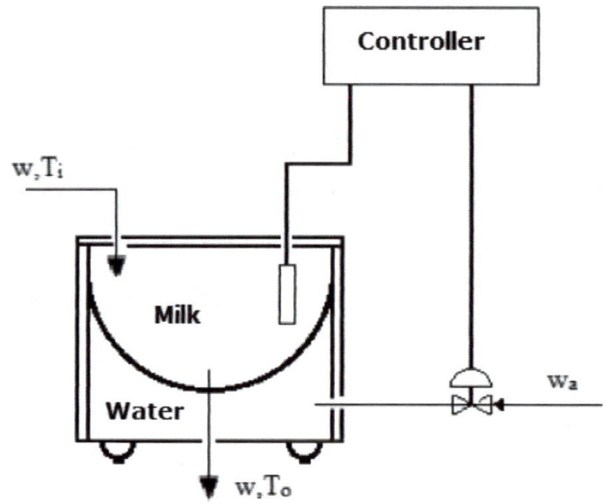

Figure 4. Continuous milk cooling system

The data for milk is as follows:

$$C_p = \text{heat capacity of milk} = 3.93\ \frac{kJ}{kg.K} \quad (2)$$

$$\rho = \text{density of milk} = 1027\ \frac{kg}{m^3} \quad (3)$$

Table 1 shows data at steady state.

Table 1. Steady state data

Parameter	Value	Unit
V_{milk}	2500	L
T_i	36	°C
T_0	4	°C
T_1	-13	°C
T_2	0	°C

Tyreus – Luybean Tuning Method

PI or PID parameters can be determined by using equation as shown in Table 2 [3-4], where:

k_{cu} = ultimate gain

P_u = ultimate period

P_u and k_{cu} are similar to those in Ziegler – Nichol method and detail can be read elsewhere [5-7].

Table 2. Tyreus – Luyben settings [3,4]

Controller	k_c	τ_I	τ_D
PI	$\frac{k_{cu}}{3.2}$	$2.2P_u$	-
PID	$\frac{k_{cu}}{2.2}$	$2.2P_u$	$\frac{P_u}{6.3}$

IOP Conf. Series: Materials Science and Engineering **703** (2019) 012004 doi:10.1088/1757-899X/703/1/012004

Hagglund – Astorm Tuning Method
Setting PI controller for Hagglund – Astorm is using equation below [8]:

$$k_c = \frac{1}{K}\left(0.14 + 0.28\frac{\tau}{\theta}\right) \quad (4)$$

$$\tau_I = \theta\left(0.33 + \frac{6.8\tau}{10\theta + \tau}\right) \quad (5)$$

Where:

K = process gain
θ = process time delay
τ = process time constant

3. Results and Discussion

Tyreus-Luyben is quite similar to Ziegler Nichols method. There are two methods in obtaining k_{cu} (ultimate gain) and P_u (ultimate periode). For simplicity, we chose plotting Bode diagram of open loop system (which depends on parameters of the process) [5]. Therefore, both Tyreus-Luyben and Hagglund-Astorm were basically model based control methods. The values for time delay were assumed ranging from 1 to 5 min so that both tuning methods could be applied. Table 3 shows the values for controller parameters obtained.

Table 3. Controller parameters

Time Delay	Controller Parameters	Tyreus - Luyben	Hagglund - Astorm
1	k_c	-10209	-2834
	τ_I [min]	4.2	5.4
2	k_c	-6200	-1441
	τ_I [min]	6.9	8.7
3	k_c	-4894	-977
	τ_I [min]	8.8	10.9
4	k_c	-3674	-745
	τ_I [min]	11.7	12.6
5	k_c	-2954	-606
	τ_I [min]	5.5	14.0

Most commercial PID controllers use a controller gain, k_c (or proportional band, PB) that is expressed as a standard dimensionless %/%. But k_c actually has units of (% of CO signal)/(% of PV signal). In a precise mathematical world, these units do not cancel [10]. However, as most process control paper do [5,6,11], we also did not show the unit.

In this research, direct acting controller was applied, but as we can see from its transfer function: $\frac{T_o}{Q}$ was negative, therefore the controller gain should be negative. Tuning parameter was based on servo problem but tested for regulatory problem. Hence the process model : $\frac{T_o}{Q}$ was still used.

At certain times, the temperature of inlet milk (T_i) was disturbed in the form of step function of -2°C (form its steady state value) at t = 35 min and at 3°C at t = 60 minutes. When interference was given, the output would move away from its setpoint but would be back immediately. These disturbances may affect the output milk temperature and the amount of heat released (Q).

Figure 5 shows profile of output milk temperature (T_0) and its setpoint when such disturbances of inlet milk temperature (T_i) occurred. Profile of heat released (Q) was also shown. This figure was for one min time delay assumption and using Tyreus-Luyben. Profile for T_0 and Q under similar disturbances are shown in Figure 6 for Hagglund–Astorm tuning method. Again, one minute time delay assumption was applied. Simulations were then continued for other time delay values (i.e time delay = 2 to 5 minutes). The figures similar to figure 5 and 6 could also be generated. The sum squared of errors (SSE) values for both method and all time delay assumptions are listed in Table 4. Both methods were successful in term of disturbance rejection, but Tyreus – Luyben gave smaller values of SSE.

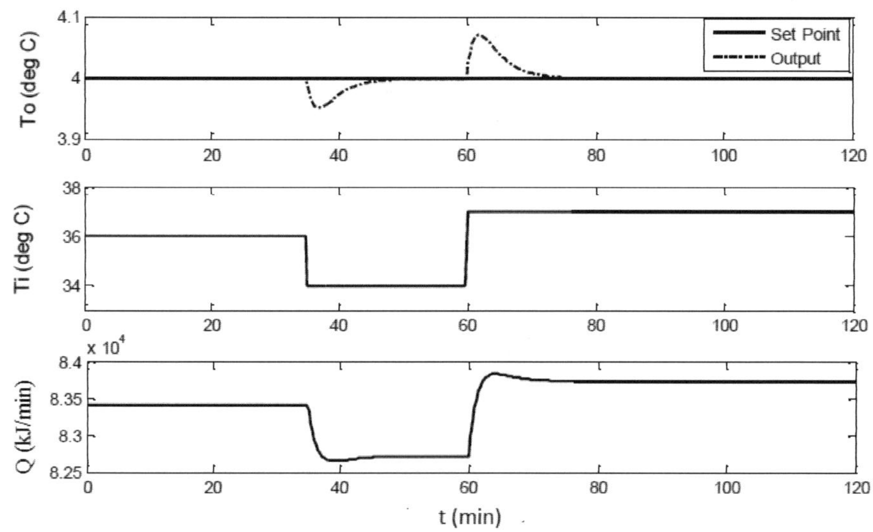

Figure 5. Profile of outlet milk temperature (T_0) and heat released (Q) for 1 min time delay assumption using Tyreus - Luyben tuning method

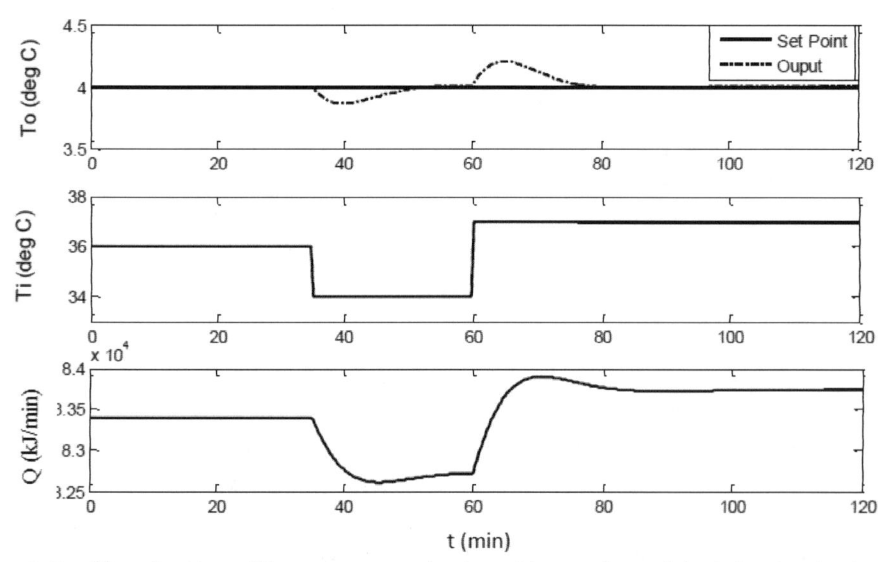

Figure 6. Profile of outlet milk temperature (T_0) and heat released (Q) for 1 min time delay assumption using Hagglund – Astorm tuning method

IOP Conf. Series: Materials Science and Engineering **703** (2019) 012004 doi:10.1088/1757-899X/703/1/012004

Table 4 also shows that small value of time delay assumption were preferred since the higher the value, the higher their SSE values (other error definitions for comparing the performance of different tuning methods such as IAE, ITAE, ISE could also be applied [6]). Looking at the controller parameters between the two methods in Table 3, it seems Tyreus – Luyben has higher values of k_c and lower values of τ_I compared with one of Hagglund - Astorm. Therefore PI controller based on Tyreus - Luyben should be better than that of Hagglund - Astorm.

Table 4. SSE (sum squared of errors) values of outlet milk temperature

Time Delay	SSE	
(min)	Tyreus - Luyben	Hagglund - Astorm
1	0.0953	1.4872
2	0.4117	8.1972
3	0.8057	17.5491
4	1.7406	27.2457
5	3.0064	36.5148

4. Conclusions

Feedback control system using PI controller has been applied for continuous milk cooling process. Two methods of tuning have been investigated (i.e. Tyreus – Luyben and Hagglund – Astorm) and both gave satisfactory results for regulatory performance. However, it seems that Tyreus- Luyben was slightly better because it gave smaller SSE values than Hagglund Astorm. It is also recommended that small time delay assumption is used instead of higher values.

References

[1] Moffat F, Khanal S, Bennet A, Thapa T B and George S M 2016 *Technical and investment guidelines for milk cooling centres* (Rome: Food and Agriculture Organization of the United Nations)

[2] Agustriyanto R, Febriyanto A and Widiawati P S 2018 *Jurnal Rekayasa Proses* **12** p 10

[3] Shahrokhi M, Zomorrodi A, Department of Chemical and Petroleum Engineering and Sharif University of Technology, https://pdfs.semanticscholar.org/116c/e07bcb202562606884c853fd1d19169a0b16.pdf.

[4] Ibrahim O, Yahaya N Z and Saad N 2016 *Int. Journal of Power Electronics and Drive System* **7** p 294

[5] Coughanowr D R and LeBlanc S E 2009 *Process Systems Analysis and Control*, 3rd ed McGraw Hill

[6] Marlin T E 2000 *Process Control*, 2nd ed McGraw Hill

[7] Riggs J B and Karim M N 2006 *Chemical and Bio-Process Control*, 3rd ed Pearson

[8] Haugen and Finn 2010 *Modeling, Identification and Control* p 31, 79

[9] O'Dwyer A 2009 *Handbook of PI and PID Controller Tuning Rules*, 3rd ed Imperial College Press

IOP Conf. Series: Materials Science and Engineering **703** (2019) 012005 doi:10.1088/1757-899X/703/1/012005

The solubility correlation of azobenzene derivatives in supercritical carbon dioxide: a short review

R S Alwi and A S Iryani

Department of Chemical Engineering, Fajar University, Makassar, Indonesia

E-mail: ratnasya@gmail.com

Abstract. This short review highlights the solubility and correlation of four azobenzene derivatives, which have similar chemical structures, available in the literature. At the same phase equilibrium (at the temperature of 353.15 K), the solubilities of four azobenzene derivatives in supercritical carbon dioxide followed in the order of Disperse Red 82 > Disperse Red 13 > Disperse Blue 79:1 > Disperse Red 1. Five different kinds of density-based models were used to correlate the solubility data. Moreover, the solubilities of four azobenzene derivatives were also correlated thermodynamically by the model of the regular solution with the Flory–Huggins theory. It was found the Sung-Shim model (four parameters) has the lowest median AARD of 12.02 % of the solubility data over the entire conditions.

1. Introduction

Carbon dioxide has unique properties compared to other substances. It has unique properties of critical temperature (Tc = 304.15 K) and pressure (Pc = 7.383 MPa), respectively. Currently, in the textile industry, Supercritical carbon dioxide ($scCO_2$) dyeing process has been in attention. This technology has many advantages compared to conventional dyeing technology because carbon dioxide is nontoxic, non-flammable, recyclable, dyestuffs can be reused, and no additional chemical is required in the process. In addition, consumption energy can be saved because drying process is unnecessary so that dyeing time is short compared to the traditional dyeing technology. To the application and development of the Supercritical carbon dioxide ($scCO_2$) dyeing process requires specific knowledge of the phase equilibrium between dyestuffs and the supercritical carbon dioxide as a solvent, and a number of authors have discussed the dependence of solubility of dye in $scCO_2$ on both temperature and pressure, and have been presented in the literatures. Some investigators have found that the solubility of anthraquinones and azo derivatives are higher, static and dynamic methods have been used to measure the solubility of dyestuffs in supercritical carbon dioxide. We have measured and correlated the solubilities of anthraquinone derivatives in supercritical carbon containing the substituent groups —NO_2 , —OH, —NH_2 [1-4] because some of the dyestuffs used in dyeing processes are configurated by these functional groups. In this short review, we highlight the solubility correlation and investigated the effects of functional groups of four azobenzene derivatives, which have similar chemical structures containing the substituent groups of —NO_2 available in the literature [5-6], namely, C.I. Disperse Red 1, C.I. Disperse Red 13, C.I. Disperse Red 82, and C.I. Disperse Blue 79:1. The solubility data obtained from the literature were correlated with the density based models, the Chrastil [7], the Sung–Shim [8], Bartle et al [9], Kumar-Johnston [10], and the Mendez-Santiago–Teja [11] models. Moreover, the solubilities of four azobenzene derivatives were also correlated thermodynamically by the model of regular solution with the Flory–Huggins theory [12].

2. Solubility of four azobenzene derivatives in supercritical carbon dioxide

In this study, we were used the solubility data available in the literature measured by Shinoda at al.[5] and Lin at al.[6]. Table 1 presents the Physical properties of C.I. Disperse Red 1, C.I. Disperse Red 13, C.I. Disperse Red 82, and C.I. Disperse Blue 79:1.

Table 1. The physical properties of C.I. Disperse Red 1 and C.I. Disperse Red 13, C.I. Disperse Red 82, and C.I. Disperse Blue 79:1.

	Red 1	*Red 13*	*Red 82*	*Blue 79:1*
CAS number	2872-52-8	3180-81-2	30124-94-8	3618-72-2
Mw	314.34	348.78	439.43	625.39
T_m / (K)	431.15[a]	406.15[a]	402.3[b]	422.15[c]
Δh_2^m / (kJ · mol^{-1})[d]	36.93	37.77	52.15	59.55
v_2^s / (m^3 · mol^{-1})[e]	$2.23 \cdot 10^{-4}$	$2.53 \cdot 10^{-4}$	$3.04 \cdot 10^{-4}$	$3.86 \cdot 10^{-4}$
T_c/ (K)[e]	946.28	951.96	901.49	952.37
P_c/ (MPa)[f]	2.025	1.924	1.3899	1.252
ω[f]	1.231	1.251	1.171	0.931

[a] Observed by Kozak et al.[13]
[b] Observed by Suesat et al. [14]
[c] Observed by Thiel et al.[15]
[d] Estimated by Jain et al. method [16]
[e] Estimated by Fedors method [17]
[f] Estimated by Edmister method [18]

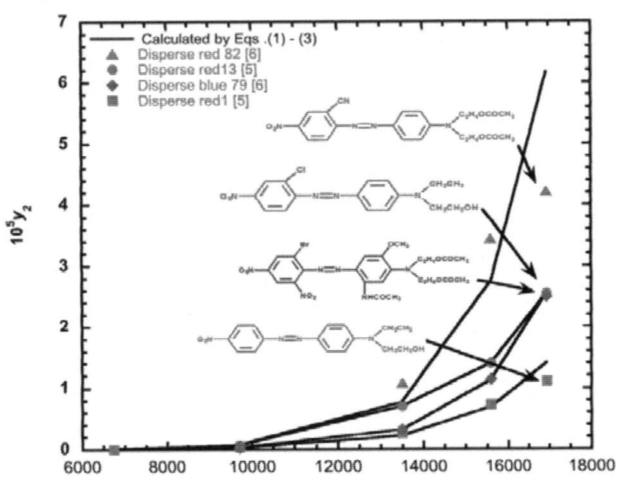

Figure 1. The solubility at temperature of 353.15 K of C.I. Disperse Red 82, C.I. Disperse Red 13, C.I. Disperse Blue 79:1, and C.I. Disperse Red 1 against density of CO_2 ρ /(mol·m^{-3}).

The same equilibrium condition (the temperature of 353.15K) is shown in the figure 1. It was found the C.I. Disperse Red 82 is the most soluble in the supercritical carbon dioxide over the entire pressure range. The magnitudes of the solubilities of four azobenzene derivatives in supercritical carbon dioxide are in the order of Disperse Red 82 > Disperse Red 13 > Disperse Blue 79:1 > Disperse Red 1. The results showed that the additional —CN group on the Disperse Red 82 cause the solubility was much better in carbon dioxide than other structures. It was found that presence of —CN group significantly decreased the polarity of dye molecule so the solubility of dye was higher because supercritical carbon dioxide is nonpolar. In addition, the solubility of azobenzene derivatives in supercritical carbon dioxide were independent of the molecular weight as shown in the table 1, the level of the molecular weight of four azobenzene derivatives were Disperse Blue 79:1 > Disperse Red 82 > Disperse red 13 > Disperse Red 1.

3. Correlations

The solubility data obtained from literature were correlated using several types of semi-empirical equations, expressed in terms of pressure, temperature, and density of CO_2, proposed previously by the Chrastil [7], the Sung–Shim [8], Bartle et al. [9], Kumar-Johnston [10], and the Mendez-Santiago–Teja [11]. These equations have three or more parameters to be adjusted to the solubility values. In the Bartle et al. model, ρ_{ref} was set 700 kg/m^3 used as a reference density, P_{ref} is set 0.1 MPa as a standard pressure. Table 2 summarizes the density-based models and equations used. Figure 3 summarizes the average absolute relative deviation (*AARD*) for four azobenzene derivatives. It was found that Sung-Shim model has lower average absolute relative deviation (*AARD*) than that of other models as illustrated in the figures 2-3. However, the correlation of Mendez-Santiago–Teja obtained the higher average absolute relative deviation (*AARD*), about 40 %. Table 3 shows the parameters of the azobenzene derivatives system used and average absolute relative deviation (*AARD*) between the solubility data and calculated values. Furthermore, by assuming that no considerable solubility of the sc-CO_2 in the solid phase, the solubility of four azobenzene derivatives were correlated by the thermodynamic model of solid-liquid equilibria. A detailed explanation of the calculation procedure of the thermodynamic criteria has been reported elsewhere [1-4]. We briefly explain the calculation procedure in this study. We used the regular solution model combined with the Flory – Huggins equation [2],[12]:

$$ln\, y_2 = \frac{\Delta h_2^m}{RT}\left(\frac{T}{T_m} - 1\right) - \frac{v_2}{RT}(\delta_1 - \delta_2)^2 - ln\left(\frac{v_2}{v_1}\right) - 1 + \frac{v_2}{v_1} \quad (1)$$

where T_m and Δh_2^m are the melting temperature and melting enthalpy of solid solute, respectively. The v_1 and v_2 are the molar volume of sc-CO_2 and solid solute. The solubility parameters of sc-CO_2 were calculated by the method of Giddings et al.[19]

$$\delta_1/(MPa)^{0.5} = 8.0325(\mathrm{Pc/MPa})^{0.5}\left(\frac{\rho_R}{2.66}\right) \quad (2)$$

The ρ_R ($= \rho / \rho_c$) is the reduced CO_2 density, ρ_c the critical density, and P_c is the critical pressure of CO_2. The solute solubility parameter was assumed to be dependent on CO_2 density, so can be expressed as:

$$\delta_2 = a + b\rho_{CO_2} \quad (3)$$

The average absolute relative deviation (AARD %) for the systems of N_D data points in this study can be calculated by:

$$AARD(\%) = \frac{100}{N_D}\sum_{n=1}^{N_D}\left|\frac{y_{2,n}^{exp} - y_{2,n}^{cal}}{y_{2,n}^{exp}}\right| \quad (4)$$

Table 2. Density-based models and equations used.

Models	Equations	Ref.
Mendez-Santiago – Teja	$T\,ln\,(y_2 P) = A + B\rho + CT$	[11]
Chrastil	$ln\,y_2 = A + \frac{B}{T} + C\,ln\,\rho$	[7]
Bartle et al.	$ln\left(\frac{y_2 P}{P_{ref}}\right) = A + B(\rho - \rho_{ref}) + \frac{C}{T}$	[9]
Kumar – Johnston	$ln\,y_2 = A + \frac{B}{T} + C\,\rho$	[10]
Sung – Shim	$ln\,y_2 = A + \frac{B}{T} + \left(C + \frac{D}{T}\right) ln\,\rho$	[8]

Table 3. Parameters of the Azobenzene derivatives system used and average absolute relative deviation (*AARD*) between the solubility data and calculated values.

System	Parameters				*AARD*
	A	*B*	*C*	*D*	/(%)
		Chrastil			
C.I. Disperse Red 1 + CO_2	-43.323	-9437.5	6.0339		20.09
C.I. Disperse Red 13 + CO_2	-46.76	-8203.7	6.1067		19.58
C.I. Disperse Red 82 + CO_2	-68.912	-4359.3	7.3209		18.06
C.I. Disperse Blue 79:1 + CO_2	-65.355	-5337.1	7.1501		13.38
		Mendez-Santiago – Teja			
C.I. Disperse Red 1 + CO_2	-14443	0.24584	35.114		59.96
C.I. Disperse Red 13 + CO_2	-13281	0.24884	32.454		54.92
C.I. Disperse Red 82 + CO_2	-9488.4	0.28638	20.967		33.19
C.I. Disperse Blue 79:1 + CO_2	-10457	0.28262	23.024		22.52
		Bartle et al.			
C.I. Disperse Red 1 + CO_2	26.167	-11391	0.015806		41.94
C.I. Disperse Red 13 + CO_2	23.501	-10179	0.015994		38.89
C.I. Disperse Red 82 + CO_2	12.409	-6016.3	0.017429		29.02
C.I. Disperse Blue 79:1 + CO_2	14.517	-7063.3	0.017204		14.19
		Kumar – Johnston			
C.I. Disperse Red 1 + CO_2	7.1483	-10008	0.00060037		40.71
C.I. Disperse Red 13 + CO_2	4.3513	-8796.7	0.00060866		39.19
C.I. Disperse Red 82 + CO_2	-7.4703	-4740.8	0.00067208		30.14
C.I. Disperse Blue 79:1 + CO_2	-5.2058	-5787.8	0.0006622		13.39
		Sung – Shim			
C.I. Disperse Red 1 + CO_2	21.718	-32853	-0.89301	2489.4	12.98
C.I. Disperse Red 13 + CO_2	26.789	-34682	-1.7264	2815.1	8.86
C.I. Disperse Red 82 + CO_2	19.446	-37588	-2.1001	3540.7	15.05
C.I. Disperse Blue 79:1 + CO_2	-10.905	-25814	1.3445	2181.9	11.20

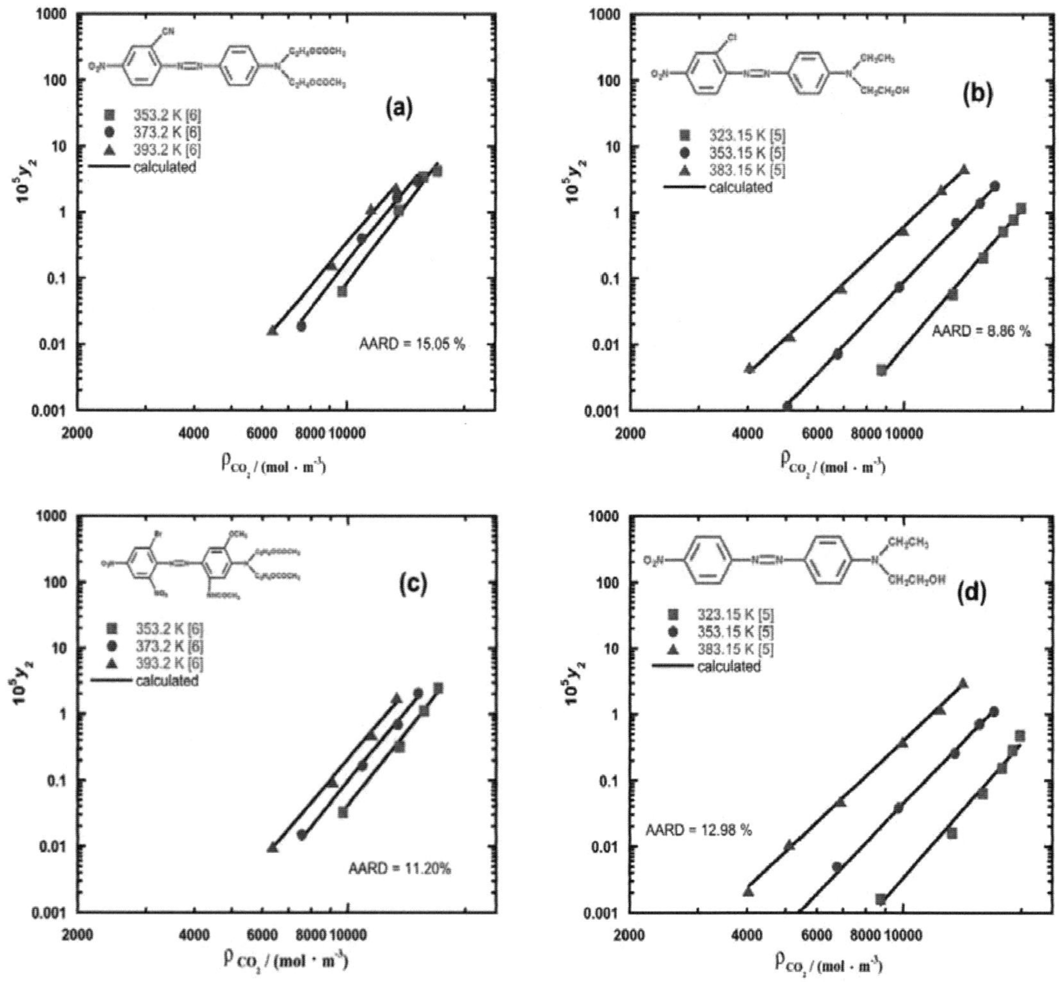

Figure 2. Plot of mole fraction 10^5y_2 against density of ρ CO_2 /$(mol \cdot m^{-3})$ to correlate results for C.I. Disperse Red 82 (a), C.I. Disperse Red 13 (b), C.I. Disperse Blue 79:1 (c), and C.I. Disperse Red 1 (d) from Sung-Shim model

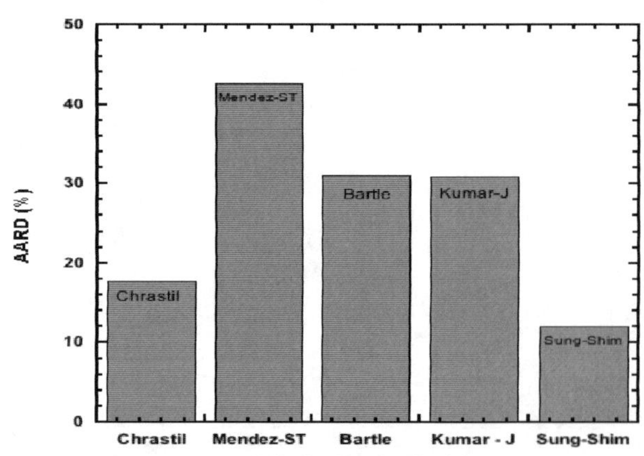

Figure 3. Summarizes the average absolute relative deviation (*AARD*) for four azobenzene derivative

IOP Conf. Series: Materials Science and Engineering **703** (2019) 012005 doi:10.1088/1757-899X/703/1/012005

4. Conclusions

The solubility correlation of four azobenzene derivatives which have similar chemical structures taken from available literature has been studied. At the same phase equilibrium (at the temperature of 353.15 K), the solubilities of four azobenzene derivatives in supercritical carbon dioxide followed the order of Disperse Red 82 > Disperse Red 13 > Disperse Blue 79:1 > Disperse Red 1. Five different kinds of density-based models were used to correlate the solubility data. Moreover, the solubilities of four azobenzene derivatives were also correlated thermodynamically by the model of the regular solution with the Flory–Huggins theory. It was found that the Sung-Shim model (four parameters) has the lowest median AARD of 12.02 % of the solubility data over the entire conditions.

Acknowledgments

This work was supported by the Directorate of Higher Education of the project on "Fundamental Research of DRPM DIKTI".

References

[1] Alwi R S, Tanaka, and Tamura K 2014 *J. Chem. Thermodynamics* **74** 119–25
[2] Tamura K and Alwi R S 2015 *Dyes and Pigments* **113** 351–56
[3] Tamura K, Alwi R S, Tanaka T, and Shimizu K 2017 *J. Chem. Thermodynamics* **104** 162–68
[4] Alwi R S and Tamura K 2015 *J. Chem. Eng. Data* **60** 3046–52
[5] Shinoda T and Tamura K 2003 *Fluid Phase Equilib.* **213** 115–23
[6] Lin H, Ho C C, and Lee M J 2004 *J. Supercrit. Fluids* **32** 105–14
[7] Chrastil J 2002 *J. Phys. Chem.* **86** 3016–21
[8] Sung H D and Shim J J 1999 *J. Chem. & Eng. Data* **44** 985–89
[9] Bartle K D, Clifford A A, Jafar S A, and Shilstone G F 1991 *J. Phys. Chem. Ref. Data* **20** 713–56
[10] Kumar S K and Johnston K P 1988 *J. Supercrit. Fluids* **1** 15–22
[11] Méndez-Santiago J and Teja A S 1999 *Fluid Phase Equilib.* **158–160** 501–10
[12] Iwai Y, Koga Y, Fukuda T, and Arai Y 1992 *J. Chem. Eng. Japan* **25** 757–60.
[13] Kozak G D, Vasin A Y, and D'Yachkova A V 2008 *Combust. Explos. Shock Waves* **44** 579–82
[14] Suesat J, Mungmeechai T, Suwanruji P, Parasuk W, Taylor J A, and Philips A S *Coloration Tech.* **127** 217–22
[15] Thiel W, Mayer R, Jauer, E A, Modrow H, and Dost H 2004 *J. fr. Prakt. Chemie*. **328** 497–514
[16] Jain A, Yang G, and Yalkowsky S H 2004 *Ind. Eng. Chem. Res.* **43** 7618–21
[17] Fedors R F 1974 *Polym. Eng. Sci.* **14** 147–54
[18] Reid R C, Prausnitz J M, Sherwood, T K *The Properties of Gases and Liquids* (2001) 5th ed (McGraw-Hill: New York)
[19] Giddings J. C, Myers M N, McLaren L, Keller R A (1968) *Science* **162** 67–73

IOP Conf. Series: Materials Science and Engineering **703** (2019) 012006 doi:10.1088/1757-899X/703/1/012006

Container storage tariff policy analysis using combining game theory and system dynamics approach

A G Budianto and B Wirjodirdjo

Department of Industrial Engineering, Faculty of Industrial Technology, Institut Teknologi Sepuluh Nopember, Surabaya 60111, Indonesia

E-mail: budisantoso.wirjodirdjo@gmail.com

Abstract. This paper examines the storage tariff setting policy. The main purpose of this paper is to give consideration to optimal tariffs based on alternative strategies that reduce dwelling time. The main problem is the basic tariff for container storage in Container Yard is too cheap. This low-cost tariff of the storage container is used by customers so that containers are piled too long in Container Yard and cause dwelling time to be high. This has an impact on the congestion effect and the decline in utilities in the container terminal. This problem has complexity and suitability with System Dynamics. The linkage system between one cause and another cause will be modelled with System Dynamics so that the estimated output of container flows can be obtained. Furthermore, with the Non-Cooperative Game approach, a payoff matrix table will be obtained which contains the total profit obtained by the container terminal management and the total cost incurred by the customer. The main factors considered in the payoff matrix are the estimated container flows, the container storage tariff and the length of time accumulated in Container Yard. Finally, the optimal tariff from both sides was found (Nash Equilibrium), from the point of view of terminal container management and customer. This will help terminal management to make the right decision.

1. Introduction

The port has an important role to support an operational supply chain system. It can be considered as a part of supply chain systems and several entities involved such as terminal management, shipper, consignee, carrier, and forwarder. The port can also be seen as a gate of the flow of goods in and out. The problem that is almost experienced in every container terminal is limited storage space and stagnation. The limited storage space of the container has caused a bottleneck of the container that can be handled by the container terminal.

Martin et al [1] state that in general, terminal operators do not derive large profits from storage charges because their main activity is container transshipment between different modes of transportation, but they would like to satisfy certain targets by introducing storage pricing, as follows : (a) to prevent customers storing containers in container yards for a long periods, (b) to ensure the efficiency of terminal performance and greater profits from the storage area, (c) to provide additional services to customers which are currently in high demand, especially for users who do not have warehouse facilities. The storage duration of the container can be considered as one of the performance indicators on the container terminal. In case the longer dwelling time indicates the lower performance of the container terminal [2].

The policy to determine the tariff of container storage had been widely discussed by the researchers. Those are: Castilho and Daganzo [3] had to focus their main objective only to maximize the benefits obtained by terminal management. Then Kim and Kim [4] introduced a storage tariff policy with progressive rates, where progressive tariffs are used as a penalty rate imposed on customers. So customers are not using the container yard facility too long. In the last research, Qiu and Lam [5] used the game theory approach with the Stackelberg game model. The main objective is to maximize the benefits of terminal management while minimizing the expenditure of customer costs in container yards. The use of exact methods to determine container storage tariff has a weakness where the mathematical approach used usually has limitations in modelling the relationship of one variable to other variables. Even though in the real world, the determination of container storage tariff has many links between one variable to other variables.

The model that has many interrelationships between one variable and other variables can be solved using a simulation approach. The simulation approach has the advantage that the model used can be probabilistic; accommodate variables that have interdependence with other variables and systems that have complexity. In this paper, the researcher uses a system dynamics simulation approach to fill the payoff matrix from the non-cooperative game theory approach. The system dynamics approach is a field of study of the structure and behaviour of socio-technical systems to guide effective decision making, learning, and policy in a world full of dynamic complexity [6]. While the use of the game theory approach has the same characteristics where the game consists of two to several people or groups have a strategy that is made to maximize their own profits or minimize the opponent's profits [7].

In this case, the researcher aims to develop a strategy where there is a balance between two players, the container terminal management and customers. So, container terminal management can have strategic considerations before making the right managerial decisions.

2. Literature study

2.1 Game Theory

Game theory is a branch of applied mathematics that is used in the social sciences, most notably in economics, as well as in biology (particularly evolutionary biology and ecology), engineering, political science, international relations, computer science, social psychology, philosophy, and management. Game theory attempts to mathematically capture behaviour in strategic situations, or games, in which an individual's success in making choices depends on the choices of others [8]. The theory breaks naturally into two parts. There is a non-cooperative theory in which the players, if they may communicate, may not form binding agreements. An example of this situation is the interaction between companies in an industry in the environment where antimonopoly laws make it illegal for companies to reach agreement on prices or production quotas or other forms of behaviour. On the other hand, in the cooperative theory, the player is allowed to form binding agreements, and so there is a strong incentive to work together to receive the largest total payoff [9].

Based on the problem-solving strategy, game theory is also divided into pure and mix strategies. The example shows how each player can choose a strategy for rock, paper, and scissors. The value in the matrix is the payoff or consequences of each decision making. In this game, each of the two players can choose rock, paper, or scissors. If both players choose the same action, there is no winner and the utilities are zero. Otherwise, each of the actions wins over one of the other actions and loses to other remaining action. Whereas when viewed from the payoff matrix in the form of zero-sum games, the type of game that describes the amount of profit received by player 1 will amount to the loss felt by player 2 and vice versa [10].

Non-zero sum games are types of games that are not totally counted because some result shave a net result that is greater or less than zero. The most famous example is the case of prisoner dilemma shows in Figure 1.Non zero-sum games case, where prisoners have the choice to confess or evade when asked by the police. In this game, player 1 can see that no matter which column player II chooses, he will be better off if he chooses row 2. For if player I choose row 2 rather than row 1, he wins 4 rather than 3 if player II chooses column 1, and he wins 1 rather than 0 if she chooses column 2. In other words, player I's second strategy of choosing the second row strictly dominates the strategy of choosing the first. On the other hand, the game is symmetric. Player II's second column strictly dominates her first. However, if both players use their dominant strategies, each player receives 1, whereas if both players use their dominated strategies, each player receives 3 [9].

$$\begin{array}{c} \\ \text{cooperate} \\ \text{defect} \end{array} \begin{array}{c} \begin{array}{cc} \text{cooperate} & \text{defect} \end{array} \\ \begin{pmatrix} (3,3) & (0,4) \\ (4,0) & (1,1) \end{pmatrix} \end{array}$$

Figure 1. Non zero-sum *games case*

2.2 System Dynamics

The system dynamic concept was developed by Forrester in the 1950s in the form of Industrial Dynamics books. The term Industrial Dynamics then changes to a system dynamics to suppress the use of this methodology in fields other than business [11]. In system dynamics, the system is conceptualized as physical state variables and information (stock) that can accumulate, run out and or increase by variable rates (flow), all of which interact through closed loops of cause and effect (feedback loops). The use of a system dynamics approach is more precisely modelled on the problem as follows:

1) Has dynamic properties (changes with time)
2) The structure of the model contains at least one feedback structure

System dynamics can also be used in reviewing a policy regarding price. In a previous study of pricing in electricity transmission services [12], the simulation results showed that price determination has a positive effect in managing transmission congestion, transmission expansion and selecting cargo locations. Price points are very sensitive to operational conditions and network characteristics. It can also encourage electricity providers and users to avoid the cost of expensive transmission congestion and dramatic electricity price fluctuations.

In addition, in a recent study of the determination of Real-Time Charge (RCP) price on Electric Vehicle (EV) [13], the models made were quite extensive including EV power consumption, a generator sets, cost pricing, user response, evaluation of benefits of all stakeholders interest and net life cycle of the electric charging station. This paper proves that increasing RCP will increase profits on charging service operators. This can happen if and only if the number of EV is large enough. While the increase in the number of EV is quite large, it only occurs if the government holds a gradual subsidy to users of EV. Besides being able to increase profits from charging EV service operators, it can also increase savings on non-renewable energy and reduce emissions.

3. System under discussion

Dwelling time is the length of the average container time in the container yard of a container terminal [14]. The consignee's pattern of container collection in the delivery process is also erratic, causing variability in container stacking time in the Container Yard.

Martin [1] proposes a mathematical model for the optimal tariff scheme on Container Yard assuming container arrivals by sea are stochastic (the number of containers unloaded per ship type is a random variable) and several vessels for two different purposes namely maximizing terminal operator profits and minimizing the cost of the whole system. The rate for storing containers based on flat-rate time can be seen in Figure 4 as follows:

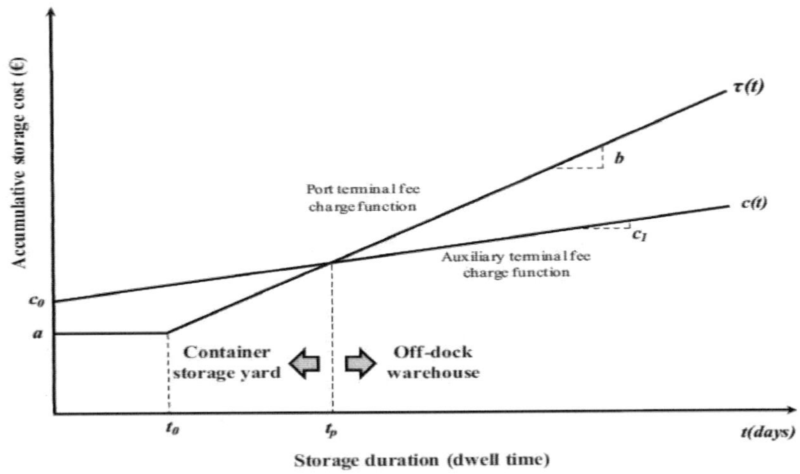

Figure 2. Container storage tariff scheme at container terminal [1]

where:

α	: zero or non-zero flat rate
b	: storage rates that are proportional to time
t0	: flat rate duration (in days)
t	: the length of time the container is in the terminal area
c(t)	: total customer expenditure at time t
tp	: threshold time

Figure 2.Container storage tariff scheme at container terminal [1] assumes that customers minimize their own storage costs, assumptions taken from Kim and Kim [4] and the customer does not reschedule the time taken to take the container according to storage costs. Based on the price structure, the threshold time will divide into two categories:

1) Customer who has a delivery time (shorter than the time limit)
2) Customer with delivery times greater than the time limit

For the last category that must be minimized because it would be better to use their own warehouse facilities because the storage costs would be lower and not cause congestion effect in the container terminal.

4. Container storage system causal loop model and payoff matrix

4.1 Container storage system causal loop model

Dwelling time and container terminal utilities are influenced by many factors including economic and social factors. The causal loop model can be seen in Figure 3.Causal Loop Diagram of Container Yard, which is divided into 3 sub-factors. The main factors include activities at container terminals that influence each other. The second factor is what activities will directly affect the total revenue of container terminal management. The last factor is any activity that will affect the total expenditure of the customer in the storage of temporary containers.

In the causal loop model, the length of accumulation in temporary facilities will be the main key of container terminal utilities. The length of accumulation in temporary facilities will be the decision of the customer where the higher the cost of warehouses outside compared to the rate of accumulation in temporary facilities, the longer the accumulation time will be. The length of container stacking will directly affect the capacity of containers that can be accommodated at container terminals. If the national

IOP Conf. Series: Materials Science and Engineering **703** (2019) 012006 doi:10.1088/1757-899X/703/1/012006

container flow is in peak season, then there will certainly be congestion in the container terminal area. Congestion will have an impact on container terminal utilities and waiting times on ships that will dock.

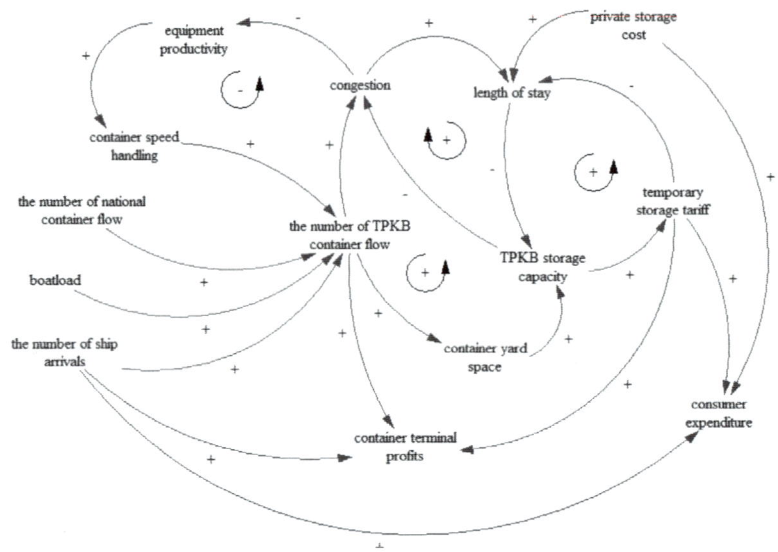

Figure 3. Causal Loop Diagram of Container Yard

In this paper, the strategy for determining stacking rates will provide additional benefits to the temporary container storage activities for container terminal management. In addition, the number of ship arrivals and incoming container flows will also add to the benefits of container terminal management. On the other hand, customers will consider the difference between the container stacking rates at container terminal facilities and the cost of storing containers in their own warehouses.

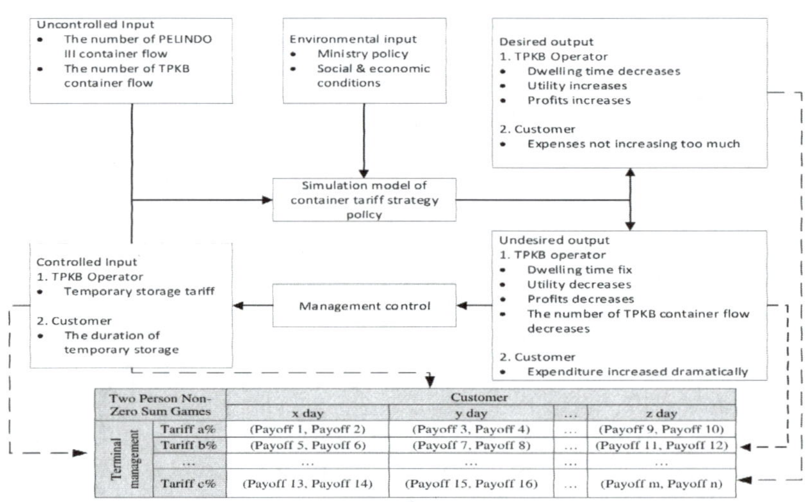

Figure 4. Input-output diagram relation to payoff matrix

In Figure 4.Input-output diagram relation to payoff matrix, policy analysis regarding storage tariff with a system dynamic approach. With the simulation of the container tariff strategy model the strategy is expected to result in a decrease dwelling time, increased utility and increased profits.

5. Future Work

Validation of the causal loop model is currently underway with a case study of the Banjarmasin container terminal. The purpose of the validation of the causal loop model is to ensure all factors and valid and reasonable factor relationships. The model of the causal loop will be updated and modified based on recommendations and feedback received from the results of validation. In the end, the researcher will help decision-makers to understand the advantages and expenses of the game results related to the increase in basic rates for container storage, dwelling time and container terminal utilities.

The advantage of this model is the model can capture the effect of increasing tariff on equipment utilities, container terminal revenue and the flow of goods in the container terminal. In previous studies only considered the effect of increasing rates on dwelling time. While the lack of this model is the model has not considered inflation and the value of profits from the flow of goods in the container terminal.

6. Conclusion

Validation of the causal loop model is currently underway with a case study of the Banjarmasin container terminal. The model of the causal loop will be updated and modified based on recommendations and feedback received from the results of the validation. In this case, the results of the validated model will be used to simulate the scenarios that are in the two-person non-zero sum game. In the end, the researcher will help decision-makers to understand the advantages and expenses of the game results related to the increase in basic rates for container storage, dwelling time and container terminal utilities. So, container terminal management can have strategic considerations before making final managerial decisions.

References

[1] Martín E, Salvador J and Saurí S 2014 Storage pricing strategies for import container terminals under stochastic conditions *Transp. Res. E Logist. Transp. Rev.* **68** 118–37
[2] Chung K 1993 Port Performance Indicators *Transp. Water Urban Dev. Dep. World Bank* 1–5
[3] De Castilho B and Daganzo C 1991 Optimal pricing policies for temporary storage at ports
[4] Kim K and Kim K 2007 Optimal price schedules for storage of inbound containers *Transp. Res. B Methodol.* **41(8)** 892–905
[5] Qiu X and Lam J 2014 Optimal storage pricing and pickup scheduling for inbound containers in a dry port system *Conf. Proc.-IEEE Int. Conf. Syst. Man. Cybern.* 2959–64
[6] Sterman J 2002 All models are wrong: reflections on becoming a systems scientist *Syst. Dyn. Rev.* **18(4)** 501–31
[7] Morgenstern O and Von N J 1953 *Theory of games and economic behavior* (Princeton university press)
[8] Roger B 1991 *Game theory: analysis of conflict* (Pres Fellows Harvard Coll USA)
[9] Ferguson T 2014 Games in coalitional form In: Game Theory (UCLA)
[10] Dutta P 1999 Strategies and games: theory and practice (MIT press)
[11] Forrester J 1989 The Beginning of System Dynamics. In: Banquet Talk at the international meeting of the System Dynamics Society Stuttgart, Germany p. 1–16
[12] Liuguo S, Shijing Z, Jianbai H. Pricing Simulation Platform Based on System Dynamics. Syst Eng Procedia [Internet]. 2012;5(932):445–53. Available from: http://dx.doi.org/10.1016/j.sepro.2012.04.067
[13] Zhang X, Liang Y, Liu W. Pricing model for the charging of electric vehicles based on system dynamics in Beijing. Energy [Internet]. 2017;119:218–34. Available from: http://dx.doi.org/10.1016/j.energy.2016.12.057
[14] UNCTAD. Port Management: Linking Performance Indicators to Strategic Objectives. United Nations Conf Trade Dev [Internet]. 2016;15(9):46. Available from: http://search.ebscohost.com/login.aspx?direct=true&db=bth&AN=19801353&site=bsi-live

IOP Conf. Series: Materials Science and Engineering **703** (2019) 012007 doi:10.1088/1757-899X/703/1/012007

Effect of NR-*g*-cellulose coupling agent into NR-cellulose composite dispersibility and its physical properties

H Handayani[1], A Cifriadi[1], A S Handayani[2], M Chalid[3], S Savetlana[4], and M Christwardana[2]

[1] Indonesian Rubber Research Institute, Salak Street No. 1, Bogor, 16151, Indonesia
[2] Department of Technical Chemistry, Institut Teknologi Indonesia, Puspiptek Serpong, Tangerang, 15320, Indonesia
[3] Department of Material Science, Universitas Indonesia, Depok, 16424, Indonesia
[4] Department of Technical Mechanic, University of Lampung, Bandar Lampung, 35145, Indonesia

E-mail: hani.ppkbogor@gmail.com

Abstract. Natural rubber and cellulose are two materials with very different compatibility, so it is difficult to produce homogenous composite without using coupling agent. Natural rubber is nonpolar while cellulose is polar. The combination of natural rubber with cellulose is expected to produce composite with higher strength. The reinforcement concept of polymeric materials, such as natural rubber with cellulose as filler, results from increased rubber-filler interactions. This study aims to determine the effect of NR-***g***-Cellulose as coupling agent on the dispersibility of the composite and its physical properties. Hexamine/resorcinol and silane were used as a reference for a commercial coupling agent. Cellulose material was mixed in two-roll open mill together with natural rubber and other chemicals using three types of coupling agents, hexamine/resorcinol, silane, and NR-***g***-Cellulose. The FESEM result of the surface of composite indicated that composite with NR-***g***-Cellulose as coupling agent showed homogenously dispersed compared to two other types of coupling agent but it is lower in physical properties, especially in tensile strength, tear strength, compression set, and rebound resilience of the composite. It may be caused by the NR-***g***-Cellulose coupling agent not working optimally and therefore it needs a high temperature to optimize the coupling agent to react with natural rubber and cellulose.

1. Introduction

Composites, the wonder materials are becoming an essential part of today's materials due to the several advantages such as low weight, corrosion resistance, high fatigue strength, and faster assembly [1]. Polymer composites are the name given to composite materials in which different polymers (binders) are used as the matrix; here, the second strengthening component (the filler) can be of any nature, including polymeric. The components of composite materials possess a clear phase boundary, while the properties of composite materials are formed by the volume combination of components and differ considerably from the properties of the individual components [2].

Natural rubber (NR) contains the high molecular weight, which is based on cis-1,4-polyisoprene. It has excellent mechanical and physical properties such as tensile strength, tensile modulus, tear strength, and hardness. However, NR has poor chemical resistance and processing ability. The combination of its unique morphology, physical properties, cost-effectiveness, and environmental friendliness make natural rubber an appealing constituent for many materials and applications [3].

Natural fibres (NF) have been studied as reinforcement in composites due to its availability, environmentally benign, and renewable [4]. One of NF that widely applied in many applications is palm oil Empty Fruit Bunch (EFB). Empty Fruit Bunch (EFB) is a biomass waste produced by oil palm industries (about 25-26%). The empty fruit bunch of palm oil contains cellulose up to 36.67% [5]. The cellulose could be extracted using the pre-treatment process. The purified cellulose would then undergo acid hydrolysis to break the cellulose structure into nano-sized. Nano-sized cellulose that has been formed is fibrous and nanofibrillated [6]. The nanosize of the cellulose (Cellulose Nano Fibre, CNF) provides excellent properties such as high crystalline content, good optical properties, and excellent mechanical properties [7].

In composite preparation, natural rubber (NR) is a material commonly used as matrix [8]. The mechanical properties of bio filler filled rubbers depend on several factors of which most important ones are dispersion level, interfacial strength, and type of filler used [9]. The combination of NR and cellulose is good to improve tear strength. The reinforcement concept of polymeric materials, such as natural rubber with cellulose as filler, result from increased rubber-filler interactions [10]. To improve interfacial bonding and mechanical properties, compatibilizer was commonly added [5].

NR and MFC cannot disperse well because of the differences in polarity. To disperse better, an appropriate coupling agent must be used. One kind compatibilizer or coupling agent for natural rubber (NR)-cellulose composite is synthesized by grafting copolymerization between cellulose and natural rubber resulting in NR-***g***-cellulose; this is the novelty of this research. This research aims to study the effect of its synthesized coupling agent and two commercial coupling agents as comparison into NR-cellulose composite dispersibility and its physical properties. The composite was produced using two roll open mill. The composite samples were vulcanized, and its physical properties were analyzed. The effects of coupling agent on sample morphology were observed using Scanning Electron Microscopy (SEM).

2. Materials and Methods

2.1 Materials

The materials used in this study include Natural rubber (NR), *Standard Indonesia Rubber* (SIR) 20 obtained from local suppliers, and also technical quality rubber compounding chemicals from local suppliers. The rubber compounding chemicals include Sulphur as vulcanization agent, stearic acid and zinc oxide as an activator, hexamine/resorcinol and silane Si-69 as commercial coupling agent, and TBBS as an accelerator.

2.2 Preparation of microfibrillated cellulose (MFC)

Microfibrillated Cellulose (MFC) and coupling agent NR-**g**-MFC were produced by ITI. The initial microcrystalline cellulose (MCC) sample was prepared from oil palm Empty Fruit Bunches (EFB) by grinding and crushing to resulting powder form. The MFC was obtained chemically from MCC by delignification process using sodium hydroxide with high-pressure and bleaching using magnesium oxide and hydrogen peroxide. The dry nano-powder was obtained by being evaporated in vacuum at 80 °C and finally dried at 50 °C up to constant weight [11]. The characterization of MFC was analyzed using Particle Size Analyzer (PSA), Scanning Electron Microscopy (SEM), and Simultaneous Thermal Analysis (STA).

*2.3 Preparation of NR-**g**-MFC*

The coupling agent between natural rubber and MFC was synthesized via copolymerization reaction by Atomic Transfer Radical Polymerization (ATRP) method using macroinitiator MCF-Bromide in an autoclave [12]. The characterization of the copolymer was analyzed using Fourier Transform Infrared Spectroscopy (FTIR).

2.4 Synthesized of natural rubber/MFC composite
Natural rubber and MFC are mixed with other ingredients in the two-roll open mill according to the compound formula. The addition of chemicals uses a unit of phr (per hundred rubbers). This is used for calculating formulas where other chemicals are calculated as parts per 100 weight of rubber. Compound was formulated with variations in 3 types of coupling agent, as seen in Table 1. Natural rubber/cellulose composites were made without coupling agent as a control. The compounds were then tested for physical properties.

Table 1. The formula of natural rubber/cellulose composite with varied kinds of coupling agents

Materials	Dosage (per hundred rubber, phr)			
	A	B	C	D
NR	100	100	100	100
MFC	25	25	25	25
NR-*g*-MFC	-	-	3	-
Silane Si-69 (*bis-triethoxy silyl propyl-tetrasulfide)*	1.5	3	-	-
Hexamine (hexamethylenetetramine)	1.5	-	-	-
Resorcinol (1,3-isomer from benzenediol)	1.5	-	-	-
Zinc Oxide	5	5	5	5
Stearic Acid	2	2	2	2
TBBS (N- tert-butyl-benzothiazole sulphonamide)	0.7	0.7	0.7	0.7
Sulphur	2.25	2.25	2.25	2.25

3. Results and Discussion

3.1 Preparation of microfibrillated cellulose (MFC)
The morphology of the MFC structure was analyzed using Scanning Electron Microscopy (SEM); the results showed a cellulose structure in the form of fibrils with a relatively wide diameter (Figure 1). Based on the PSA analysis, MFC has polydispersity 2.174 with size 4.625.5 ± 18.862.9 nm. These results indicate that the synthesized MFC still has various particle sizes that need to be refined and homogenized again to reach smaller size particles. Meanwhile, the thermal properties of MFC based on the STA diagram analysis (Figure 2) show good thermal resistance as evidenced by high degradation temperatures (> 300 ° C).

*3.2 Preparation of NR-**g**-MFC*
The NR-**g**-MFC that were successfully synthesized can be proven through FTIR analysis. Figure 3 shows the results of the FTIR spectrum of MFC before and after the synthesis of MFC graft copolymers with natural rubber. Based on Figure 3, it can be seen that the aromatic ring at wave number 1593.12 has disappeared which indicates that the aromatic ring breaks up and binds to the double bond carbon group (C = C) and with the side groups of natural rubber form a double bond and carbonyl group. A Double bond carbon and carbonyl group form can be seen at wavenumbers 1661.77 and 1737.47 which are the characteristic for the absorption of carbon groups C = C and C = O.

3.3 Synthesized of natural rubber/MFC composite
An amount of four NR/MFC composite formulas according to Table 1 have been made compound and cured on a press machine to produce vulcanized rubber as can be seen in Figure 4. Visually based on Figure 4, it can be seen that mixing of natural rubber with MFC and other materials in the two rolls open mill did not result in homogeneous dispersion, the particles of cellulose are still visible on the surface, not well dispersed with natural rubber. This is evidenced by the results of SEM analysis for three composites A, B, and C while SEM results for D composites are still tested (Figure 5).

From the results of SEM analysis, it can be seen that the mixing process has not produced a composite with uniform dispersion. In composite A, cellulose particles still appear on the surface, which is characterized by the appearance of white spots. This indicates that the MFC particles have not completely filled in the natural rubber matrix. In composite B, white spots appear more than composite A, while in composite C, white spots look less than composites A and B. This indicates that the mixing of materials in composite C using NR-*g*-MFC as coupling agent is relatively more homogeneous compared to the other two composites.

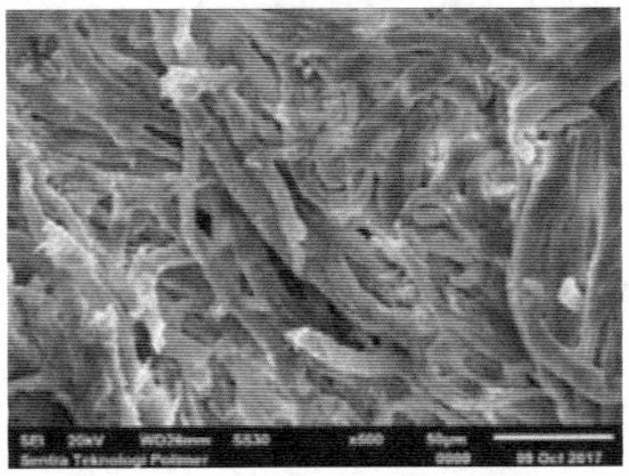

Figure 1. SEM micrograph of microfibrillated cellulose (MFC)

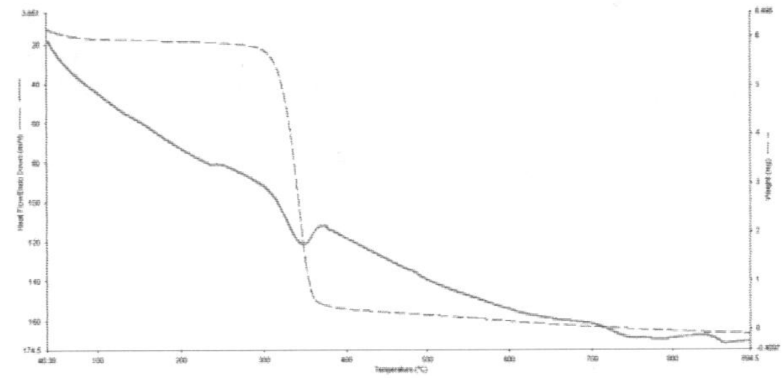

Figure 2. Thermal analysis of microfibrillated cellulose (MFC)

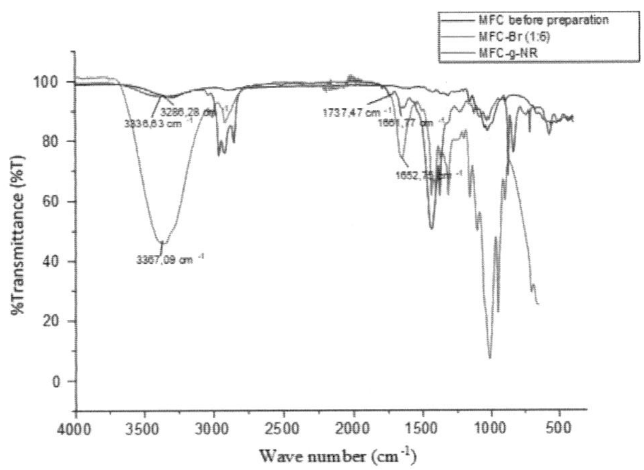

Figure 3. FTIR spectra of MFC before and after synthesis of graft copolymers with natural rubber

3.4 Physical properties of the composite.

The addition coupling agent into natural rubber/MFC composites affected the physical and mechanical properties of the composites. The results of physical properties testing for all vulcanized composites can be seen in Table 2.

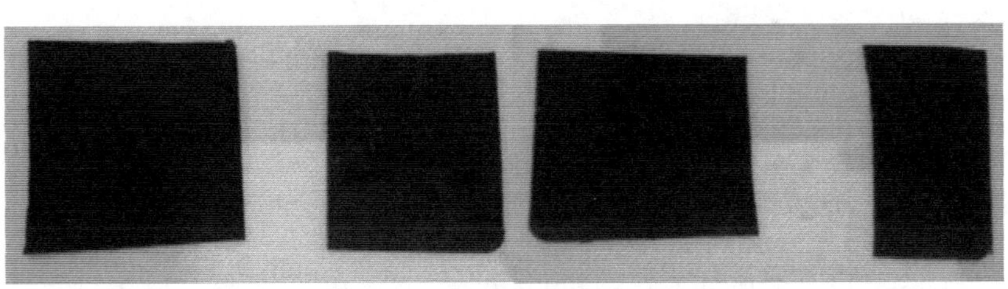

Figure 4. Visualization of composite A-D (from left to right)

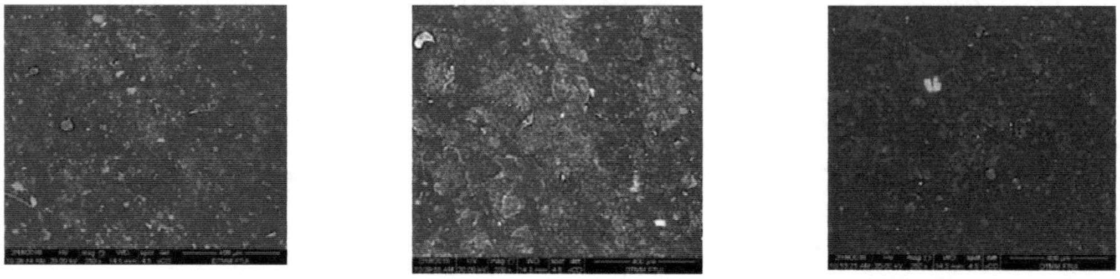

Figure 5. SEM micrograph of composite A, B, and C (from left to right)

Table 2. Physical properties of natural rubber/MFC composite with varied kinds of coupling agent

Physical Properties	Vulcanized			
	A	B	C	D
Hardness (Shore A)	75	67	65	63
Tensile strength (MPa)	4.4	4.3	2.4	5.1
Elongation at break (%)	350	520	420	560
Tear strength (MPa)	35.8	27.3	22.6	27
Compression set, temp. 27±2 °C (%)	9.89	12.64	43.23	13.64
Compression set, temp. -5±2 °C (%)	2.61	5.08	23.14	7.52
Rebound resilience (%)	69	58	51	64

3.5 Effect on hardness

The addition of combinations coupling agent, hexamine/resorcinol and silane, increases the hardness up to 12 shores A compared to D composites, which did not use coupling agent. Meanwhile, the addition of a single coupling agent, silane, increases the hardness by 4 shore A and the addition of NR-***g***-MFC increases the hardness by 2 shore A. This can be caused by the coupling agent that increases the interaction of natural rubber-filler (NR-MFC interaction) so that the composite structure becomes compact and solid as a result of increased composite hardness.

3.6 Effect on tensile strength

Composites without coupling agent (composite D) provide better tensile strength compared to other composites that using coupling agent. This is probably due to the fact that the coupling agent used is not suitable with the NR/MFC composite so that the addition of the material weakens the NR-MFC interaction which result to the decrease of the elasticity of the NR/MFC composite.

3.7 Effect on elongation at break

Composites without coupling agent (composite D) provide better elongation at break than other composites that using coupling agent. The decrease in elongation at break can be caused by the function of coupling agent to increase NR-MFC interaction so that the intermolecular interactions of natural rubber decrease. It caused decreasing in the elasticity of natural rubber, so the composites with coupling agent become easier to break than composite without coupling agent.

3.8 Effect on tear strength

Composite A, which uses a combination coupling agent, hexamine/resorcinol and silane, gives better tear strength than other composites. The increase in tear strength can be caused by the addition of coupling agent to strengthen the NR-MFC interaction, thus increasing the strength of the NR/MFC composite. Composite B, which uses a single coupling agent, silane, has better tear strength compared to composite D (without coupling agent). However, the tear strength of composite C using NR-***g***-MFC as coupling agent is lower than other composites. It can be caused that NR-***g***-MFC is not suitable with NR/MFC composites so that the addition of material weakens the NR-MFC interaction resulting to the decrease of the strength of NR/MFC composite.

3.9 Effect on compression set

Composite A, which uses a combination coupling agent, hexamine/resorcinol and silane, provides better compression set than other composites. The greater compression set value shows that after compressed, rubber elasticity is decreased [13]. Composite B, which uses a single coupling agent, silane, provides better compression set than composite D (without coupling agent). Whereas composite C, which uses NR-***g***-MFC as coupling agent, has the worst compression set than other composites. This can be caused the NR-***g***-MFC is not suitable with NR/MFC composites so that the filler-filler interaction of the MFC used is greater than NR-MFC interaction. This shows that the intermolecular interaction of MFC is stronger than the NR-MFC interaction so that when the composite is compressed in high pressure at certain temperature and time, its elasticity will drastically decrease.

3.10 Effect on rebound resilience

Composite C, which uses NR-***g***-MFC as coupling agent, provides better rebound resilience than other composites. This can be caused by the addition of NR-***g***-MFC increasing intermolecular interactions of natural rubber so that the elasticity of composites is better than other composites. The lower value of rebound resilience indicates that the composite is more resistant to impact.

4. Conclusions

The uses of NR-***g***-MFC as a coupling agent for NR/MFC composite could increase the dispersibility of MFC into natural rubber matrix in process compounding. On the other hand, the addition of NR-***g***-MFC decreases almost all physical properties except rebound resilience. To optimize the function of NR-***g***-MFC as coupling agent, the temperature of mixing should be increased so the reaction of the coupling agent with natural rubber and MFC will be optimum.

Acknowledgment

This work was carried out with the financial support of Ministry of Research, Technology, and Higher Education of the Republic of Indonesia through The National Strategic Research Consortium program with contract numbers: 34 / KP / LPKT-ITI / III / 2018.

References

[1] Jasmin P J, Sant K M, Sabu T, Kuruvilla J, Koichi G, and Meyyarappallil S S 2012 *Polymer Composites* (Germany: Wiley-VCH)

[2] Yu A M, Kerber M L, and Gorbunova I Y 2002 *International Polymer Science and Technology*

[3] Chin H C, Jithin J, Hanna J M, and Sabu T 2013 *Natural Rubber Materials* (Cambridge: Royal Society of Chemistry)

[4] Ujianto O, Noviyanti R, Wijaya R, and Ramadhoni B 2017 *Proc. Int. Conf. Series: Mater. Sci. and Eng.* 223 (Indonesia: IPST 2016)

[5] Dwi S, Uju, Neli M, Isroi, Nyoman B S, and Ami A N 2018 *Proc. Int. Conf. Series: Earth and Environ. Sci.* 141 (Indonesia: Bogor)

[6] Mohd H S, Eraricar S, Siti N H M, and Ida I M 2013 *Int. Tech. Symp. Adv. Food Process. Qual. Manage.* (China: Guangzhou)

[7] Abraham E, Deepa B, Pothan L A, Jhon M, Narine S S, Thomas S, and Anandjiwala R 2012 *Cellul.* **20** 417-27.

[8] Geethamma V G 2005 *Composites, Part A* **36** 1499-506.

[9] Mohd K M and Ruslimie C A 2016 *Adv. Mater. Res.* **1133** 593-7.

[10] Hani H, Adi C, Aniek S H, Mochammad C, Riana H, and Shirley S 2018 *J. Penelit. Karet* **36** 173-82.

[11] Michael I 2012 *J. Nanosci. Nanotechnol.* **2** 9-13.

[12] Mohammad C, Aniek S H, and Emil B 2014 *Adv. Mater. Res.* **1051** 90-4.

[13] Hani H, Mohammad I F, and Indri K 2011 *J. Penelit. Karet* **29** 49-62.

IOP Conf. Series: Materials Science and Engineering **703** (2019) 012008 doi:10.1088/1757-899X/703/1/012008

Formulation and characterization of chitosan-alginate freeze dried matrices loaded with oleoresin extract of red ginger

E A Krisanti, A Safiya, and K Mulia

Chemical Engineering Department, Faculty of Engineering, Universitas Indonesia, Depok 16424, Indonesia

E-mail: elsakm@che.ui.ac.id

Abstract. A simple method of encapsulation of bioactive compounds that does not involve washing, heating, and degradation of active substance is desired. The encapsulation of red ginger oleoresin in the chitosan-alginate matrix was developed using the freeze drying method to achieve higher matrix yield and loading capacity in the matrix. Bioactive substances that are encapsulated in the chitosan-alginate matrix aim to be released in the colon. Oleoresin red ginger extract was obtained from maceration of dried rhizome powder in ethanol. A mixed solution consisting of chitosan, oleoresin extract, Tween 80 as a surfactant and alginate was lyophilized at -50 C at a vacuum pressure of about 20 atm for 48-72 hours. The X-ray diffraction pattern of the formed matrix shows that the crystalline chitosan and alginate peaks disappear in the chitosan-alginate matrix which shows compatibility. In the Infra Red matrix spectrum the interaction between chitosan and alginate in the matrix is shown by the change of peak height at 1520 and 1385 cm^{-1}. Release assay showed that the matrix with a weight ratio of chitosan: alginate of 1: 0.5 slightly released phenolic compounds from oleoresin in the simulated gastric fluid but highly released in the simulated intestinal and colon fluids.

1. Introduction

Zingiberofficinale, generally known as ginger, originated from South-East Asian and introduced to many parts of the globe, has been cultivated for thousands of years as a spice and also for medicinal purposes. The rhizome of this plant has been used as a medicine in Asian, Indian, and Arabic herbal traditions since ancient times also as herbal medicine for the treatment of arthritis, rheumatological conditions and muscular discomfort [1,2]. Some phenolic substances present in ginger, in general, have strong anti-inflammatory and anti-oxidative properties. It also exerts substantial anti-carcinogenic and anti-mutagenic activities. It has been found that ginger supplementation suppresses colon carcinogenesis in the presence of procarcinogen [3,4].

Encapsulation of active compounds in matrices or microparticles with various polymers has been widely known for its use in protecting and acting as carriers of drug or active substance. Chitosan is a common polymer used and known as a biodegradable, biocompatible and non-toxic biopolymer [5] which has a good mucoadhesivity [6]. Alginate is another common biopolymer in oral drug delivery known for its stability in acidic conditions of the stomach [7]. The combination of chitosan and alginate have been previously used in oral controlled drug delivery formulations [8,9].In this research, the active compounds of ginger oleoresin were loaded into chitosan-alginate matrices using the freeze drying or also called lyophilization method, which converts a solution into a solid by taking out water molecules through the process of sublimation and desorption in vacuum condition [10]. The concentration of alginate and oleoresin content in the matrices were varied. The matrices had been evaluated for its release characteristics in the *in vitro* assay using simulated gastrointestinal fluids. The total phenolic content, matrix yield, loading capacity and the morphology of matrices were also evaluated and reported.

2. Materials and methods

2.1. Materials

Fresh red ginger (*Zingiberofficinale var. Rubrum*) was purchased from a local market in East Jakarta, Indonesia. Pure grade of ethanol 96%, 2,2-Diphenyl-1-picrylhydrazyl (DPPH), acetic acid, Folin-ciocalteu's (FC) reagent, gallic acid, Tween 80, Sodium carbonate (Na_2CO_3), hydrochloric acid (HCl), potassium chloride (KCl), monopotassium phosphate (KH_2PO_4), and sodium hydroxide (NaOH) were procured from Sigma-Aldrich, Inc. and Merck.The chitosan of MW of 80-120 kDa, particle size of 200-300 mesh, viscosity of 100-200 mPas, was purchased from Chemultiguna, Indramayu, Indonesia.

2.2. Formulation of chitosan-alginate freeze-dried matrices

The red ginger oleoresin extract were resulted from maceration of dry rhizome powder in ethanol 96% for seven days using similar procedure as reported on other publication[11]. The chitosan-alginate matrices were prepared with six variations, i.e. one blank matrix without oleoresin extract and five matrices with variation of alginate and extract composition. All of the matrices were prepared using the freeze drying method reported previously with some modification [12]. A certain volume of chitosan solution (0.1g/50 mL acetic acid 2.5%)was mixed with oleoresin extract. The mixture was stirred using magnetic stirrer (IKA C-Mag HS 7) at 1500 rpm for 10 minutes to form a homogeneous emulsion. For stabilizing the emulsion, 1 mL of 2 % Tween 80 solution was added drop by drop and emulsion was stirred at1000 rpm for 5 minutes. The alginate aqueous solution of 0.8% (g/mL) was added into the chitosan mixture emulsion and stirred at 1500 rpm for 20 minutes to form a homogeneous emulsion. The emulsion was lyophilized using EYELA FDU-2100 freeze dryer, at -50°C and at vacuum pressure about 20 atm for 48h-72 h, depending on the volume of solution. The formed matrices were grinded into powder (AISHUKA KS-168) and sieved with a 45 mesh metal filter to get particles with size <100 μm.

2.3. Matrix morphology test

Powder X-Ray Diffraction (XRD) of chitosan, alginate, and all samples of freeze dried matrices were obtained using SHIMADZU XRD 7000 MAXIMA-X with Cu tube. All samples were analyzed between 2θ angles of 10° and 80° using the voltage of 40 kV, current of 30 mA, and time per step of 1 second. The crystalline or amorphous character of chitosan, alginate, and obtained matrices were determined through the diffractogram. The Fourier Transform Infra-Red (FTIR) analysis was performed (THERMO SCIENTIFIC NICOLET iS5Spectrophotometer with optic iD5 ATS) using wavelength ranging from 4000 to 400 cm^{-1}. The FTIR analyzed the morphology of matrix, that was based on the absorbance peaks of functional groups vibrations present in oleoresin-chitosan-alginate matrices.

2.4. Total yield and loading capacity of matrices

The yield and loading capacity of the matrices were quantified based on the mass of dried matrices and the oleoresin extract used using the following formulas:

$$\text{Yield (\%)} = \frac{\text{Mass of dried matrix (mg)}}{\text{Total mass of materials used for matrix (mg)}} \times 100\% \quad (1)$$

$$\text{Loading Capacity (\%)} = \frac{\text{Mass of oleoresin used in matrix (mg)}}{\text{Total mass of matrix (mg)}} \times 100\% \quad (2)$$

2.5. Total Phenolic Content determination

The total phenolic content (TPC) was determined using spectrochemistry analysis and Folin-ciocalteu's (FC) reagent, the procedure was described in previous publication[11]. The absorbance of sample with FC reagent was measured at 765 nm using a UV/VIS-spectrophotometer of Spectroquant® Pharo 300. Phenolic content in samples was expressed as gallic acid equivalent (mg GAE/g sample) and analyzed triplicate.

2.6. In-vitro drug release test in simulated gastrointestinal fluids

The release profile of phenolic compound from the oleoresin-loaded chitosan-alginate matrices was obtained from the sequential series release testing using dialysis membrane tube (Pur-A-LyzerMega dialysis kit Mega 12000, PURG12020, Sigma-Aldrich) in simulated gastrointestinal fluids. Three release media were used: *simulated gastric fluid* (SGF; pH 1.2), *simulated intestinal fluid* (SIF; pH 7.4) and *simulated colonic fluid* (SCF; pH 6.8). The procedure for the preparation of simulated gastrointestinal fluids is as described in a previous publication[5]. 20 mg of matrix particles were immersed in 60 mL of simulated fluids and incubated at 37°C. A sample volume of 5 mL of each fluid media was taken periodically in 2, 4, 6, 8, and 24 hours, to analyze the total phenolic content.

3. Result and discussion

3.1. Formulation of matrices, yield, and loading capacity

Six different matrices have been made. All compositions contain a fixed amount of chitosan, but the ratio of alginate to chitosan and the ratio of oleoresin to chitosan vary from 0.1: 1 to 1: 1. Data obtained from the matrix are shown in table 1. The matrix yields are quite high, with values ranging from 81 % to 96%, which is calculated using equation (1). The matrix yield using the freeze drying method is higher than that formed using other techniques: 29% - 43% are reported for the spray drying method [13] and 34% - 76% are reported for the ionotropic gelation method [14]. This shows that the freeze drying method effectively form chitosan-alginate matrix.

Table 1. Matrices formulation, Yield and Loading Capacity

Matrix	Chitosan:Alginate:Oleoresin (w)	Matrix Yield (%)	Loading Capacity (%)
A1E0	1:0.1:0	91.91	0
A1E1	1:0.1:0.1	86.59	6.54
A5E1	1:0.5:0.1	81.51	5.33
A10E1	1:1:0.1	92.65	3.50
A5E3	1:0.5:0.3	95.76	11.59
A5E5	1:0.5:0.5	94.01	17.79

As predicted, the loading capacity, which is calculated using equation (2), increases with the decreasing of alginate amount at a constant amount of oleoresin extract, as seen in matrices of A1E1, A5E1, and A10E1. The higher alginate content, the more matrix structure is formed, hence the loading capacity decreases. This result is in agreement with the result reported by Soliman et al. [15]

3.2. Characteristics and morphology of matrices

Figure 1 shows the XRD patterns of pure alginate, chitosan and chitosan-alginate matrices. The XRD pattern of chitosan showed two typical diffraction peaks at $2\theta= 15.45°$ and $24.49°$, whereas the for alginate there was only one peak at $2\theta= 20.47°$. These typical diffraction peaks of chitosan and alginate are known as semi-crystalline, which some of it are crystalline and some are amorphous [12]. The XRD pattern of the A5E1 and A5E3 matrices in figure 1 had no diffraction peaks. This indicates that chitosan and alginate in the matrices had high compatibility, hence the XRD pattern of matrices had a weak broad profile without peaks of individual component. The disappearance of crystalline peaks concluded that the freeze dried matrices formed had amorphous structures. This result is in agreement with the study reported by Wang et al. [16]. Cheng et al.explained the decreased of crystallinity in the matrix chitosan-alginate was attributed to the broken of hydrogen bonding between amino groups and hydroxyl groups in chitosan. When the chitosan-alginate matrixformed, the amino groups of chitosan and hydroxyl groups of alginate form hydrogen bonds, that results in an amorphous structure of matrices [17].

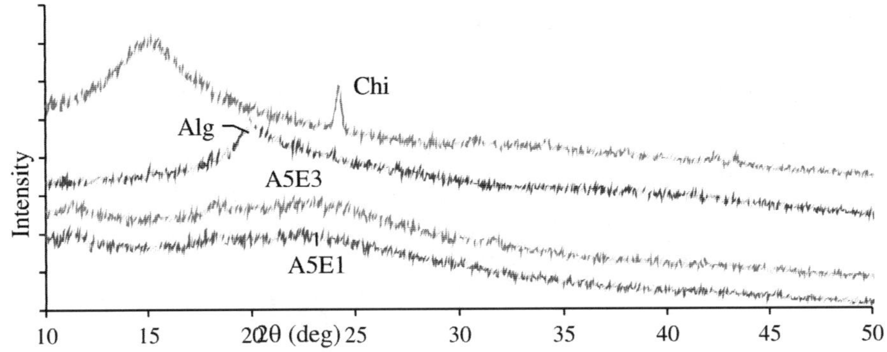

Figure 1. XRD graphs of chitosan (Chi), alginate (Alg), and matrices of A5E1 and A5E3 (formulation is shown in table 1).

The results of samples analysis using Fourier Transform Infra-Red spectroscopy or FTIR is shown in figure 2. Matrices of A5E5, A5D3 and A5E1 have the same alginate and different oleoresin, while A1E1, A5E1, and A10E1 have the same oleoresin and the different amount of alginate. The specific broad peaks of chitosan-alginate matrix are on the wavenumber ranged 3244-3433 cm^{-1} corresponding to the stretching vibration peaks of hydroxyl groups (O-H) and amine (N-H). The peaks at 1520 and 1385 cm^{-1} are belong to stretching of carbonyl (C=O) group and bending of secondary amide (-N-H), respectively, as the consequence of interaction between chitosan and alginate. The peak at 1107 cm^{-1} is due to the stretching vibration of -C-O-C groups in chitosan and alginate. The vibration of hydroxyl (OH) phenolic groups in oleoresin is shown in peak at wavenumber of 595 cm^{-1}. Similar peaks of absorption of chitosan-alginate matrices and ginger oleoresin are reported by other researchers [8,18,19].

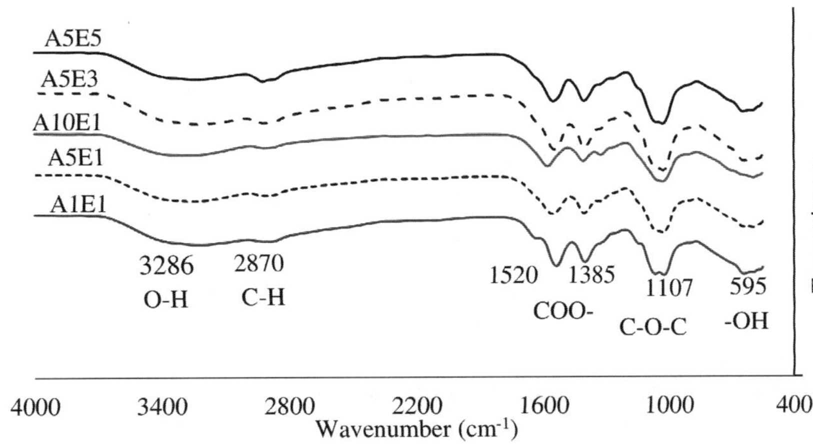

Figure 2. FTIR spectra of matrices (formulation is shown in Table 1)

The typical vibration peaks at 1520 cm^{-1} and 1385 cm^{-1} in each matrix sample confirm the interaction between chitosan and alginate that affect the stretching vibration of carbonyl group and the bending vibration of secondary amine groupof chitosan. The interaction between

However, when the alginate increased the height of both peaks reduced. This might corresponds with the interaction occurred in the matrices between chitosan and alginate, which encapsulate the ginger oleoresin

3.3. Total phenolic content and invitro release study

The total phenolic content (TPC) of oleoresin extract was reported previously as 40.6 mg GAE/g dry dried ginger powder, while the antioxidant activity determined as the IC_{50} had a value of 17 ppm was reported previously[11]. The *in vitro* release profile of oleoresin out of the chitosan-alginate matrices are shown in figure 3.After two hours in buffer solution with pH 1.2 (SGF) the cumulative release of TPC was low. As reported by Wen et al. alginate in the matrix can form alginic acid at low pH condition, which trigger forming gel or hydrocolloid layer with high viscosity[20]. The hydrogel layer could prevent the degradation of chitosan matrices hence the active compounds, such as oleoresin, slightly release out of the matrices [5].

The release rateis faster in SIF of pH 7.4 and slower from 8 hour to 24 hour in SCF of pH 6.8. When the content of alginate is high, as in matrix A10E1 with weight ratio of chitosan to alginate 1:1, the alginate hydrogel in alkaline condition might swell that cause the polymer (matrix) surface becomes softer. This could be followed by water uptake by matrices that cause massive diffusion and high release of active compounds [9,21]. Zhang et al. also found out that in high pH solution the carboxyl groups of alginate are more ionized and the electrostatic repulsion between ionized groups increases swelling rate [22]. The release profile of matrices A10E1from 3 hours to 7 hours period was inline with the previous explanations. However, in long period from 8 hours until 24 hours the release rate was low, andreach cumulative release of 90% at 24 hours.

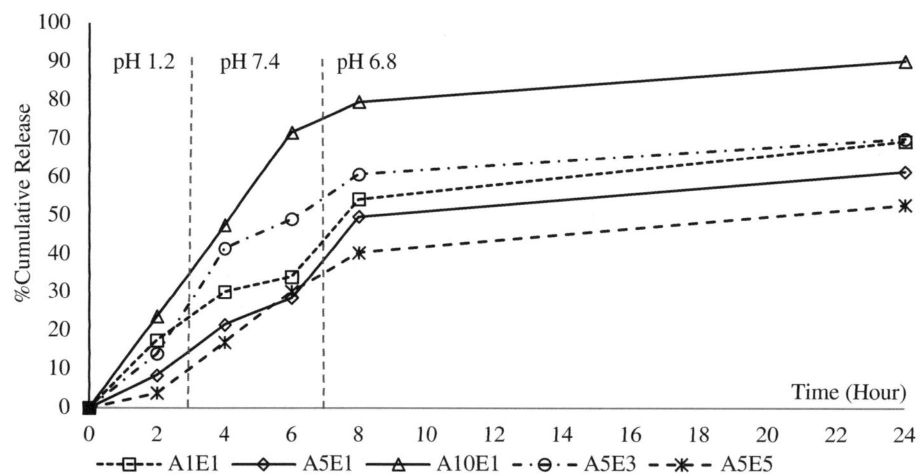

Figure 3. Oleoresin release profile from chitosan/alginate matrices. Effect of alginate and oleoresin in matrices of A1E1; A5E1; A10E1 in SGF, SIF and SCF.

The release profile shows that matrices formulations will affect the release characteristics of bioactive compounds in gastrointestinal fluids (digestive system). The chitosan-alginate matriceswith weight ratio of chitosan: alginate 1:0.5, that were formed using freeze-drying method, showed slightly release in gastric condition (pH 1.2) as low as 8.5 %. This formulation of matrices releasedthe phenolic compounds of oleoresin extract up to 61% in colonic condition (pH 6.8). The result indicates that chitosan-alginate matrices prepared by freeze-drying method, has the potential to be used as a carrier to deliver phenolic compounds of oleoresin extract to target areasin colon via oral administration.

4. Conclusion

The freeze-drying method successfully had been used to form chitosan-alginate matrices that encapsulated oleoresin extract with a relatively high yield (90%) and loading percentage (18%).

The XRD patterns of matrices showed that lyophilization process reduced the crystallinity of chitosan and alginate by forming the matrices with amorphous structures. The FTIR spectrum showed the specific absorbance peaks of various functional group for each matrix that support the presence of strong interaction between chitosan and alginate in the matrix, as it is shown by the absorbance peaks at 1520 cm^{-1} and 1385 cm^{-1}. The *in vitro* release profile of oleoresin from chitosan-alginate matrix showed the dependency of pH conditions. The presence of alginate in the matrix suppresses the release of oleoresin in acidic conditions but increase the release in higher pH conditions. Certain composition of matrix, such as weight ratio of chitosan to alginate of 1:0.5, resulted in low release of oleoresin in gastric condition (around 8%) and high release in neutral or high pH condition (around 61%). Therefore, the formulation of chitosan-alginate matrix can be designed to be used as a carrier for extended release of oleoresin extract to the sites in the colon via oral administration.

Acknowledgment

The authors are grateful for the financial support received from the Ministry of Research, Technology and Higher Education of the Republic Indonesia through the PDUPT Research Grant contract NKB-0747/UN2.R3.1/HKP.05.00/2019

References

[1] Tanaka K, Arita M, Sakurai H, Ono N and Tezuka Y 2015 *Biomed Res. Int.* **2015** 1-7
[2] Al-Nahain A, Jahan R and Rahmatullah M 2014 *Arthritis.* **2014** 1-8
[3] Wang C-Z, Qi L-W and Yuan C-S 2015 *Am. J. Chin. Med.* **43** 7 1-13
[4] Prasad S and Tyagi A K 2015 *Gastroenterol. Res. Pract.* **2015** 1-11
[5] Mulia K, Halimah N and Krisanti E 2017 *AIP Conf. Proc.* **1823** 020010
[6] Krisanti E, Aryani S D and Mulia K 2017 *AIP Conf. Proc.* **1817** (1) 020014
[7] Zhou M, Hu Q, Wang T, Xue J and Luo Y 2018 *Int. J. Biol. Macromol.* **120** 859-864
[8] Li P, Dai Y, Zhang J, Wang A and Wei Q 2008 *Int. J. Biomed. Sci.* **22** 1-8
[9] Siepmann J, Siegel R A and Rathbone M J 2012 *Fundamentals and Applications of Controlled Release Drug Delivery*. New York: Springer 107-229
[10] Sowasod N, Nakagawa K, Tanthapanichakoon W and Charinpanitkul T 2013 *Food Sci. Tech. Res.* **19** 633-640
[11] Mulia K, Risqi U Y, Pane I F and Krisanti E A 2019 *J. Phys. Conf. Ser.* **1198** 062008
[12] Ruiz-Caro R and Veiga-Ocha M D 2009 *Mol.* **14** 4370-86
[13] Mulia K, Putri T, Krisanti E A and Handayani N A 2019 *AIP Conf. Proc.* **2092** 030005
[14] Suskamran T, Opanasopit P, Rojanarata T and Ngawhirunpat T 2011 *Trop. J. Pharm. Res.* **10** (3) 317-324
[15] Soliman E A, El-Moghazy A Y, El-Din M S M and Massoud M A 2013 *J. Encapsulation Adsorpt. Sci.* **3** 48-55
[16] Wang G, Wang X and Huang L 2017 *Biotechnol. Biotechnol. Equip.* **31** 776-773
[17] Cheng M, Deng J, Yang F, Gong Y, Zhao N and Zhang X 2003 *Biomater.* **24** 2871-2880
[18] Jayanudin and Rochmadi 2017 *Int. J. Pharm. Sci.* **9** 29-34
[19] Purnomo H, Jaya F, and Widjanarko S B 2010 *Int. Food Res. J.* **17** 335-47
[20] Wen H, and Park K, (Eds.) 2011 *Oral controlled release formulation design and drug delivery: Theory to practice* John Wiley and Sons. Hoboken, Hoboken, New Jersey, USA
[21] Tonnesen H and Karlsen J 2002 *Drug. Dev. Ind. Pharm.* **28** 621-630
[22] Zhang H, Alsarra A and Neau S H 2002 *Int. J. Pharm.* **239** 197-205

IOP Conf. Series: Materials Science and Engineering **703** (2019) 012009 doi:10.1088/1757-899X/703/1/012009

The effects of electroculture on shoot proliferation of garlic (*Allium sativum l.*)

Von Louie R Manguiam[3], Ashley Marie N. Margate[1], Rose Danielle G Hilahan[1], Harold Gian L Lucin[1], Kristopher Ray S Pamintuan[2,3], and Adonis P Adornado[2,3]

[1] Mapúa Senior High School Department, Mapúa University, Manila 1002, Philippines
[2] School of Chemical, Biological, and Materials Engineering and Sciences, Mapúa University, Manila 1002, Philippines
[3] School of Graduate Studies, Mapúa University, Manila 1002, Philippines

E-mail: adonisadornado@yahoo.com

Abstract. The Philippines is an archipelago that always experiences yearly devastating typhoons which hinders continuous agricultural crop production. Using an old horticulture technique called electroculture, supplied electric voltage can stimulate plant growth, improve crop quality, and increase crop yields as long as the contributing factors are present – dormancy and essential nutrients. Using garlic (*Allium sativum* L.), three set-ups were prepared: control, 6 V, and 12 V electrocuted systems. Then, the effects of the supplied voltage in terms of plant height were visualized. The reaction of the garlic (*A. sativum* L.) to the application of voltage gave unfavorable results – the average height is relatively higher for the controlled system having 78.2 cm compared to the 6 V and 12 V set-ups of 55.7 cm and 57.1 cm, respectively. However, from the data of the 12th, 17th, 18th, 19th, 20th, 21st, 25th, and 26th day of the experimental period, the electricity applied somehow assisted the plant's growth. Drying and decrease in height were also visible during the 29th day of the experiment. Further experimentation and optimization can be done to substantiate the results of this present study.

1. Introduction

The Philippines is a tropical country where farming is the staple source of food with principal crops growing such as rice (*Oryza sativa*), corn (*Zea mays*), coconut (*Cocos nucifera*), banana (*Musa*), pineapple (*Ananas comosus*), coffee (*Coffea*), and mango (*Mangifera indica*) [1,2]. As one of the major exporters of agricultural products, the rate of production of each crop must be continuous. However, the country always experiences yearly devastating typhoons and natural calamities that hinders ceaseless production [3].

In order to compensate for this loss, electroculture, an overlooked planting method, can be a promising technique to use. Electroculture is a method of applying electric currents to a plant to assist its seed germination process [4]. The earliest experiments on the subject recorded appear to be those of Dr. Maimbray of Edinburgh in 1746 where he electrified two myrtle plants which in turn gave a positive result [5]. Based on the observations of Palafox (2013) [6], plants that grew beside power lines are healthy. In this context they concluded that the plants utilize the electric current produced by the power lines to speed up its growing process. They also considered the contribution of thunderstorm to the plant's germination and growth process, hence, gave a favorable result. Though, some researchers have found out that applying electrical current to a plant decreases its growth rate [7,8] while others stated that electric current application does not affect the plant's growth [9]. Previous experiments on electroculture use root crops and monocotyledon seed and showed satisfactory results.

Yet, the influence of electricity to garlic (*Allium sativum* L.) plant, which is considered to be a monocotyledon bulb, has never been studied. Garlic (*A. sativum* L.) is an important dry season crop that usually takes about seven to nine months for it to be ready for harvest [10] and the Philippines need about 46,000 metric tons to meet the local garlic (*A. sativum* L.) demand [11].

Thus, this study aims to determine the effects of applying varying electrical voltages (6 V and 12 V) for 30 days on shoot proliferation of garlic (*A. sativum* L.). The influence of electricity will only be based on the observable changes in the height of the garlic (*A. sativum* L.) shoots. The results may lead to an early emergence, a sign of higher produce yields, or a prolonged germination process that will hinder continual harvest. Nevertheless, these collections of data can contribute to the understanding of the electroculture technology in relation to garlic (*A. sativum* L.) and other crop farming. It can potentially alleviate the Philippines' dependence on importation and uplift the agricultural sector in terms of crop production.

2. Materials and methods

The garlic (*Allium sativum* L.) cloves bought in a local nursery was planted in plastic pots (5.4 in. x 7 in.) with loam soil in an upright position, 2 inches deep from the surface then covered with compost. Watering schedule was set every two days at 12:00:00 PM with 200 mL of water to maintain the dampness of the soil. Also, exposure to sunlight was maintained. A measuring stick was installed until outgrowths are visible.

Electrocution was performed with 6 V and 12 V from a nail buried 1 inch beside where the garlic (*A. sativum* L.) clove was planted attached to battery sources via alligator wire clip. The electricity applied was scheduled same time twice a day (morning and evening) for 5 minutes. A total of three set-ups were prepared – set-up A for the control, set-ups B and C with 6 V and 12 V applied, respectively. Only the favorable conditions for planting garlic (*A. sativum* L.) were maintained. All measurements performed in this study were carried out in three replicate runs and the mean values were reported.

3. Results and discussion

Figure 1 shows a month-long progress measurement of the garlic (*Allium sativum* L.) plant's average height, expressed in centimetrers(cm), for the controlled and electrocuted systems. The average weather conditions recorded during this specific experimental period were as follows: temperature of 27°C, 0.79% relative humidity, and atmospheric pressure of 1,009.50 mbar.

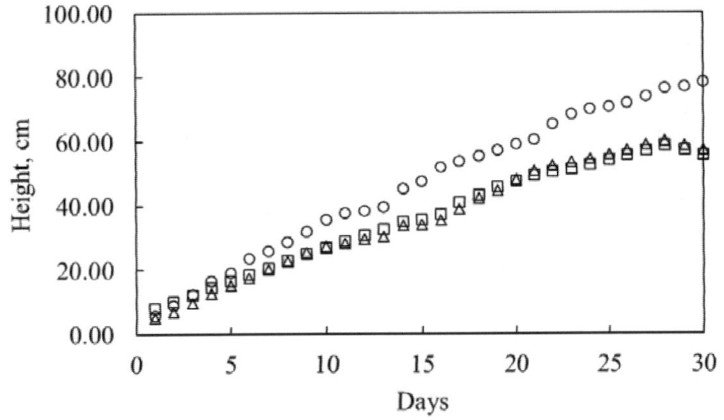

Figure 1. Garlic (*Allium sativum* L.) shoot's height daily progress – Control (○), 6 V (□), and 12 V (△).

As shown in figure 1, the garlic (*A. sativum* L.) plant's average height in the controlled system is relatively higher with value of 78.20 cm compared to the two electrocuted systems – 6 V and 12 V set-ups having only 55.70 cm and 57.10 cm, respectively. However, the tallest outgrowth first became visible under 6 V having a value of 8.10 cm. Based on the gathered data, application of electricity gave an unfavorable result. Uneven sprouting occurred during the period of the application of electricity. A continuous growth was visible until the 29th day where the outgrowth drying occurred for 6 V and 12 V electrocuted set-ups. These results are comparable to the study conducted by Murr (1964) [7] where leaf tip damage was evident from grain sorghum (*Sorghum bicolor*), orchard grass (*Dactylis glomerata*), yellow bush and wax beans (*Phaseolus vulgaris*), and corn (*Z. mays*). These phenomena were caused by the damages gained by the important cells for metabolism.

Nonetheless, looking at figure 2, based on the daily change in height of garlic (*A. sativum* L.) plant, data on the 12th, 17th, 18th, 19th, 20th, 21st, 25th, and 26th day of the experimentation somehow exhibited enhancement in the garlic (*A. sativum* L.) plant's shoot proliferation.

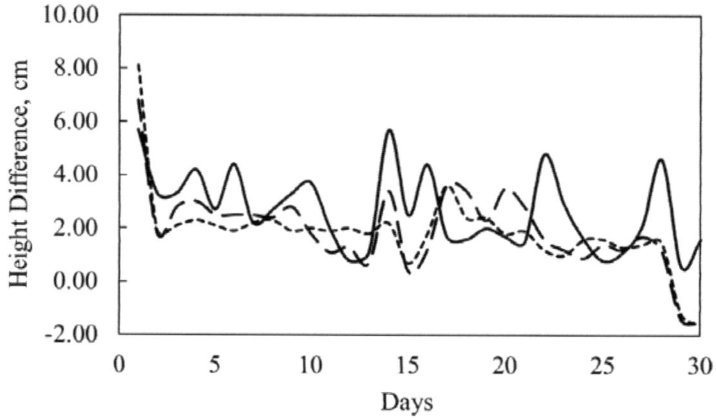

Figure 2. Garlic (*Allium sativum* L.) shoot's daily change in height – control (—), 6 V (− −), and 12 V (- - -).

As theorized by Dudgeon (1915), in the presence of electrical liberation, essential nutrients of the soil tend to be more soluble for easy assimilation, the formation of starch and sugar was increased, and all processes were accelerated, thus, an increase activity in the plant [12]. Somehow, the effects of adopting electroculture farming techniques will help to increase the yield unless the planted crop did not undergo dormancy period. According to a study conducted by Woldeyes (2016), three months of dormancy period must be maintained for the crop in order for it to adjust to its environment [13]. In this case, simultaneous application of electricity was performed during this period for visualizing the effects of an accelerating parameter to the plant's growth.

Nutrients, another essential factor for the plant's growth must be present in the soil. As reported by Goldsworthy (2006), cells only respond to artificial current if there are calcium ions available in the environment [14]. Calcium ions act as an accelerator together with the supplied electric current in a process of channel opening. Once a plant sensed a change in voltage, calcium channels (voltage gated) flip open to let calcium ions enter. Also, these ions follow electrical stimulation. The connection of the two parameters to plant growth is the main reason why sometimes, electroculture experiments are not successful.

Another source of disparity is the type of plant used for this experiment. Early experiments utilized plants of different classification that falls under root crops – carrots (*Daucus carota*), potato (*Solanum tuberosum*), and radish (*Raphanus raphanistrum*); and grains – mostly barley (*Hordeum vulgare*), corn (*Z. mays*), and wheat (*Triticum*). Almost all root crops that were subjected to electroculture are all dicotyledon and all grains are monocotyledon. However, garlic (*A. sativum* L.) is considered to be a crop that grows underground having a monocotyledon type of seed. The feasibility of this study to use garlic (*A. sativum* L.) for electroculture planting gave disapproving results. It seems that garlic (*A.

sativum L.) plant's properties, together with the applied electricity inhibit the metabolic capacity of the plant.

4. Conclusions

The success of electroculture technique depends mainly on the following parameters: dormancy and essential nutrients. Dormancy is a period wherein a crop adjusts to the environment's condition in order for it to survive drastic changes. On the other hand, essential nutrients are needed for the plant's metabolic processes. Comparison between the controlled system vis-à-vis electrocuted systems – 6 V and 12 V set-ups, showed that the electricity supplied inhibits the maximum growth of the garlic (*Allium sativum* L.) plant within a month of experimental period with values of 78.20 cm, 55.70 cm, and 57.10 cm, respectively. During the 29th day of electrocution process, garlic (*A. sativum* L.) experienced dryness and decrease in height for electrocuted systems – 6 V and 12 V set-ups because of the damage received by the important cells for metabolism and growth. Still, based on the data of the 12th, 17th, 18th, 19th, 20th, 21st, 25th, and 26th day of the experimental period, the electricity applied somehow assisted the plant's growth. The electrocution process would be successful if there were calcium ions present in the soil based on the records and observations of past researches.

References

[1] Macaraeg P 2017 What are the Most Profitable Farm Products to Grow in the Philippines? (Retrieved from: https://www.entrepreneur.com.ph/news-and-events/what-are-the-most-profitable-farm-products-to-grow-in-the-philippines-a1673-20170915)
[2] Nation Encyclopedia 2010 *Philippines – Agriculture* (Retrieved from: https://www.nationsencyclopedia.com/economies/Asia-and-the-Pacific/Philippines-AGRICULTURE.html)
[3] Rice D 2013 Why is Philippines a hot zone for typhoons? (Retrieved from: https://www.usatoday.com/story/weather/2013/11/09/philppines-typhoon-hot-zone/3483837/)
[4] Doorne Y 2012 Electroculture, Good vibes for Agriculture (Retrieved from: http://www.electrocultureandmagnetoculture.com/)
[5] Solly E 1845 *On the influence of electricity on vegetation* (London: W. Clowes and Sons)
[6] Palafox K, Villangca L, and Asuncion H 2013 Rapid Rice Plant Augmentation by means of Electroculture Technique (Retrieved from: https://www.scribd.com/document/184430981/Thesis-Document)
[7] Murr L E 1964 *Pennsylvania Academy of Science Proc.* **38** 44–46.
[8] Russell E J 1920 *J. of the Board of Agric.(Great Britain)* **26** 971–81.
[9] Hendrick J 1918 *Scottish J. of Agric.* **1** 160–70.
[10] Donahue M 2009 How long does it take for garlic to grow & mature? (Retrieved from: http://homeguides.sfgate.com/long-garlic-grow-mature-72059.html)
[11] Olchondra R 2014 Garlic prices come down, imports tightened (Retrieved from: http://business.inquirer.net/174695/garlic-prices-come-down-imports-tightened)
[12] Warner C 2016 Early Reading on Electroculture (1892-1922) (Retrieved from: https://hatch.kookscience.com/wiki/Early_Reading_on_Electroculture_(1892_-_1922))
[13] Woldeyes F, W/tsadik K and Tabor G 2016 Emergence of Garlic (Allium sativum L.) as Influenced by Low Storage Temperature and Gibberellic Acid Treatments (Retrieved from: http://www.journalrepository.org/media/journals/JAERI_37/2016/Dec/Woldeyes1022016JAERI29843.pdf)
[14] Goldsworthy, A. 2006 Electroculture, Good vibes for Agriculture. (Retrieved from https://www.electrocultureandmagnetoculture.com/history.html)

IOP Conf. Series: Materials Science and Engineering **703** (2019) 012010 doi:10.1088/1757-899X/703/1/012010

Preparation and characterization of polyvinyl alcohol-chitosan-tripolyphosphate hydrogel for extended release of anti-tuberculosis drugs

K Mulia[1], S A Chadarwati[1], A J Rahyussalim[2], and E A Krisanti[1]

[1]Department of Chemical Engineering, Universitas Indonesia, Depok 16424, Indonesia
[2]Department of Orthopedic and Traumatology, Faculty of Medicine, Universitas Indonesia, Cipto Mangunkusumo Hospital, Jakarta 10430, Indonesia

E-mail: kmulia@che.ui.ac.id

Abstract. Polyvinyl alcohol (PVA)-chitosan-tripolyphosphate hydrogels were prepared using a solvent casting-evaporation method to obtain anti-TB drugs extended release formulations. X-ray diffraction and scanning electron microscope results indicated that chitosan reduced the crystallinity of PVA hydrogels and the polymers formed homogeneous mixtures. The added chitosan reduced the encapsulation efficiency, and, the rate of drug release. The profile release of the four drugs indicated that the formulation of 80% PVA-20% chitosan hydrogel matrix had the highest drug release rate in a short period of time in phosphate buffer solution of pH 7.4. The maximum drug loading efficiency was obtained in a matrix formulation contain only PVA, while the addition of sodium tripolyphosphate increased the drug loading efficiency and also reduced the amount of drug released. The hydrogel matrix formulation of PVA-chitosan-tripolyphosphate showed its potential to be used as an extended delivery system for anti-TB drugs.

1. Introduction

Tuberculosis (TB) is an infectious disease caused by Mycobacterium tuberculosis that could affect various organs, such as lung, pericardium, meninges, and also backbone [1,2]. One of the most common types of TBC is osteoarticular tuberculosis, an extrapulmonary tuberculosis affecting bones, including backbones. The current treatment method is by surgery to remove the infected part, and continued by oral drug treatment of isoniazid (INH), pyrazinamide (PZA), ethambutol (ETH), and rifampicin (RFP) for two months, followed by isoniazid and rifampicin treatment for four more months to kill remaining active and dormant bacteria [3]. Unfortunately, this treatment could potentially produce discomfort and damage to the liver of the patients, as well as, drug resistant bacteria. These problems call for the development of an extended drug release formulation, in which drugs are accumulated in a depot to be later released slowly in the infected site. TB treatment using an extended drug release system has several advantages, such as decreased drug dosage, reduced side effect to the patients' liver, and drastically reduced drug consumption frequency. Xue et al. [4] showed that TB drugs encapsulated in poly-lactide co-glycolic acid (PLGA) could be released in 25 days. A study on alginate-PLGA core-shell for slow release of TB drugs indicate that only 0.4% of the drugs were released in 25 days [5]. Polyvinyl alcohol (PVA) is a synthetic polymer vastly used for biomedical application having biodegradable, hydrophilic, non-toxic, and biocompatible characteristics [6], as well as, a relatively high drug encapsulation efficiency of hydrophilic drugs [7].

However, PVA hydrogels in aqueous media swollen easily facilitating the release of drugs. Rogojanu et al. [8] showed that 50% of salicylic acid was released in only 10 h, while Takamura et al [9] demonstrated that 45% of flurbiprofen was released in 3 h. The addition of chitosan (CS) to the PVA matrix the slowed the bovine serum albumin release 60% of the total amount in 25 h [10]. It is also known that chitosan could form crosslinking in the presence of sodium tripolyphophate (STPP) [11]. In this research, the previously prepared PVA-chitosan matrix was immersed in the tripolyphosphate solution in order to form cross linking the chitosan in the matrix. The aim was to produce PVA hydrogel matrices with a high encapsulation efficiency and a slow release of anti-TB drugs to be used as an implant material for spinal tuberculosis treatment.

2. Methods

2.1. Material

Polyvinyl alcohol (degree of hydrolysis of 99%), chitosan (15-75 kDa, degree of deacetylation of 84%), acetic acid, and sodium tripolyphosphate were purchased from Sigma-Aldrich. Anti-TB drugs isoniazide (INH), pyrazinamide (PZA), ethambutol (ETH), and rifampicin (RFP) were supplied by Phapros.

2.2. Preparation of PVA-CS-drug-STPP hydrogel matrices

A casting solvent evaporation method was used in the preparation of the matrices. Chitosan was dissolved in 2% acetic acid solution to obtain 4%-w/v chitosan solution. PVA was dissolved in distilled water by heating at 90 °C and stirred until PVA was completely dissolved to obtain a 4%-w/v PVA solution. The PVA solution was mixed thoroughly with the chitosan solution at room temperature resulting in mixtures with PVA to chitosan mass ratios of 100:0, 80:20, 60:40, and 40:60. Each of the TB drugs (2 mg) was added to the PVA-chitosan solution (50 ml), the mixture was agitated using a sonicator until they become homogenous, and allowed to stand at room temperature to remove the bubble from the agitated solution. The mixture was poured into 5 cm petri dishes (20 ml), and then evaporated in an incubator at 37 °C until a constant weight was obtained. The crosslinking process was carried out by immersing the hydrogels in 2 or 4% STPP solution for 30 min. The crosslinked hydrogels were dried in the incubator at 37 °C until the weight was constant.

2.3. Drug encapsulation efficiency

The drug encapsulation efficiency was determined based on mass of drugs encapsulated in the matrices and mass of drugs dissolved in the STTP solution during the crosslinking process. The STPP solution after the crosslinking process was collected for analyzing the non-encapsulated drug using high performance liquid chromatography (HPLC) analysis. The drug encapsulation efficiency was calculated using equation (1):

$$encapsulation\ efficiency\ (\%) = \frac{initial\ drug\ weight - drug\ loss\ to\ STPP}{initial\ drug\ weight} x100\% \quad (1)$$

2.4. In vitro drug release test

In vitro drug release test was conducted in PBS of pH 7.4. The drug-matrix samples of 15 mg were placed in a dialysis membrane containing 20 ml of PBS. The dialysis membrane was immersed in test tube containing 40 ml of PBS. The test tube was placed in a water bath at 37 °C and stirred at 250 rpm. For assaying the drug released, the PBS solution samples (4 ml) were withdrawn several times up to 72 h. The same volume of the fresh PBS solution was added to keep a constant volume. The drug (isoniazid, pyrazinamide, ethambutol) contents in the sample solution were assayed using an HPLC-UV detector at a wavelength of 250 nm and mobile phase composed of 8% acetonitrile and 92% 20 mM NaH_2PO_4 solution of pH 6.8. Mobile phase composed of 50% acetonitrile and 50% 20 mM NaH_2PO_4 with a pH of 6.8 was used for rifampicin content determination at a wavelength of 333 nm.

IOP Conf. Series: Materials Science and Engineering **703** (2019) 012010 doi:10.1088/1757-899X/703/1/012010

2.5. Morphology crystallinity of hydrogels

The surface morphology of hydrogel matrices was determined using the scanning electron microscope (SEM) SEI JEOL-JSM 6510LA with the secondary electron image mode of 20kV acceleration voltage and 500-10,000× magnification. X-Ray Diffraction (XRD) pattern of hydrogel matrices was obtained using SHIMADZU XRD 7000 MAXIMA-X with Cu tube. All samples were analyzed between 2θ angles of 10° and 80° using the voltage of 40 kV, current of 30 mA, and time per step of 1 second.

3. Results and discussion

3.1. Drug encapsulation efficiency

Table 1 gives the formulation code, PVA to chitosan mass ratio in the hydrogel matrices, the concentration of the STTP solution used, and, the drug encapsulation efficiency for each anti-TB drug as determined using equation (1).

Table 1. Compositions of PVA-chitosan-STPP matrices.

Formula	PVA:chitosan mass ratio	[STPP] (%-w/v)	Encapsulation efficiency (%)			
			INH	PZA	ETH	RFP
100A	100:0	2	98.6	57.7	62.9	99.6
100B	100:0	4	98.4	40.8	78.8	99.3
80A	80:20	2	99.7	28.1	45.0	99.9
80B	80:20	4	99.3	24.5	37.9	99.9
60A	60:40	2	98.8	29.7	55.2	99.9
60B	60:40	4	98.9	35.1	53.9	99.9
40A	40:60	2	97.8	27.2	49.1	99.8
40B	40:60	4	97.4	18.1	54.0	99.8

The encapsulation efficiencies of the anti-TB drugs follow the order of: RFP~INH>ETH>PZA. This trend could be correlated to water solubilities of these drugs in the order of: RFP<INH<PZA<ETH or 0.0003<42<363<649 mg/mL, respectively. Although PZA is slightly less soluble in water than ETH, the encapsulation efficiency values of this drug were the lowest among the four drugs, indicating that PZA is leached out of the chitosan-PVA hydrogels more extensively. This might be due to the interaction between the more positively charged amide functional group of PZA, stabilized by the amide resonance structures, and the more negatively charged tripolyphosphate anion.

3.2. The SEM pictures and XRD patterns of the hydrogel matrices

Figure 1 shows SEM pictures (500x magnification) that show surface morphology of three PVA-CS matrices: (left) PVA:CS=100:0, 2%-STPP, pre-immersion; (center) PVA:CS=40:60, 4%-STPP, pre-immersion; (right) PVA:CS=40:60, 4%-STPP, post-immersion of 72 h. Sample 100A (PVA only) showed a smooth and even surface of the matrix. PVA thin film has a good structural integrity, smooth, even, and no crack [12]. Sample 40B-new (40% PVA, 60% chitosan in 4% STPP) also shows a homogenous, although a less smooth surface compared to sample 100A. This shows that PVA and chitosan were compatible polymers.

Figure 1. SEM pictures of Morphology Anti-TB drugs release profiles from PVA-chitosan-tripolyphosphate matrices.

Sample 40B-72 (60% chitosan in 4% STPP after 72 hours release test) shows an uneven and rough surface, due to degradation and erosion of the polymer in the release media. Figure 2 shows an XRD pattern with a high intensity peak at 2θ of 20° due to the crystalline of the PVA-only matrix. The intensity of this peak was diminished after chitosan was incorporated into the hydrogel matrices due to the reduced crystallinity of the PVA matrices. An XRD pattern could be related to crystal size where a wider peak is a result of imperfect crystals [13]. The SEM and XRD data indicate that the amine or hydroxyl functional groups in chitosan were able to form bonds with the hydroxyl groups in PVA.

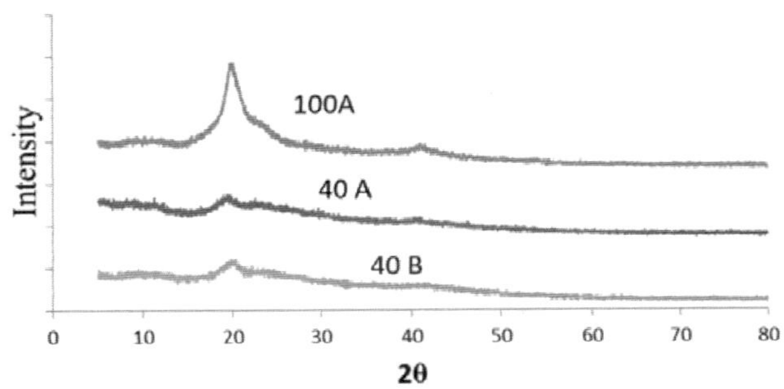

Figure 2. XRD patterns for samples 100A, 40A, and 40B.

It is reported earlier that combining PVA with chitosan for preparing scaffold was occurring due to the formation of hydrogen bonding between hydroxyl groups of PVA and chitosan [14]. By immersing the matrix of PVA-CS-drug into the solution of STTP, the matrix seemed hardened because of the formation of networking between the positively charged amine group in chitosan and the negatively charged phosphate group of STPP [15]. The schematic of hydrogel matrix formation between PVA, chitosan and STPP, loaded with anti-TB drugs is shown in figure 3.

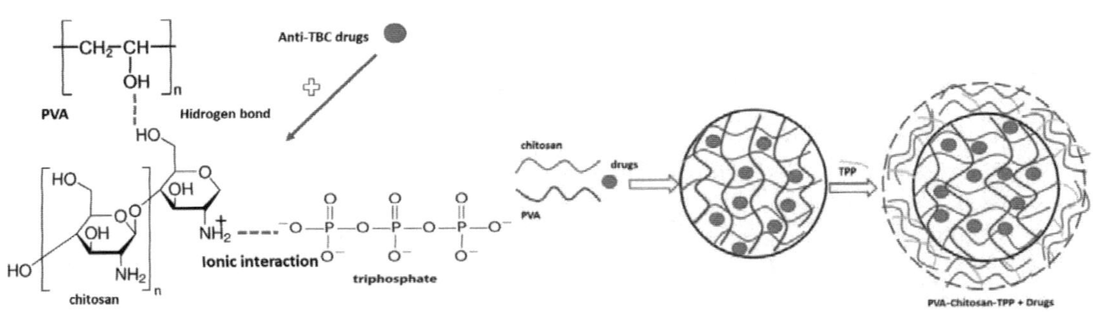

Figure 3. Schematic of anti-TB drug-loaded PVA-chitosan-STPP hydrogel matrix.

3.3. Drug release profile

The samples taken at a certain period of time were assayed for the drug content and the cumulative release plots are shown in figure 4 for anti-TB drugs INH, PZA, ETH, RFP. Although these drugs showed different release profiles in buffer solution of pH 7.4, the release profiles of INH and PZA followed a similar pattern. Less than 20% of INH were released from the PVA-only matrices (100A &100B) due to the high crystallinity of the matrices (see figure 2). On the other hand, it took only about 4 h to achieve a 65% cumulative release of INH from the matrix with 20% chitosan (80A). A similar burst release of PZA from a matrix with similar composition (80A), reached 100% cumulative release in about 60 min. These drugs were readily released after the matrix was swollen with the buffer solution in a short period of time, probably due to the disrupted PVA network in the presence of chitosan. Matrices with increasing chitosan to PVA mass ratios (60A and 40A) showed even lower cumulative release of INH and PZA because of the more extensive crosslinked between chitosan and tripolyphosphate, and also, due to the formation of hydrogen bonds between the carbonyl functional group of INH and PZA and the amine functional group of chitosan [16]. Ethambutol showed slightly longer release time, probably because the secondary amine functional groups do not interact as strongly as the primary amine functional group of INH and PZA.

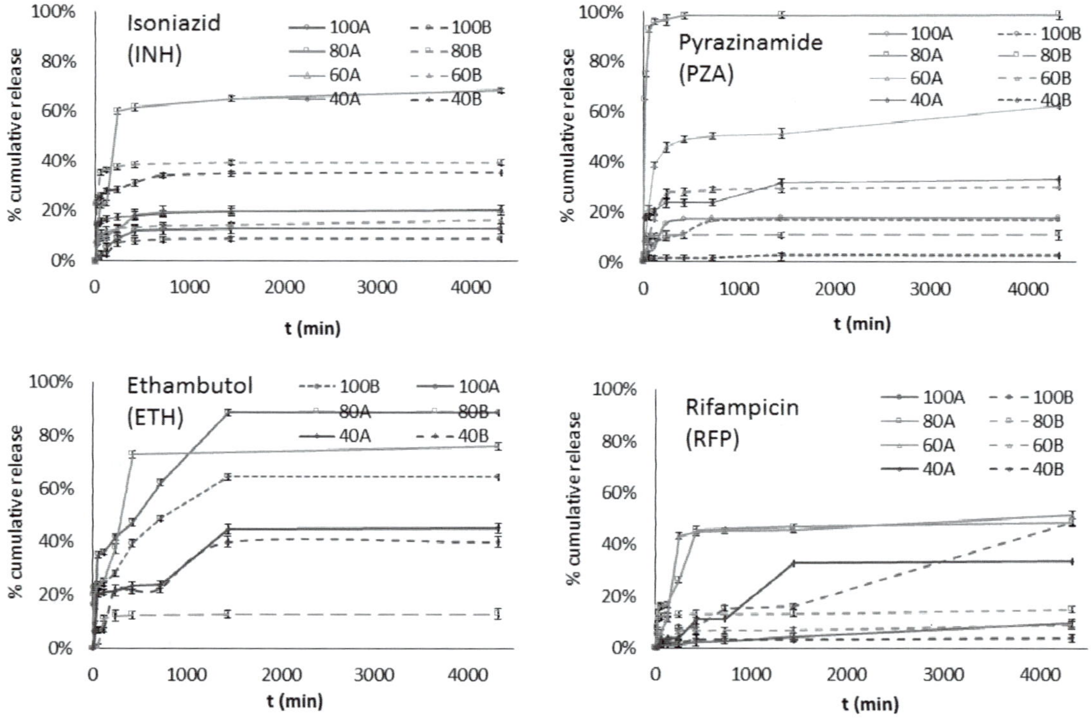

Figure 4. Release profiles of anti-TB drugs from PVA-CS-STTP matrices.

The release profile of the RFP in buffer solution was quite different from the other three drugs. It was only about 45% released at 4 h when matrices contained chitosan of 20% and 40%. More importantly, the release of RFP in a matrix of 100% PVA was almost linear for 3 days, with the cumulative release of 5 and 45% from matrices prepared with 4 and 2%-w/v STTP, respectively. Based on its size, it is likely that RFP molecules would be entrapped in the PVA matrix, allowing for the desired zero-order release.

4. Conclusion

The encapsulation efficiencies of the anti-TB drugs follow the order RFP~INH>ETH>PZA that could be related to their water solubility values. The release profiles of the four drugs showed significant burst release up to 500 min (INH and PZA) and to 500 min (ETH and RFP). In general, the effect of the added chitosan was more drugs were released from hydrogels with PVA to chitosan ratio of 80:20, while hydrogels with higher content of chitosan hindered the release of the drugs. The hydrogels with the lower concentration of tripolyphosphate produced higher cumulative release of all drugs. In contrast to the burst release of the hydrophilic drugs, approximately linear release profiles of the RFP from PVA-only matrices were observed for 3 days. As a large molecule, it is likely that RFP would be entrapped in the PVA matrix, allowing for the desired extended zero-order release. The formulation of the PVA hydrogel matrix containing tripolyphosphate should be evaluated further in order to obtain longer release time of the drug.

Acknowledgment

The authors are grateful for the financial support received from the Ministry of Research, Technology and Higher Education of the Republic of Indonesia through the PDUPT Research Grant contract 410/UN2.R3.1/ HKP05.00/2018.

References

1. Fogel N 2015 Tuberculosis: A disease without boundaries *Tuberculosis* **95** pp 527-531
2. Centers for Disease Control and Prevention Tuberculosis 2016 [cited 7th May.2017] Available from: https://www.cdc.gov/tb/topic/basics/default.htm
3. World Health Organization Guidelines for Treatment of Tuberculosis 2010 [Internet] [cited 21 st Nov. 2017] Available from: http://www.who.int/tb/publications/2010/9789241547833/en/
4. Xue M, Hu H, Jiang Y, Liu J, HE H, and Ye X 2012 Biodegradable Polymer-Coated, Gelatin Hydrogel/Bioceramics Ternary Composites for Antitubercular Drug Delivery and Tissue Regeneration *J. Nanomaterial* **2012**
5. Lio D, Yeo D and Xu C 2016 Control of Alginate Core Size in Alginate – Poly (Lactic-Co-Glycolic) Acid Microparticles *Nanoscale Res Lett.* **11** pp 9
6. Paradossi G, Cavalieri F, Chiessi E, Spagnoli C and Cowan M K 2003 Poly(vinyl alcohol) as versatile biomaterial forpotential biomedical applications *J. Mater. Sci.: Mater. Med.* **14** pp 687–691
7. Li J and Mooney D J Designing hydrogels for controlled drug delivery. *Nat Rev Mater.* **1**(12)
8. Rogojanu A, Rusu E, Olaru N, Dobromir M and Dorohoi D O 2011 Development and characterization of poly(vinyl alcohol) matrix for drug release. *Dig. J. Nanomaterials and Biostructures* **6** pp. 809 – 818
9. Takamura A, Ishii F and Hidaka H 1991 Drug release from poly(vinyl alcohol) gel prepared by freeze-thaw procedure *J. Controlled Release* **20** pp 21 – 28
10. Wang Q, Du Y, and Fan L 2004 Properties of chitosan/poly(vinyl alcohol) films for drug controlled release *J. Appl. Polym. Sci.* **96** pp. 808 – 813
11. Shu X Z and Zhu K J 2002 Controlled drug release properties of ionically cross-linke chitosan beads: the influence of anion structure *Int. J. Pharm.* **233** pp 217-225
12. Liu Y, Wang S, Lan W and Qin W 2017 Fabrication and Testing of PVA/Chitosan Bilayer Films for Strawberry Packaging *Coatings* **7(8)** 109
13. Cho A R, Chun Y G, Kim B K and Park D J 2014 Preparation of Chitosan–TPP Microspheres as Resveratrol Carriers *J. Food Sci.* **79** pp 4
14. Alhosseini S N, Moztarzadeh F, Mozafari M, Asgari S, Dodel M, Samadikuchaksarei A, Kargozar S and Jalali N 2012 Synthesis and characterization of electronspun polyvinyl alcohol nanofibrous scaffolds modified by blending with chitosan for neural tissue engineering *Int. J. Nanomedicine* **7** pp 25-34

15. Liang S, Liu L, Huang Q and Yam K L 2009 Preparation of single or double-network chitosan/poly(vinyl alcohol) gel films through selecively cross-linking method *Carbohydr. Polym.* **77** pp 718-724
16. Takahashi H, Chen R, Okamoto H and Danjo K, Acetaminophen particle design using chitosan and a spray-drying technique *Chem. Pharm. Bull.* **53(1)** pp 37-41

IOP Conf. Series: Materials Science and Engineering **703** (2019) 012011 doi:10.1088/1757-899X/703/1/012011

The surface roughness analysis using sound signal in turning of mild steel

Anayet U Patwari, A A Zamee, M H Bhuiyan and S M Sakib

Department of Mechanical and Chemical Engineering, Islamic University of Technology (IUT), Dhaka, Bangladesh

E-mail: apatwari@iut-dhaka.edu

Abstract. Among every other parameters of production process, surface roughness holds its ground as one of the most crucial factors for the quality analysis. Good surface finish is a major criterion in almost every machining process. If the surfaces aren't smooth, many kinds of mechanical, thermal, frictional, vibrational problems may occur. So minimizing the surface roughness should be the top priority. Surface roughness depends on many factors. But it has a direct relation with the tool conditions and overhang length of the tool. With the variations of tool conditions and overhang length the machining sound level also varies. In this present study, an analysis has been made based on captured sound signal to correlate the surface roughness parameters with sound level at different tool and overhang length conditions. As the cutting tool wears out due to continuous usage, surface roughness also develops on the tool surface. It has been observed that with the increase of overhang length the sound generated level within the cutting zone varied and surface roughness in the job also varied due to the effect of vibration and friction. A correlation factor has been investigated with the sound level variation to analyze the surface roughness condition with different tool wear and overhang length.

1. Introduction

Lathe machines are one of the most widely used metal shaping machines for cylindrical job pieces. For effective and multipurpose uses they are being used in almost every workshop. Many research works has already been conducted to improve the surface roughness of machined work-piece in turning operations using different types of process parameters, coolant, hot machining, cryogenic machining etc. G.M. Sayeed Ahmed et al [1] conducted an experiment using AISI 1050 material of different diameters of 20, 30, and 40 mm in which the surface roughness of the work piece was determined through experiments using constant cutting speed and feed rates with different depth of cuts (DOCs) and tool overhang lengths. Safeen Y. Kassab et al. [2] experimented to find relation between surface roughness and cutting tool vibration in lathe dry turning of medium carbon steel. They found that vibration of cutting tool depends strongly on cutting tool overhang length and with the increasing feed rate the surface roughness of work piece increase. H.H. Habeeb et al. [3] conducted machining of nickel based alloys 242 included using four different cutting tool materials under wet condition. Flank wear modes are noticed as acceptable results at lowest cutting depth with high cutting speed and moderate feed rate. Optimum surface roughness results were also recorded with decreasing of cutting depth. In a similar type of experiment by Anayet U. Patwari et al. [4] investigated the effects of overhang length on surface roughness, chip morphology to determine the suitability of over-hang length with coated carbide inserts so that with the proper selection of the overhang length the productivity will increase with the increased process parameters. Vladimir Aleksandrovich Rogov et. al. [5] used the Taguchi method to optimize the surface roughness with vibration in turning of Aluminum AA2040 under dry condition where Spindle speed, feed rate, depth of cut and tool overhang were chosen as cutting parameters. Kong Lingguo et al.[6] in their study found that within a certain range, the cutting speeds increased will first make the amount of vibration and surface roughness increase, then decreased, and more than this range, the surface roughness goes up along with the vibration increase. The cutting edge of the worn tool acting like un-uniformly larger nose radius which produced better surface was found in an experiment by S. Thamizhmanii et al. [7]. In this present study, several experiments have been conducted with different types of tool wear and overhang length and generated sound were captured at different cutting conditions in turning process. The captured sound level was analyzed to correlate the surface profile with the sound level in machining process and find out the relationship with sound level, surface roughness, tool wear and overhang length for conventional turning processes.

2. Experimental details

Many mild Steel bars of 200 mm length and 32 mm diameter were cut from the same bar using LET LHM-280 power saw machine for conducting lathe operation using Pinacho Mod. L-1/180 Lathe machine. Three same types of coated carbide insert have been used with different tool wear. In the experiment, one new insert, one moderate wear insert and large wear insert has been used for machining purposes. During machining audio recorder was used to capture the sound for sound analysis and an open source software was used for sound analysis which was captured by audio recorder. After the machining process, Profilometer was used for measuring surface roughness. A detail of the experimental setup is shown in Figure 1.

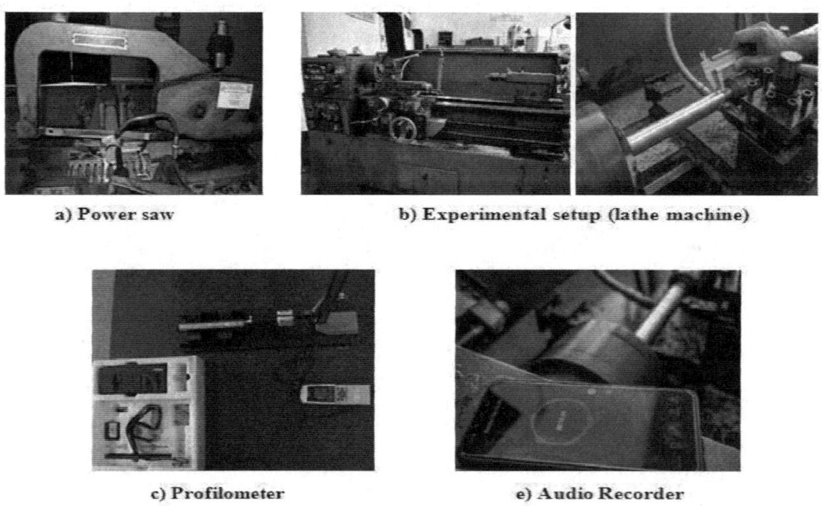

Figure 1. Details of the Experimental setup used in this study.

For this experiment, same type of insert with different tool wear namely no wear-new insert, medium wear and high tool wear as shown in Figure 2 have been used for machining operation of different overhang length keeping all other process parameters constant. Conventional turning machining was done in the following manner:

- 1st work piece: High wear insert for 1st 10 mm, Moderate wear insert for 2nd 10 mm, no wear insert for 3rd 10 mm were used.

- 2nd work piece: No wear insert for 1st 10 mm, high wear insert for 2nd 10mm, moderate wear insert for 3rd 10 mm were used.
- 3rd work piece: Moderate wear insert for 1st 10 mm, no wear insert for 2nd 10mm, high wear insert for 3rd 10 mm were used.

(i) High Wear (ii) Moderate Wear (iii) No Wear

Figure 2. Different types of Insert used in this study

2.1 Experimental conditions:

This experiment was conducted by using lathe machine where 3 types of (no wear, moderate wear, high wear) coated carbide insert were used for machining. In both machining process following parameters were applied Cutting speed, Depth of cut, Feed rate was kept constant throughout the whole experiment while overhang length became variable.

Constant parameter: Cutting Speed: 875 rpm; Depth of Cut: 1 mm; feed rate: 0.1 mm/rev;

Variable parameter: Overhang length: 160 mm, 170 mm and 180 mm.

3. Results and discussions

While measuring the surface roughness, different surface roughness was found using different insert for each work piece. Comparing work pieces, insert 3 (no wear insert) had the best surface finish at various overhang length as shown below.

Surface Roughness **Surface Roughness Graph**

- For work piece 1: 160 mm Overhang, No wear insert

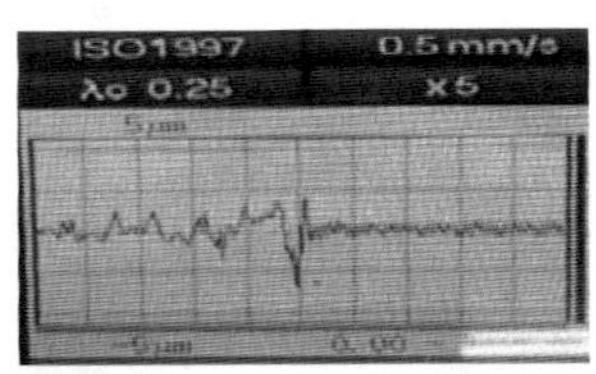

- For work piece 2: 180mm overhang, No wear insert

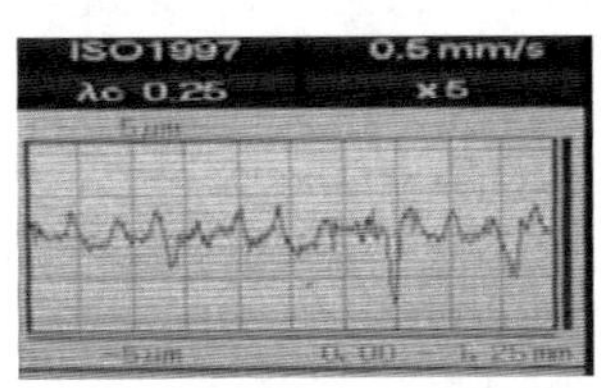

- For work piece 3: 170 mm overhang, No wear insert

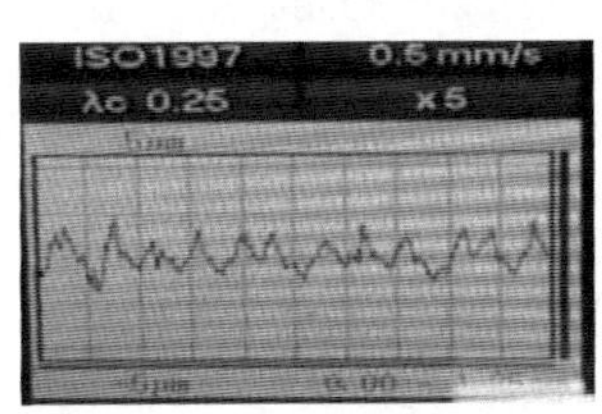

In this experiment, using different insert for each job piece while varying the overhang length, it is found that 3rd insert (No wear) has the better surface finish for conventional machining. For job piece 1, in 160 mm overhang length for insert 3 (No wear) has the better surface finish of 0.360 micro-m. Similarly, for job piece 2 & 3 better surface finish is obtained using 3rd insert that has no wear for varying overhang length 180 mm & 170 mm as shown in Figure 3. It is clearly observed that with the increase of overhang length the surface roughness increased in case of no wear insert. In case of sound level analysis, it is also found that no wear insert generated less sound level compared to other two inserts. It is also observed that at overhang length 170 mm and 180 mm for moderate and high wear insert the surface roughness varied considering the frictional and vibration effect.

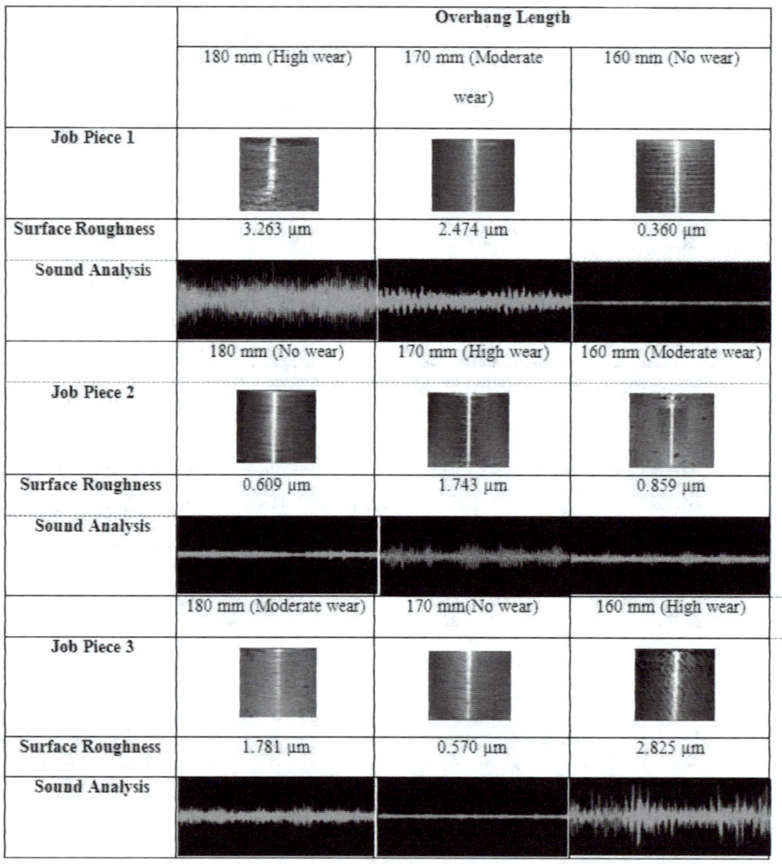

Figure 3. Surface profile with average roughness and sound level variation at different overhang length.

While machining the job piece, different types of chips were formed as shown in figure 4 using different insert for varying overhang length. For 1st job piece, using high wear insert for 180 mm overhang length discontinuous chip was formed which was black in color (because of high friction between tool and job piece). For 1st job piece, during machining discontinuous chips were formed for other insert (Moderate & No wear) for varying overhang length (170 mm & 160 mm). But the chips which were formed during machining became bluish (because of friction between tool and job piece it slightly burnt) for insert 2 (Moderate wear) and silver for insert 3(No wear).

For 2nd job piece continuous chip was formed using insert 3(No wear) where overhang length was 180 mm and using insert 1(High wear) with overhang length 170 mm formed continuous chips with build up edge due to excessive friction and vibrational effect between tool and job piece and it was coil type shape. For overhang length of 160 mm insert 2 (moderate wear) formed discontinuous chip

IOP Conf. Series: Materials Science and Engineering **703** (2019) 012011 doi:10.1088/1757-899X/703/1/012011

which was brownish in colour. For job piece 3, continuous chips were formed using 2nd (Moderate wear) & 3rd insert (No wear) for varying overhang length at 170 mm.

	Overhang Length		
	180 mm(High wear)	170 mm(Moderate wear)	160 mm(No wear)
Job Piece 1	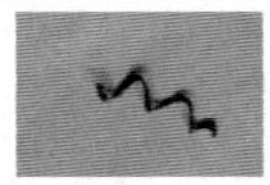	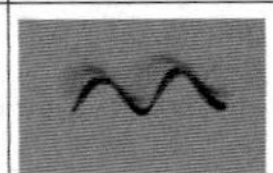	
	180mm (No wear)	170 mm(High wear)	160 mm (Moderate wear)
Job Piece 2	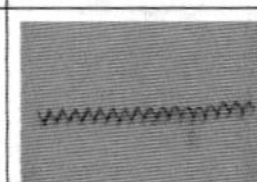	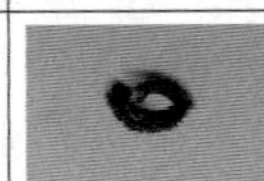	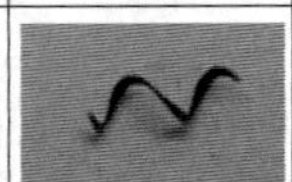
	180mm (Moderate wear)	170 mm(No wear)	160 mm (High wear)
Job Piece 3	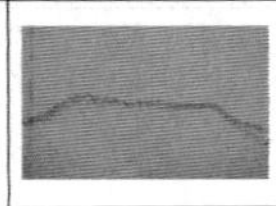	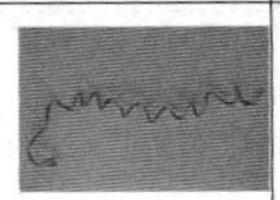	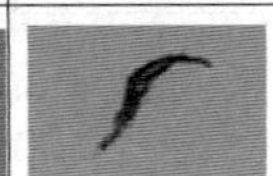

Figure 4. Different types of chip formed in machining processes.

But the chips which were formed during machining became bronze type color (because of friction between tool and job piece it slightly burnt) for insert 2 (Moderate wear) and silver for insert 3 (No wear). Beside this continuous chips with build up edge was formed using insert 1(High wear) (overhang length 160mm) due to excessive friction between tool and job piece.

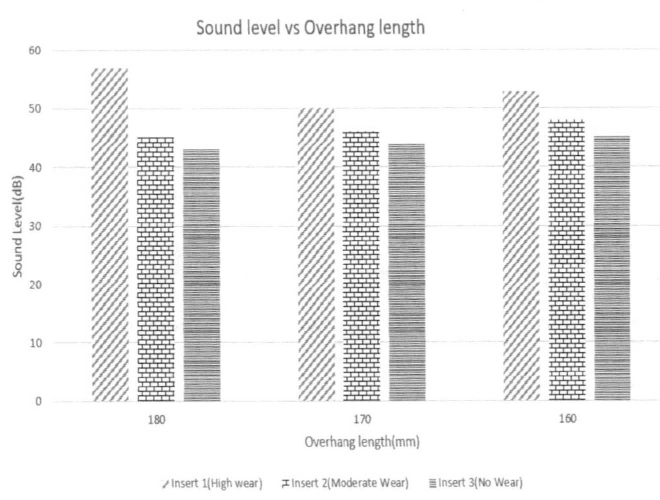

Figure 5. Sound level variation at different overhang length for different insert.

Figure 5 shows the visual demonstration of sound analysis vs. overhang length, where machining is done for this experiment under same feed rate, depth of cut and cutting speed but overhang length is variable for each insert. From this histogram, it can be seen that sound level is decreasing with decrease of overhang length. When overhang length is decreasing less vibration occurs, that's why sound level decreases and surface improvement is also achieved. Correlating figure 3 and figure 5, it is found that for insert 1 (High wear) surface roughness first decreasing then increasing with the increase of overhang length and same tendency is reflected in sound level variation. For insert 2 (Moderate wear) surface roughness first increasing then decreasing with increase of overhang length. For insert 3 (No wear) surface roughness gradually decreases with the decrease of overhang length. Generally surface roughness decreases with decrease of overhang length. Similar tendency has also been reflected in captured sound level variation. According to the sound analysis, it is showed that maximum sound is recorded [57db] for overhang length 180 mm in case of high wear insert and minimum [43db] is found in case of no wear insert.

4. Conclusions

In this study, surface roughness has been compared with the sound level variation at different overhang length for three different types of insert. Generally, it has been observed that with the increase of overhang length, the average surface roughness value is going to be increased. Overhang length variation results the cutting process stability and insert status reflects the frictional behaviour in between the tool and the work-piece. The performance of machining is related to the stated factors and sound captured during machining will be varied at various combinations of the process and system parameters. So, in this study, an attempt was made to investigate and correlate the output response in terms of average surface roughness and generated sound signal during cutting process. Surface profile response, tool life, vibration may be predicted from the sound level variation.

References

[1] Ahmed G M S, Ahmed H and Samad S S 2013 Experimental investigation of effect of tool length on surface roughness during turning operation and its optimization *IOSR J. Mech. Civ. Eng.* **7** pp 2320-34

[2] Kassab S Y and Khoshnaw Y K 2007 The effect of cutting tool vibration on surface roughness of work piece in dry turning operation *Eng. Tech.* **25**

[3] Habeeb H H, Hossein K A A E, Mohamad B, Ghani J A and Kadirgama K 2008 Investigating of tool wear, tool life and surface roughness when machining of nickel alloy 242 with using of different cutting tools *Asian J. Sci. Res.* **1** pp 222-30

[4] Patwari A U, Alfakih Y M and Bamnjo I 2018 Investigation of tool overhang length effects on machining responses during turning operation annals of faculty engineering Hunedoara *Int. J. Eng.* Tome XVI

[5] Rogov V A and Siamak G 2013 Optimization of surface roughness and vibration in turning of aluminum alloy AA2024 using Taguchi technique world academy of science, engineering and technology *Int. J. Mech. Mechatronics Eng.* **7**

[6] Lingguo K, Guangjun C, Liang W, Dingwen W, Weidong C and Ting Z 2016 Research on vibration and surface roughness in hard cutting process of hardened steel *Key Eng. Mater* **693** ISSN: 1662-9795 pp 878-83

[7] Thamizhmanii S, Kamarudn K, Rahim E A, Saparudin A and Hassan S 2007 Tool wear and surface roughness in turning AISI 8620 using coated ceramic tool *Proc. World Congr. Eng.*

A review of a machine design of chocolate extrusion based co-rotating twin screw extruder

P Pitayachaval and P Watcharamaisakul

School of Industrial Engineering, Institute of Engineering, Suranaree University of Technology, 111 University Avenue Muang, Nakhon Ratchasima, Thailand

E-mail: paphakorn@g.sut.ac.th

Abstract. Based on innovation and competitive market for food industry, there are several food products which have been designed to attract customer. Since there is USD 39,431 millions of chocolate sales in 2018, USA [1], chocolate product shapes have been developed based on manufacturing process.This paper presents a review process of a machine design of chocolate extrusion based co-rotating twin screw extruder. A property of suitable chocolate for extruder was established. The pros and cons of machine extruder for food processing including, a screw extruder design were exposed. Since there were problems in the chocolate extruder, the process parameters such as barrel temperatures, feed rate, screw speed, motor load and melt pressure were established. These parameters would be applied to design screw extruder for chocolate processing.

1. Introduction

Several new technologies or innovations have been introduced in order to increase productivity while reducing production cost. There are several areas in food industry that have high competition, such as, the chocolate market which has market of USD 39,431 millions of chocolate sales in 2018 in USA with consumers ranging from children to adults [1]. Therefore, many technologies have been applied to increase productivity of chocolate production while enraging flavors and various shapes. Since the properties of chocolate are melted in normal temperature, flowed in a container and changed viscosity depending on the temperature, chocolate is processed via many production processes such as extrusion, printing, and so on. However, there are several problems during extrusion process such as low productivity rates, product waste and improper extrusion temperature affect to increase production costs. Therefore, the chocolate extrusion is processed with the screw extruder in order to reduce those problems, and the screw extruder has been applied to control various factors including high flexibility. This paper presented a review process of a machine design of chocolate extrusion based co-rotating twin screw extruder.

2. Chocolate

Chocolate is composed of nonfat particles (sugar, cocoa solids, and milk powder particles) dispersed in cocoa butter as a continuous phase [2]. The flow properties to control the quality of chocolate product of a chocolate flow is an important parameter since too low viscosity will cause the weight of the chocolate over the enrobed candy will also be too low. However, if it is too high, then bubbles may be formed and put chocolate. Moreover, the different chocolate viscosity provides distinguished flavor, that can be detected by consumers. The perceived taste is related to the viscosity and rate of melt. Therefore, the flavor of chocolate is different between stored in the refrigerator or in a room temperature [3]. Therefore, the designs of the machine to process chocolate are important to control the parameters that affect the flow and viscosity of chocolate. Those factors affect the chocolate extrusion process because they are involved in melting, as well as extrusion through the die.

3. Extrusion processing in food industrial

The extrusion processing for the food industry has been wildly used in production process since it provides a variety of food shapes. The products vary depending on the ingredients and the condition of the control. The Machine, called Extruder, has been incorporated with various processes. The extruder mechanism is to blend raw materials and ingredients (feed or transport section) under pressure and heat conditions to shape or meter the section from compression, and to push the product out to die. Extrusion technology has been introduced to shape a variety of food productions because there are many advantages such as saving energy, reducing labor and production costs, flexibility for several raw materials, ability for mixing ingredient, and saving space for tool installation. Those properties can also improve the quality of products as needed [4].

4. Screw extruder design

Hot-melt extrusion equipment consists of an extruder, auxiliary equipment for the extruder, downstream processing equipment, and other monitoring tools used for performance and product quality evaluation [5]. The extruder is typically composed of a feeding hopper, barrels, single or twin screws, and the die and screw–driving unit (figure 1). The auxiliary equipment for the extruder mainly consists of a heating/cooling device for the barrels, a conveyer belt to cool down the product and a solvent delivery pump. The monitoring devices on the equipment include temperature gauges, a screw-speed controller, an extrusion torque monitor and pressure gauges.

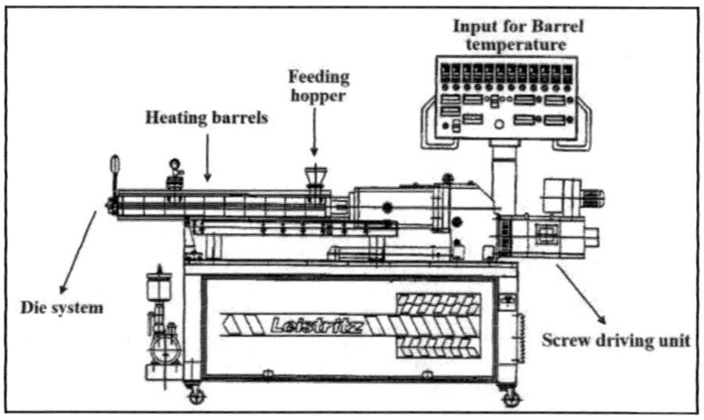

Figure 1. Micro-18 Twin screw co-rotating Leistritz extruder[6]

The theoretical approach to understanding the melt extrusion process is generally presented by dividing the process flows into four sections, as follows [7]: 1) Feeding of the extruder, 2) Conveying of mass (mixing and reduction of particle size), 3) Flow through the die, 4) Exit from the die and down-stream processing. Generally, the extruder consists of one or two rotating screw inside a stationary cylindrical barrel. The barrel is often manufactured in sections, which are bolted or clamped together. An end-plate die, connected to the end of the barrel determines the shape of the extruded product shown in figure 1 and figure 2.

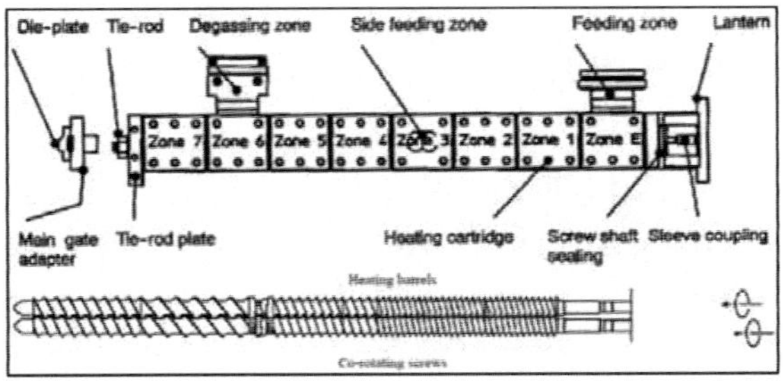

Figure 2. Heating barrels and co-rotating screws for hot-melt extruder [6]

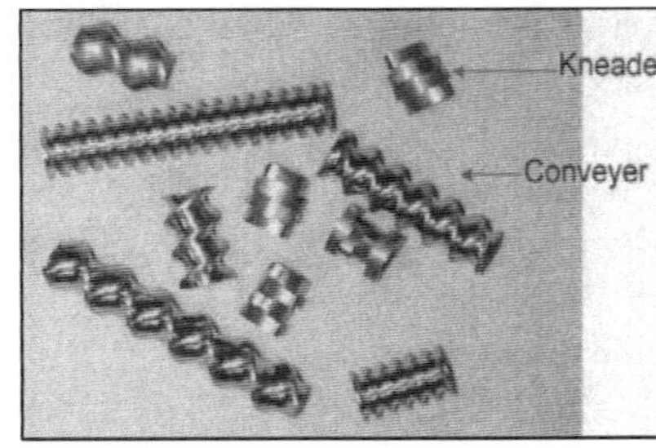

Figure 3. Screw and kneading elements [6]

The heat that is generated from friction force is required to melt or fuse the material. This force is a shear force occurring between the rotating screws and the wall of the barrel that is mounted with electric or liquid heaters [8]. Most commercial extruders have a modular design, providing a choice of screws or interchangeable sections which modify the configuration of the feed, transition, and metering zones. The modular design has been introduced to meet special process requirements, such as, increase high shear extrusion, addition more solvent and enlarge evaporate solvent from raw material. Modifying screw designs also allow the extruder to perform a mixing and reduction of particle size in addition to extrusion, so that material can be blended into the extrudate or even dissolved, as shown in figure 3.

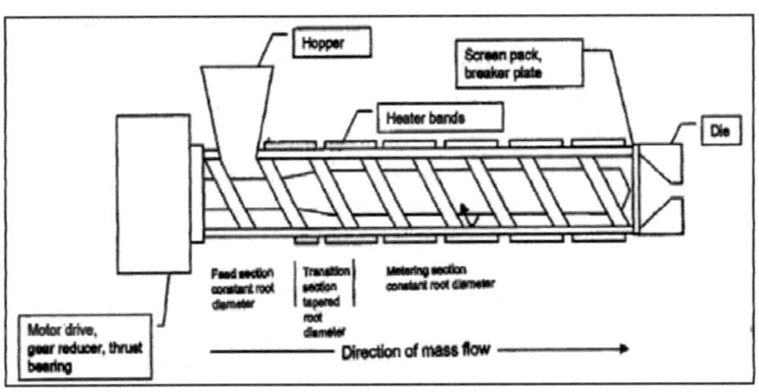

Figure 4. Component parts of single –screw extruder [6]

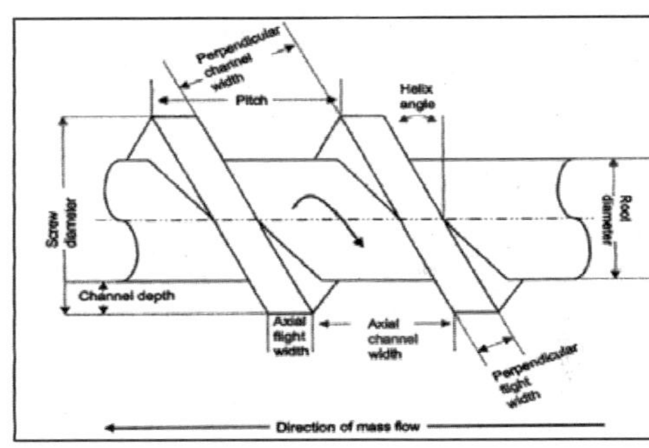

Figure 5. Extrusion screw geometry [6]

The screw and die design was available, and a practical considerations of thermoplastic extrusion were reviewed by Whelan and Dunning [9]. The extrusion channels were conventionally divided into three sections: feed zone, transition zone, and metering zone, as shown in figure 4. The starting material was fed from a hopper directly into the feed section, which has deeper flights or flights of greater pitch, as shown in figure 5. This geometry enabled the feed material to fall easily into the screw for conveying along the barrel. The pitch and helix angle determined the throughput at a constant rotation speed of the screws. The material was transported as a solid plug to the transition zone where it was mixed, compressed, melted and plasticized. Compression was developed into two types: (1) decreasing the thread pitch while keeping a constant flight depth, and (2) decreasing flight depth but keeping a constant thread pitch [10]. Both types resulted in increasing pressure as the material moved along the barrel. The melt conveyed by circulation in a helical path by means of transverse flow, drag flow, pressure flow and leakage; the latter two mechanisms reversed the flow of material along the barrel. The space between screw diameter and width of the barrel is normally in the range of 0.1-0.2 mm [8]. The material reaches the metering zone in the form of a homogeneous plastic melt suitable for extrusion. To extrude a uniform part thickness, flow must be consistent without stagnant zones right up to the die entrance. The function of the metering zone is to eliminate pulsating flow and ensure a uniform delivery rate through the die cavity [7].

5.Twin screw extrusion

The twin-screw extruder has two agitator assemblies mounted on parallel shafts. These shafts are driven through a splitter/reducer gearbox and rotate together in the same direction of rotation (co-rotating) or in the opposite direction (counter rotating) and are often fully intermeshing. In such case, the agitator element wipes both the surface of the corresponding element on the adjacent shaft, and the internal surfaces of the mixing chamber and ensures a narrow and well-defined residence time distribution. In general, co-rotating shafts have better mixing capabilities as the surfaces of the screws move towards each other. This leads to a sharp change in mass flow between the screw surfaces [7]. As the screws rotate, the flight of one screw element wipes the flank of the adjacent screw, causing material to transfer from one screw to the other. In this manner the material is transported along the extruder barrel. The twin-screw extruder was characterized by the following descriptive features [8]: 1) Short residence time: The residence time in the twin-screw extruder in a typical extrusion processes ranges from 5-10 minutes depending on the feed rate and screw speed, 2) Self-wiping screw profile: The self-wiping screw profile is the flight of the one screw wipes the root of the screw on the shaft next to it, ensures near complete emptying of the equipment and minimizes product wastage on shutdown, 3) Minimum inventory: Continuous operation of the equipment coupled with the continuous feeding of the material helps in reducing inventories of work in progress. This is important when processing valuable or potentially hazardous materials, 4) Versatility: Operating parameters can be changed easily and continuously to change extrusion rate or mixing action. The segmented screw elements allow agitator designs to be easily optimized to suit a particular application. Die plates can also be easily exchanged to alter the extrudate diameter. This allows processing of many different formulations on a single machine, leading to good equipment utilization. Polymers with a wide range of viscoelastic and melt viscosities may be processed and even fine powders may be directly fed into the system, 5) Superior mixing: The screws have various mixing elements which impart two types of mixing, distributive mixing and dispersive mixing. The distributive mixing ideally maximizes the division and recombine the material while minimizing energy. The dispersive mixing ideally breaks droplet or solid domains to fine morphologies using energy at or slightly above the threshold level needed. This mixing aids in efficient compounding of two or more materials in the twin-screw extruder.

Typical twin-screw laboratory scale machines have a diameter of 16-18 mm and length of four to ten times the diameter. Atypical throughput for this type of equipment is 0.5- 5 gm/min. As the residence time in the extruder is rather short and the temperature of all the barrels are independent and can be accurately controlled from low temperatures (30◦C) to high temperatures (300◦C), degradation by heat can be minimized [8]. Extrusion processing requires close monitoring and understanding the various parameters: viscosity and variation of viscosity with shear rate and temperature, elasticity and extensional flow over hot metal surfaces. Today, extruders allow in-process monitoring and control of parameters, such as the temperature in the extruder, head and die as well as pressure in extruder and die [11].

6. Extrusion parameters

The main monitoring and controlling parameters are barrel temperatures, feed rate, screw speed, motor load and melt pressure. Barrel temperature, feed rate and screw speed are controlling parameters and motor load and melt pressure are monitoring parameters. 1) Barrel temperatures: The glass transition or melting temperatures of polymers and drug usually determines the barrel temperature, 2) Feed rate and screw speed. The constant feeding rate and screw speed throughout the process is important as the combination of these two factors establishes the level of fill in extruder. This is critical to the process, because it governs the balance between the weak and strong mass transfer mode [8]. Due to constant feed rate and screw speed, there will be a constant amount of material in the extruder and thus the shear stress and residence time applied to material remains constant, 3) The motor load and melt pressure: These parameters depend on feed rate and screw speed. With constant feed rate and screw speed these parameters depend upon the molecular weight of polymer and drug as well as polymer miscibility in binary mixtures [7].

7. Co-rotating and counter rotating twin screw

Since there are varieties of raw materials to mix into chocolates, the co-rotating twin screw extruders has been interesting to investigate in order to improve chocolate extruder machine. The twin screw extruder has been classified into two types as following [12]:1. Based on screw rotation direction: co-rotating and counter-rotating, as shown in figure 6, 2. Based on degree of meshing: non overlap (non-intermeshing) and overlaps which are a partially intermeshing and a fully intermeshing as shown in figure 7.

The range of twin screw extruder operation is during low speed (10-40 rpm) to high speed (200-500 rpm). In most industries, counter-rotating twin screw extruders are preferred because they have better conveying performance than co-rotating twin screw extruders as well as counter-rotating twin screw extruders and the heat generated by the shear is less [12].

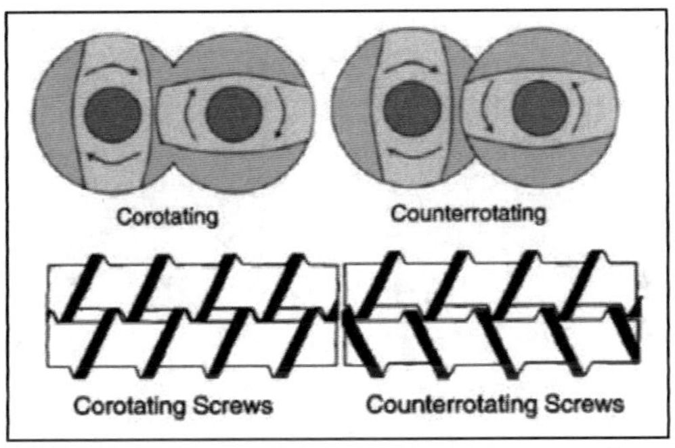

Figure 6. Screw rotation direction, co-rotating screw and counter-rotating [6]

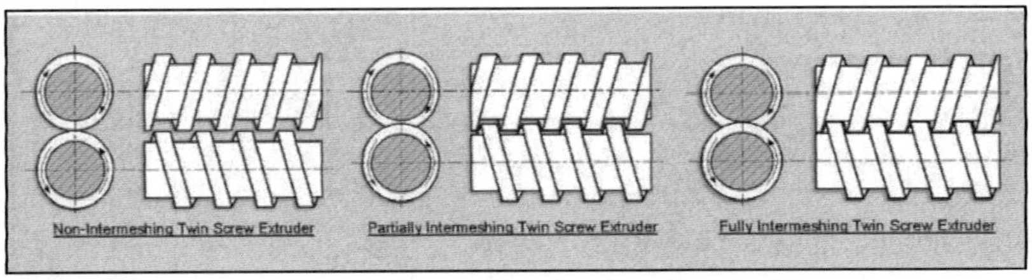

Figure 7. The amount of overlap between the screws in the twin-screw extruder [6]

Twin screw extruder offers advantages over single screw extruder in that it does not require a drag force to push the material. Less shear results in less shear heat, more efficient mixing and less flexibility for low speed applications [6]. A comparison of screw parameters was concluded as table 1.

Table 1.A review of investigating screw parameters.

Researchers	Year	Screw parameters				
		Single screw extruder	Twin screw extruder	Ram extruder	Screw speed	Material temperature
Milivoje M. Kostic [14]	2006	die design	-	-	-	-
Vivar-Vera [15]	2008	-	-	-	40–60rpm	40- 80°C
HesamAnvariArdakani [16]	2014	-	-	-	-	30- 50°C
Chris Rauwendaal [1]	2015	solving extrusion problems	-	-	50 rpm	50 - 100°C

8. Screw extrusion problems

Although there are several benefits of screw extrusion e.g. available for many chocolate formulas, flexible for process parameters and suitable for continuous process, there are many extrusion problems such as shape accuracy, appearance problems, functional product properties, high melt temperature, high motor load, wear of screw and/or barrel. Moreover, process problems had been mentioned [1] such as low yield productivity, and products defected, as shown in figure 8. A non-appropriate extrusion temperature affects to product surface that increases production cost, as shown in Figure 8(a), and to chocolate state that constructs un-uniform chocolate line, as shown in figure 8(b).

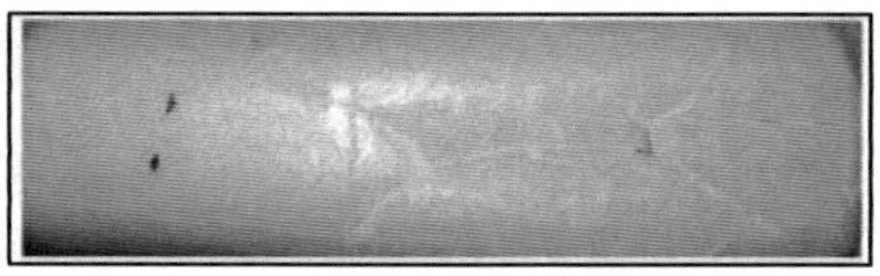
(a) The cracks surface

(b) Un-uniformchocolate line

Figure 8. Defective product [13]

9. Future experimental design

Based on the literature reviews, the temperature and screw speed parameters affected the product shape and flaw. However, there were a few studies on twin screw condition. In order to improve product development, the co-rotating twin screw extrusion will be investigated as a future work to determine suitable conditions such as screw speed and appropriate temperature for extrusion process.The co-rotating twin screw for chocolate extruder will be conducted based on two parameters that are screw speed and temperature in barrel. Then, the width of the extruded chocolate lines and the error between extruded chocolate and nozzle diameter will be measured as the results in which the design tolerance will be controlled with in ± 0.1 millimeters.

10. Conclusion

Screw extrusion is widely used in the food industry e.g. pasta and dessert. One of the desserts with high market value is chocolate, in which chocolate is also capable of melting and extrusion. The extruder also has many problems in the manufacturing process, such as low productivity and defective products. As a result, the extruder design has been investigated to improve process efficiency and reduce the production problems that occurred during extrusion process. The co-rotating twin-screw extruder has been redesigned in order to reduce product flaws while increase productivity. Since co-rotating twin-screw composes of feeder element, transition element and metering element, the advantages of co-rotating twin-screw are well mix chocolate paste and good conveying performance of paste products. However, the process parameters of co-rotating twin-screw chocolate extrusion should be investigated in order to minimize screw extrusion problems as a future work.

Acknowledgments
The authors would like to acknowledge the Suranaree University of Technology, in particular School of Industrial Engineering, Institute of Engineering for support the research

References

[1] Chris R 2015 Effective troubleshooting of extrusion problems *Proc. Reg. Conf. Graz Polym. Processing Soc. (PPS)*
[2] Schantz B and Rohm H 2005 Influence of lecithin–PGPR blends on the rheological properties of chocolate *Lebensmittel Wissenschaft und Technologie* **38** p 41-5
[3] Beckett ST 2009 *The Science of Chocolate* vol 2 (Cambridge: Royal Society of Chemistry Paperbacks) p 240
[4] The Institute of Nutrition 2017 *Extruder Design* (Mahidol University)
[5] Kruder G A 1985 *Extrusion* vol 1, 2nd ed John Wiley and Sons Inc (New York: Encyclopedia of Polymer Science and Engineering) p 571-631
[6] Rosato DV 1998 *Extruding Plastics a practical processing handbook* (USA: Chathem MA)
[7] Breitenbach J 2002 *Melt extrusion: from process to drug delivery technology* (Biopharm: EurJ Pharm) chapter 54 p 107-17
[8] Martin C 2001 *Guidelines for Operation of Leistritz Twin screw Extruder* (Somerville: American Leistritz Corporation)
[9] Whelan T and Dunning D 1988 *The Dynisco Extrusion Processors Handbook*, 1st ed (London: London School of Polymer Technology)
[10] Johnson P S 1982 *Development in Extrusion Science and Technology* (Ontario: Polysar technical publication)
[11] Shah R D, Kabadi M, Pope D G and Augsburger L L 1994 Physicomechanical characterization of the extrusions pheronization process *Pharm. Res.* chapter 11 p 355- 60
[12] Nicholas P 1998 *Cheremisinoff: Guidebook to Extrusion Technology* (New York: Potsdam Clarkson College of Technology)
[13] Arun J S, George M, Kalpana K, Janardhana S, Jaya J and Jaina G 2014 Defects identification in chocolates using neural fuzzy interface segmentation *Int. J. Eng. Sci. Inv.* **3** p 50-7
[14] Milivoje M K 2006 *Design of Extrusion Dies* (Encyclopedia of Chemical Processing)
[15] Vivar-Vera G, Torrestiana-Sanchez B, Monroy-Rivera J A and Fuente E B 2008 *Chonching Rheological and Structural Changes of Chocolate Mass (Deutsche Lebensmittel-Rundschau)* **104** p 376-82
[16] Ardakani H A, Mitsoulis E and Hatzikiriakos S G 2014 Capillary Flow of Milk Chocolate J Non-Newtonian Fluid Mech. **210** p 56-65

IOP Conf. Series: Materials Science and Engineering **703** (2019) 012013 doi:10.1088/1757-899X/703/1/012013

Tofu wastewater treatment through a combined process of coagulation-flocculation and ultrafiltration

P Prawati, A Oktariany, S S Putri, I Aditya and S Kartohardjono

Department of Chemical Engineering, Universitas Indonesia, Kampus UI Depok, Depok, 16424, Indonesia

E-mail: sutrasno@che.ui.ac.id

Abstract. Wastewater from tofu industries contain many parameters such as total suspended solid (TSS), total dissolved solid (TDS), turbidity, and chemical oxygen demand (COD). These parameters are present in high amount and can endanger the aquatic environment and also decrease water quality. Several studies have been conducted to reduce these parameters, but still require a longer time which is around 12 to 78 days. This study aims to discuss the processes and methods to reduce TSS, TDS, turbidity, and COD parameters from tofu wastewater. The processes discussed in this study include a combined process of coagulation-flocculation and ultrafiltration (UF). The results of the study show that TSS, turbidity and COD reduced to 60%, 60% and 20%, respectively, in the coagulation-flocculation process, but TDS increased to around 90% because of the neutralization process using Sodium Hydroxide (NaOH) solution. Furthermore, it was shown that the combined process of coagulation-flocculation and UF was able to reduce TSS, turbidity, COD and TDS up to 99%, 99%, 75% and 50%, respectively. All the parameters contained in the water produced from the combined process of coagulation-flocculation and UF was in line with the National Environmental Quality Standard requirements, excluding COD due to the COD content in fresh tofu wastewater is very high.

1. Introduction

Tofu is a favorite food in Asia [1, 2], especially in Indonesia because of its low-cost, and it contains good and healthy nutrition. Almost every city in Indonesia has tofu processing industries, which are usually run by small household industries, and are managed with simple technology [3]. Therefore, there is low effective and efficient use of resources such as raw materials and water [4, 5]. High amount of harmful organic and chemical compounds are present in tofu wastewater such as biological oxygen demand (BOD), chemical oxygen demand (COD) and total suspended solid (TSS) in the range of 5000 - 1000 mg/L, 7000-26000 mg/L and 6000-8000 mg/L, respectively [2, 6, 7].

The waste from the tofu industry can be in the form of liquid waste and solid waste, where a portion of liquid waste can be used for subsequent production, while solid waste can be utilized as feed for animal[5, 8]. However, because of the large amount of water used for tofu processing, the wastewater is also released in great volume, and it is generally not fully reusable; so the remaining liquid waste is released into the nearby aquatic environment. Unfortunately, discharging of tofu wastewater causes environmental pollution such as bad odors and soil pollution, increased water turbidity, decreased photosynthetic rate of phytoplankton and other aquatic plants; thereby decreasing water productivity. Several health problems such as skin diseases, stomach aches, and digestive problems can also be triggered by consuming water that has been polluted by tofu wastewater [5, 9, 10].

Several recent studies have been conducted on the utilization of tofu wastewater to produce products such as bio-ethanol [11], vitamin B12 [1, 2, 12] and cultivation of micro algae [13, 14]. Several attempts have also been made to treat tofu waste water in Indonesia through anaerobic processes in the fixed dome anaerobic and anaerobic baffled reactor [5] and the combination of anaerobic baffled and activated sludge system. However, the processes take longer time which is around 12 to 78 days [15]. Another method used is ozonation. In this process, the waste water is diluted first prior to processing [16]. The aim of the study is to assess the success of the combined process of coagulation-flocculation and ultrafiltration (UF) in treating tofu industrial wastewater. The COD, TSS, TDS, and turbidity are the parameters that will be observed to assess the success and effectiveness of the proposed process. A coagulant used in the coagulation-flocculation process is poly aluminum chloride (PAC), and it is one of the best options for reducing COD, TSS, and turbidity of wastewater [17]. Meanwhile, the UF process is expected to further reduce COD, TSS, and turbidity level of the wastewater [18].

2. Materials and methods

2.1 Materials

The experimental set up for the combined process of coagulation-flocculation and UF is schematically shown in Fig. 1. The wastewater was provided by the tofu industry in South Jakarta, Indonesia. The sodium hydroxide used for pH adjustment was analytical grade and was obtained from Merck Indonesia and the filter paper used was sized 10 μm BIPMED. Meanwhile, the PAC used was technical grade and was purchased from PT Brataco Indonsesia. There were three hollow fiber membrane modules used in this study, which were made of polysulfone, cellulose acetate, and ceramics and were supplied by CV GDP Filter Bandung Indonesia.

2.2 Methods

Prior to the coagulation process, the wastewater's pH was set to about 7.0 by addition a certain amount of 2 M NaOH as a base solution. The wastewater solution was added with 1 ml of coagulant solution in the coagulation-flocculation process, and the process was performed at 120 rpm for 2 min and then continued at 40 rpm for 10 min in the VELP JLT-6 Jar Test. After 30 minutes, the solution was filtered using a 10 μm filter paper, and the filtrate was used as feed for the UF process. In the UF process, compressed air was employed to generate the pressure driving force a cross the membrane. TSS and turbidity, COD, and TDS and pH were measured by Hach DR/890 Colorimeter, Spectrophotometer UV-Vis BEL UV-MS1 single beam and Hanna Waterproof Tester Combo pH and EC, respectively. The whole experiments were repeated for three times. All experiments to measure the observed parameters were repeated three times and the average values were taken as final data.

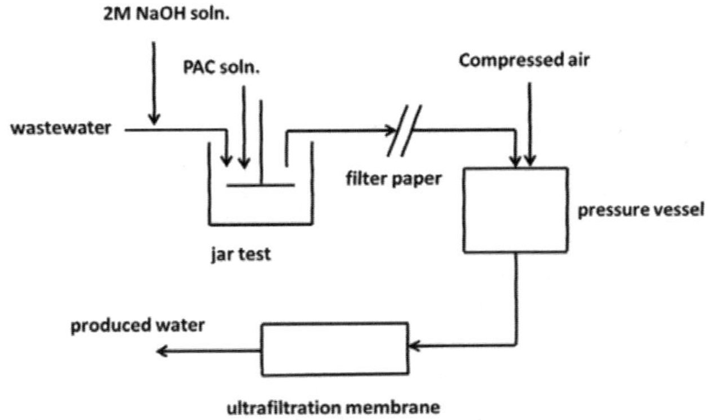

Figure 1. Schematic diagram of a combined process experiment.

3. Results and discussion

3.1 Coagulation-flocculation process

The tofu wastewater characteristic as well as the National Environmental Quality standard [19] is shown in table 1. Based on the data present in table 1, it is clear that all the observed parameters have already passed the threshold set by the government. Besides being acidic as shown by the wastewater's pH which is below 4.0, tofu wastewater also contains a high suspended solid as indicated by high turbidity, TSS and COD. The TSS as well as turbidity of the wastewater decreased, while TDS and COD increased after the addition base solution to neutralize the wastewater as shown in figure 2. The increase in TDS was mainly due to the increase in the level of dissolved ions present in the wastewater which is derived from Na^+ ions. Meanwhile, the decrease in TSS and turbidity was due to a small amount of suspended solids that precipitate during the addition base solution.

Table 1. Tofu wastewater characteristic

Observed Parameters	Concentration range	Government Regulation [19]
TSS (mg/L)	280 - 415	100
Turbidity (FAU)	370 - 525	25
COD (mg/L)	5980 - 7950	275
TDS (mg/L)	850 - 1050	2000
pH	3.4 - 3.8	6.0 - 9.0

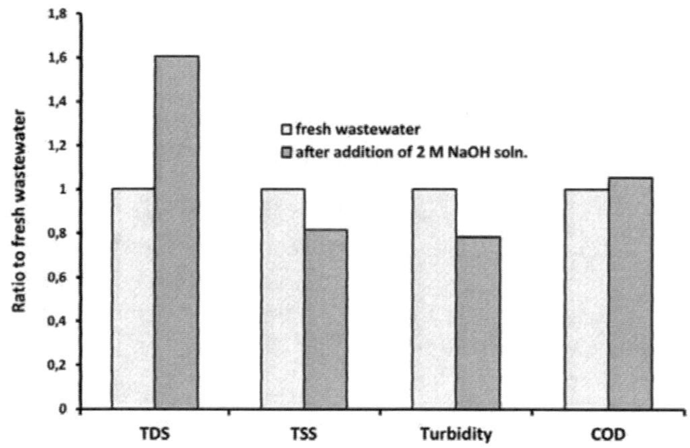

Figure 2. The Observed Parameters in the Tofu Wastewater before and after adding of 2 M NaOH Solution.

The initial pH of solution effect on the TSS, turbidity and COD in the coagulation-flocculation process is presented in figure 3. These parameters decrease in the range of pH 6 to 7, and then increases until pH 9. At low pH the hydrolyzate produced by PAC is $Al(OH)^{2+}$, $Al_2(OH)_2^{4+}$ and $Al_3(OH)_4^{5+}$. The positive hydrolyzate can easily neutralize the negative charge on the outer part of the organic molecule in the solution to destabilize the colloid. At pH between 6 and 8, positive polymer hydrolyzate and $Al(OH)_3$ are formed in the solution. These hydrolyzates have low solubility and high surface area which makes it easy for them to precipitate to the bottom of the container. When the pH is above 8, the solution will be difficult to destabilize because the hydrolyzate formed is $Al(OH)_4^-$ [20].

The TSS, turbidity, COD and TDS of wastewater after the process of coagulation-flocculation are demonstrated in figure 4. The TSS, turbidity and COD decreased until the settling time of 30 minutes, and relatively constant from 30 to 90 minutes. The decrease in TSS and turbidity reaches 60%, while the decrease in COD is around 33% after the coagulation-flocculation process. The decrease occurred due to the flock formed was broken so that it dissolves back into the wastewater. Meanwhile, TDS increases with increasing settling time due to more solids that keeps dissolving into wastewater. The increase can reach 90% as presented in figure 4.

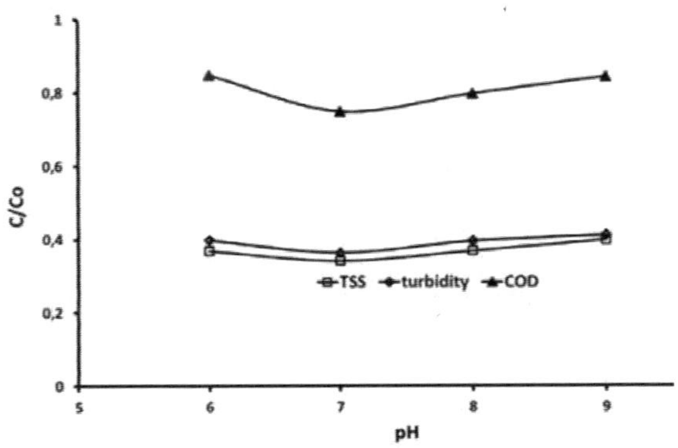

Figure 3. The effects of initial pH on TSS, turbidity, COD and TDS in the process of coagulation-flocculation.

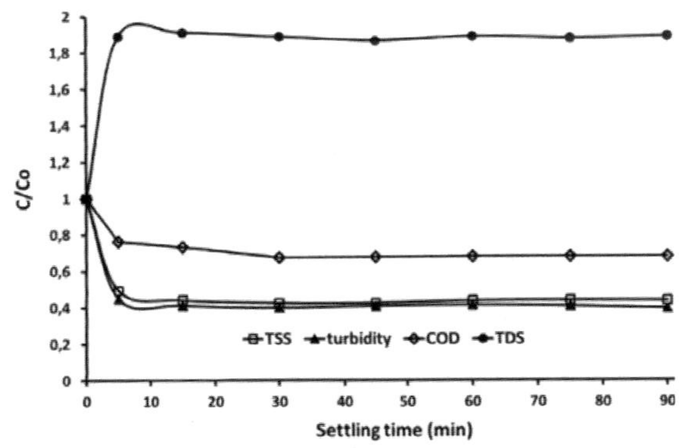

Figure 4. The settling time effects on the TSS, turbidity, COD and TDS in the coagulation-flocculation process.

The effect of PAC dose on TSS, turbidity, COD and TDS in the process coagulation-flocculation of wastewater solution at pH 7.0 is shown in figure 5. There is adecrease in the level of TSS, COD and turbidity, while TDS increase after the process of coagulation-flocculation. The pH of the solution applied in the process of coagulation-flocculation is closely associated to the coagulation process efficiency. The PAC has low solubility at a pH in the range of 6 -7, so that the coagulant can be maximally transformed into flock. The coagulation-flocculation process involves colloidal particles destabilizing by neutralizing forces that cause the colloids to stay far away from each other. Cationic coagulants provide a positive electrical charge which reduces the negative colloids charge. As a result, the particles collide with each other to produce macro particles known as flocks [21].

Generally, coagulants to form hydroxides when it dissolves in water, so that the metal hydroxide polymers formed will be of amorphous structures, positively charged, and have a large surface area. These hydroxides are hydrophobic, which can be absorbed onto the surface of organic anionic particles, therefore becoming insoluble [17]. As shown in figure 5, TSS from the wastewater has been reduced to nearly 50% after the coagulation-flocculation process due to destabilization of suspended solids, so the process runs effectively. The flock formed from the process has sufficient time to settle at the bottom of the Erlenmeyer so that the filtrate of tofu wastewater was free of particles larger than 10 microns. Similar effects also occurred for the turbidity as it can be seen that the decrease is almost equal to that of TSS, so that the turbidity decreasing curve coincides with the TSS decreasing curve. Since the turbidity of the solution is strongly influenced by the amount of suspended solids, so that the solution turbidity increases with suspended solids, while the brightness of the solution decreases with suspended solids. Figure 5 also shows that the decrease in COD only in the range of 8 to 28%. Particle deposition can indeed decrease the value of COD, especially in wastewater with high TSS content. Generally, a 1mg/L decrease in TSS will also decrease COD by about 10 mg/L[22]. In this process, TSS successfully decreased by about 175 mg/L, while the COD was decreased by 1584 mg/L. Thus, for every 1mg/L decrease in TSS, there is also a 9mg/L decrease in COD. Meanwhile, figure 5 shows that TDS tends to increase due to the addition of a base solution to neutralize the wastewater's pH. An increase in TDS can also arise due to the fact that dissolved medium solids (colloid transition) in the wastewater solution passes through the filter paper because their size is less than 0.45 microns.

3.2 Ultrafiltration process

Figure 6 shows the water fluxes in the UF process for polysulfone (PS), cellulose acetate (CA) and ceramic membranes. As expected, water flux increases with increasing trans-membrane pressure (TMP) in the ultrafiltration process. The fluxes of the PS and ceramic membranes are almost the same, while that of the CA membrane is lower due to the smaller membrane pore size.

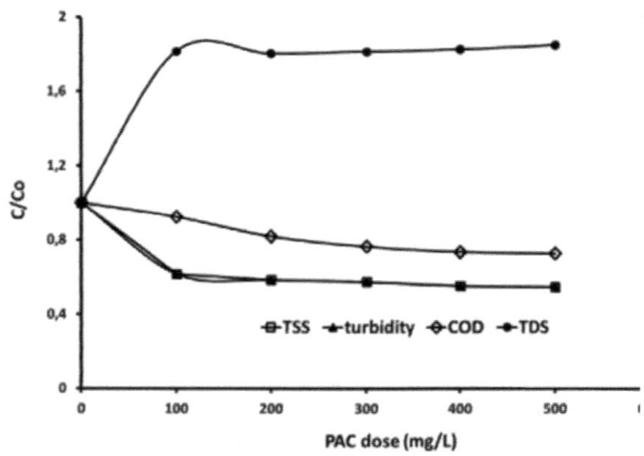

Figure 5. The effects of PAC dose on TSS, turbidity, COD and TDS in the process of coagulation-flocculation.

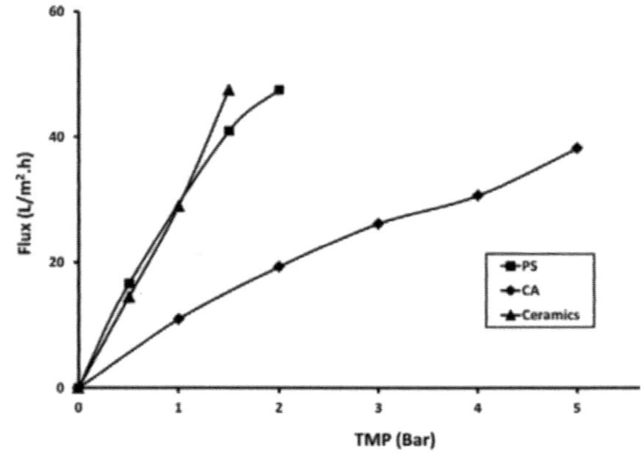

Figure 6. The Effects of Trans-membrane Pressure, TMP, on the Water Flux in the Ultra filtration Process.

The effect of TMP on TSS in the ultrafiltration process using PS, CA and ceramic membranes is shown in figure 7. It is shown in figure 7 that all the membranes used were able to produce products with very low TSS levels (<10mg / L) with a rejection of 92 to 99%. It also reveals that suspended solids cannot penetrate the membrane in the UF process. The lowest TSS of water produced in UF process was 2 mg/L, which far below the National Environmental Quality standard of 100 mg/L. This effect also occurs in turbidity where turbidity significantly decreased by more than 95% as shown in figure 8. The lowest turbidity of the water produced was 4 FAU which is far below the National Standard of 25 FAU. The effects of TMP on TDS and TDS rejection on the UF process is shown in figure 9. TDS decreases at TMP of 1.5 bar and then increases at TMP of 2.0 bar for PS membranes. This is due to the fact that at TMP of 2.0 Bar, more dissolved solids penetrate the membrane, so it increases the TDS of the water produced.

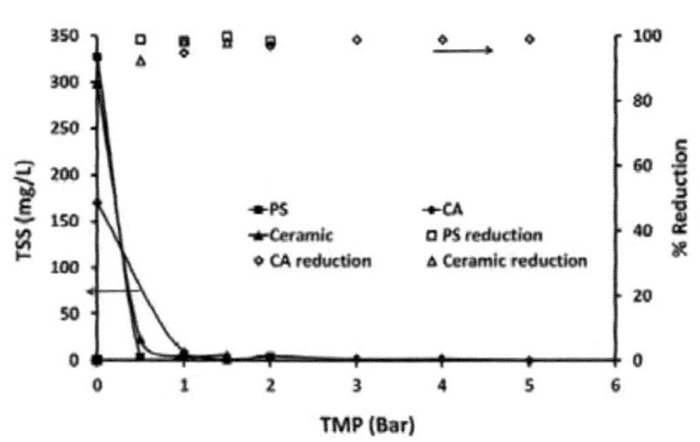

Figure 7. The effects of trans-membrane pressure (TMP) on TSS and TSS Reduction in UF process using PS, CA and ceramic membranes.

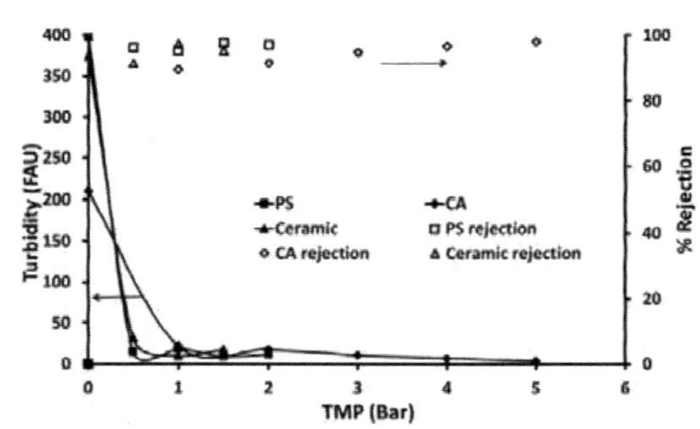

Figure 8. The effects of trans-membrane pressure (TMP) on turbidity and turbidity reduction in UF process using PS, CA and ceramic membranes.

Figure 10 shows the COD of the water produced after the UF process. The COD decreases with TMP of 1.5 and 1.0 bar and then increases at TMP of 2.0 and 1.5 bar for PS and ceramics membranes, respectively. Meanwhile, the COD decreases with TMP applied in the experiments for the CA membrane. It reveals that the UF membranes could retain the soluble and particulate organic matter in water not to cross the membrane pores up to TMP of 1.0 and 1.5 Bar for ceramic and PS membranes, respectively. The CA membrane was also able to hold soluble and particulate organic matter in water not to cross the membrane pores in the TMP applied in the experiment. The lowest COD of water that can be produced in the UF process is around 1460 mg/L, which still does not meet the National Standards. A high COD contents is mainly due to the presence of soluble and particulate organic matter in water which cannot be completely discarded by membrane pores in the UF process.Therefore, it is necessary to do a further process so that the COD of produced water complies with government regulations. The reverse osmosis (RO) process can be recommended for this purpose.

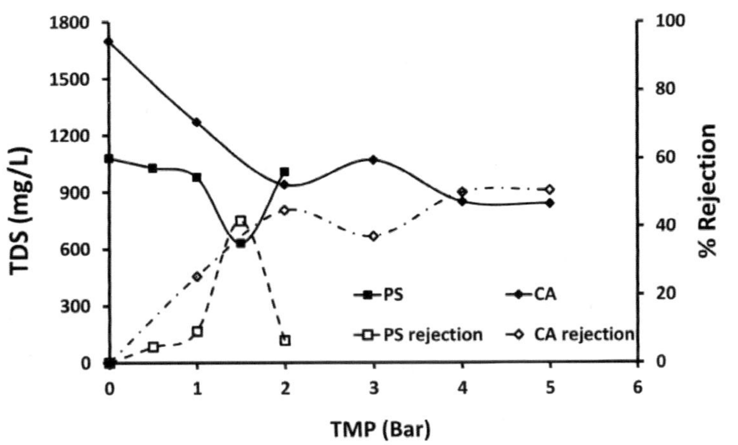

Figure 9. The effects of trans-membrane pressure (TMP) on TDS and TDS Rejection in UF process using PS, CA and ceramic membranes.

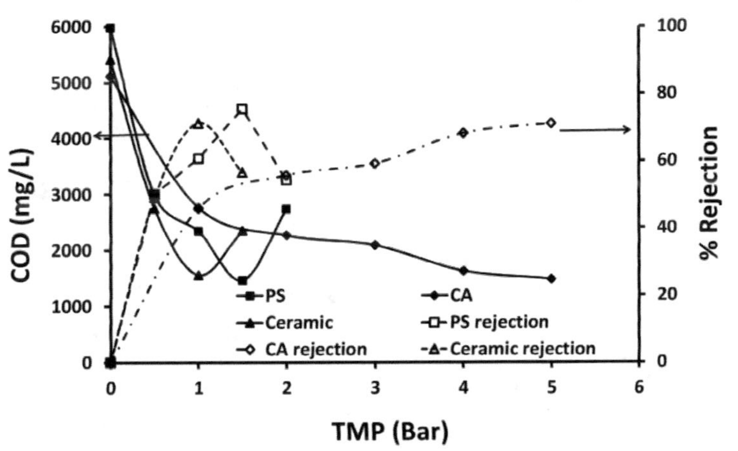

Figure 10. The effects of trans-membrane pressure (TMP) on COD and COD rejection in ultrafiltration process using PS, CA and ceramic membranes.

4. Conclusion

Tofu industries produce a lot of wastewater that cannot be discharged directly into the aquatic environment without any treatment process as they contain a high level of TSS, TDS, turbidity and COD. This study aims to reduce TSS, TDS, turbidity, and COD parameters from tofu wastewater through a combined process of coagulation-flocculation and UF. The results of experiment revealed that TSS, turbidity and COD can be reduced up to 60%, 60% and 20%, respectively, while TDS increased by about 90% in the coagulation-flocculation process. The increase in TDS was due to adding of base solution when neutralizing the wastewater. Meanwhile, TSS, turbidity, COD and TDS can be reduced up to 99%, 99%, 75%, 50%, respectively, after UF Process. All parameters in the water produced from UF process fulfill the National Standard except COD, which is still above the standard due to the high COD content in the fresh tofu wastewater. COD content in the fresh tofu wastewater can be reduced from 8000 ppm to 1460 ppm, while the requirement should not exceed 275 ppm.

Acknowledgments
Authors wish to acknowledge the financial support of this publication from PIT9 Project provided by Universitas Indonesia through Contract No. NKB-0079/UN2.R3.1/HKP.05.00/2019.

References
[1] Yu Y, Zhu X, Shen Y, Yao H, Wang P, Ye K, Wang X, and Gu Q 2015 *App. Micro & Bio.* **99** 10481
[2] Zheng G, Wang L and Kang Z 2010 *Ren. En.* **35** 2910
[3] Aditya I and Kartohardjono S 2018 *E3S Web of Conf.* **67** 04006
[4] Belén F, Sánchez J, Hernández E, Auleda J and Raventós M 2012 *J food Eng.* **110** 364
[5] Faisal M, Gani A, Mulana F and Daimon H 2016 *Asian J. of Chem.* **28** 501
[6] Anggarini S, Hidayat N, Sunyoto NMS and Wulandari PS 2015 *Agr. and Agr. Sci. Proc.* **3** 95
[7] Zhu H, Suzuki T, Tsygankov AA, Asada Y and Miyake J 1999 *Int. J. H. En.* **24** 305
[8] Putri SS and Kartohardjono S 2018 *E3S Web of Conf.* **67** 04005
[9] Sobral P and Wagner J 2009 *Inf. Tech.* **20** 65
[10] Sudiyani Y, Alawiyah S, Anita Y and Badria I 2007 Rep. UGM
[11] Febrianti F, Syamsu K and Rahayuningsih M 2017 *Int. J. Tech.* **5** 898
[12] Yuwono SD and Kokugan T 2008 *Bio. Eng. J.* **40** 175
[13] Wang S-K, Wang X, Miao J and Tian Y-T 2018 *Biores. Tech.* **253** 79
[14] Yonezawa N, Matsuura H, Shiho M, Kaya K and Watanabe MM 2012 *Biores. Tech.* **109** 304
[15] Wagiman A and Suryandono A 2006 *Agritech* **26** 39
[16] Hadiyanto H 2018 *E3S Web of Conf.* **31** 03002
[17] Irfan M, Butt T, Imtiaz N, Abbas N, Khan RA and Shafique A 2017*Arabian J. Chem.* **10** S2307
[18] Minesterial Regulation of Ministry of Environment No. 5 2011. Republic of Indonesia
[19] Zhang P, Wu Z, Zhang G, Zeng G, Zhang H, Li J, Song X And Dong J 2008 *Sep. Pur. Tech.* **3** 642
[20] Arinaldi A, Ferdian F and Widiasa IN 2013 *J. Tek. Ind.* **2** 8
[21] Tchobanoglous G, Stensel HD, Tsuchihashi R, Burton F, Abu-Orf M, Bowden G And Pfrang W 2014 *Wastewater Engineering: Treatment and Resource Recovery* (New York: McGraw-Hill Book)

IOP Conf. Series: Materials Science and Engineering **703** (2019) 012014 doi:10.1088/1757-899X/703/1/012014

Carbon emission modelling in container terminal operations planning using a system dynamics approach

D N Prayogo

Department of Industrial Engineering, University of Surabaya, Raya Kalirungkut, Surabaya, 60293, Indonesia

E-mail: dnprayogo@staff.ubaya.ac.id

Abstract. This paper discusses the model development of carbon emission cost in container terminal operations planning using a system dynamics approach. Operational planning in a container terminal consists of planning a vessel at berth process, containers loading and unloading processes by quay cranes, containers transportation process by container trucks, containers loading and unloading processes by yard cranes, and containers receiving and delivery processes by container trailers. In each stage, the process will produce carbon gas emissions. A system dynamics simulation model was applied to determine the optimal allocation of the number of used material handling equipment based on the number of handled containers to minimize the total carbon emissions in a container terminal operations planning. The design of a system dynamics simulation model considers the uncertainty in the vessel arrival time and the number of containers that should be handled on each moored vessel at the berth. The system dynamics model had been applied to a numerical example for resulting the optimal solutions of planning decisions at container terminal operations.

1. Introduction

In recent years, container terminal industries have been under great pressure to meet economic and environmental standards. The level of energy consumption and resulting emissions in this industry significantly increases the energy consumption rate and costs. Nevertheless, few container terminal industries have implemented performance measurement and strategy of energy efficiency. Container terminal performance needs to be improved to create industries that are not only competitive and productive, but also need to be more sustainable. As a consequence, performance measurement efforts that must be beyond the measurement of traditional efficiency and productivity are emerging challenges. In terms of energy consumption, there is a strong relationship between sustainability, efficiency, competitiveness and port profitability. The relationship of sustainability with the efficiency of energy consumption and port performance has not been fully understood, nor has been analyzed in details [1].

Energy consumption level is one of the important factors in port operations and economic activities related to ports. Increased energy costs need to be responded by port authorities and container terminal operators by finding ways or strategies to reduce fuel expenditure. With the growth of global container trade and port infrastructure development, ports have become significant energy consumers.

An analysis of energy consumption requires a detailed understanding of the portion of energy expenditure at container terminal operations expressed by different sizes and types of containers, so that it can determine what operating area consumes energy [2].

The remainder of this paper is structured in the following manner. Section 2 reviews the relevant literature with the main focus on energy consumptions and carbon emissions resulting from container terminal operations. A detailed description of the problem studied herein is described in Section 3. In Section 0, the model development using system dynamics approach is presented for minimizing the total carbon emissions in container terminal operations. Section 5 presents and discusses the results of simulation model and, finally, in the last section summarizes findings and outlines potential future research directions.

2. Literature review

Although the literature related to energy consumption in container terminal operations is quite limited, several research work on energy consumption in specific types of container handling equipment had been conducted by Yun et al. [3], Yang and Chang [4], Acciaro et al. [5] and Geerlings and Duin [6]. However, in general, there has been no systemic view research of the beyond impact of energy consumption on technological advancements in container terminals. Yun et al.'s work was one of findings reported on the impact of electric rubber-tired gantries on green port performance [3]. Yang and Chang's research indicates that bulbar-powered RTGs equipped with online braking can reduce energy consumption by up to 60% [4].

Sim has made energy activity clusters as follow: vertical operations (quay cranes), horizontal operations (e.g. reach-stacker (RS) cranes, rubber-tired gantry (RTG) cranes, rail-mounted gantry (RMG) cranes, etc.), lighting, buildings and cooling (reefers). Usage time is another important factor when it comes to measuring energy consumption and setting indicators for energy efficiency because of: (1) ship calling patterns, all which can trigger significant variations and peaks in energy consumption; (2) variations in the dwell time of different container types (e.g. import and export containers); and (3) the seasonality of certain types of traffic (e.g. reefers) [7]. Geerlings and Duin have classified the energy consumption per type of cargo handling equipment in terminals as shown in the following Table 1 [6].

Table 1. Energy consumption per type of equipment.

Energy	Type of equipment	Fixed consumption per container move	Variable consumption
Electric	Quay Crane [a]	6.00 kWh	
	Barge Crane [a]	4.00 kWh	
	Rail Crane [a]	5.00 kWh	
	Automated Stacking Crane [a]	5.00 kWh	
	Rail Mounted Stacking Crane [b]	7.25 kWh	
	Platform [b]	5.00 kWh	
Diesel	Automated Guided Vehicle [a]	1.10 liter	1.80 liter/km
	Straddle Carrier [a]	0.80 liter	3.50 liter/km
	Terminal Tractor [a]		4.00 liter/km
	Multi Trailer System [a]		4.20 liter/km
	Reach Stacker/Top Lifter [a]		5.00 liter/km

[a] Based on a TNO project by Oonk
[b] Based on a comparison with the ASC on the ECT Delta Terminal, in which the reach of the equipment (stack length) is taken into consideration

Carbon emissions at container terminals are produced by direct and indirect carbon emissions based on fuel consumption or electrical energy [3]. Equipment operated using fuel, such as container trucks and yard cranes, consumes fuel energy while operating such that it will produce direct carbon emissions. Equipment operated by using electricity, such as OPS, will produce indirect carbon emissions, which means that emissions are generated outside the seaport. Carbon emissions either directly or indirectly produced by each equipment (Figure 1.) can be calculated according to ISO 14064-1: 2006 standard [8] and IPCC 2006 [9], which are determined by the amount of energy consumption and the coefficient of emissions carbon from that energy, as given in Eq (1).

$$W = F.C_{energy} \quad (1)$$

where W represents the carbon emissions; F represents the energy consumption, which can be diesel (kg), LNG (kg), or electric (kWh); C_{energy} is the carbon emission coefficient of the energy, which can be obtained from IPCC 2006 [9] and the energy statistics department in corresponding region. The CO_2 emission coefficient of electricity for the port is 1073.65 lbs CO_2/MWh (1 lb =0.45359 kg), namely 0.7369 kg/kWh [10]. Therefore, once the energy consumptions are formulated, the carbon emissions can be obtained.

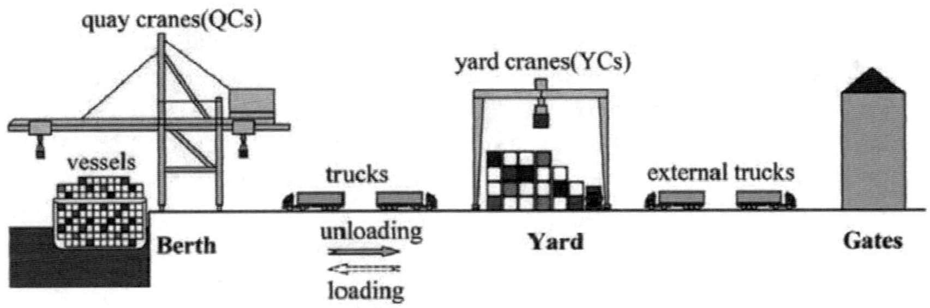

Figure 1. The equipment in container terminals [3].

The energy consumption level of ships can be divided into three components, namely: (1) energy consumption when the ship sails in the waterway, (2) energy consumption when the ship is waiting at the port, and (3) energy consumption when the ship is berthing at seaport. Based on Korean Environmental Industry and Technology Institute, 2016 in [7], this paper used emission density values published by a South Korean organization as a basis for estimating carbon emissions produced in South Korea (KLCI DB, 2016). The carbon density values of the major vessel and equipment activities are shown in Table 2.

Table 2. The carbon emission density of major vessel and equipment activities.

Activity	CO_2-e (kg/TEU)
Container ship maneuver	1.99000
Container ship at berth	0.02324
Quay crane	1.39000
Yard truck	0.04287
Yard crane	0.06873
Container trailer	0.38645

2.1. CO_2 emission calculation for vessels

The vessel emissions at the hoteling phase are calculated using the following equation from Hu et al. [11]

$$E_{vessel} = \sum_p PO.t_s.LF.EF_p.FCF_p.C_y \quad (2)$$

Where: E_{vessel} is the total vessel emission while hoteling in kg of CO_2;

PO is the rated power (kW) of the auxiliary engine of the vessel,

t_s is the makespan or completion time in which the engine is used (hours),

LF is the load factor of the engine while the vessel is moored;

EF_p is the emissions factor kg CO_2/kWh for each pollutant p considered,

FCF_p is the fuel correction factor for each emission factor, to replicate changes in the fuel properties over time

C_y is the number of cylinders of the auxiliary engines of the vessel.

2.2. CO_2 emission calculation for QCs

A QC is an electrical machine that does not emit CO_2 directly, but the electricity used, was previously generated through a process with CO_2 emissions. According to Geerlings and Van Duin [6] the QCs' energy consumption is considered fixed per handled container, with a fixed value of 6 kWh/container. Thus, the CO_2 emission during the crane operation can be calculated as:

$$E_{cranes} = \sum_k C_s . W_k . EFC_k \quad (3)$$

Where: E_{cranes} is the total crane emission in kg of CO_2 during the loading/unloading process,
C_s is the fixed energy consumption for a QC, in kWh/container,
W_k is the amount of containers handled by each QC k,
EFC_k is the emission factor per electric QCs, in kg of CO_2/kWh;
k changes from 1 to n, which is the total number QCs used.

In this paper, the system dynamics simulation model is applied to help decision makers in solving the complex systems problem to minimize the total carbon emissions in container terminals in spatial and multi-temporal integration. The steps needed for system dynamics modeling and simulating of complex systems based on systems thinking are: (1). Identify the problem. (2). Develop a dynamic hypothesis explaining the cause of the problem. (3). Create a basic structure of a causal graph. (4). Augment the causal graph with more information. (5). Convert the augmented causal graph to a system dynamics flow graph. (6). Translate a system dynamics stock-flow graph into Powersim programs and equations [12].

3. Problem description

Based on real data at the port, the number of vessel calls can be 10 calls/week up to 60 calls/week. Quay cranes can process loading and unloading containers from and to vessels between 25 to 35 TEUs/h. The yard tractor moves at an average speed of 25 km/h and the handling capacity of each block is 25 TEUs/h. The number of loading containers may differ a lot between vessels of different sizes to simulate real conditions in container ports. The number of loading containers for one service call can reach 3600 TEUs and not less than 400 TEUs. For various types of loading tasks, mooring productivity may be different.

In this paper, the sets of parameter used in the system dynamics simulation model are as follow: the vessel arrival rate is Uniform Discrete distribution, UD [3,15] vessel calls/day at container terminals that have berth capacity of 20 vessels. The number of handled containers per vessel by Normal distribution with mean of 500 TEUs and standard deviation of 150 TEUs. Quay crane productivity rate Uniform Discrete distribution, UD [500,600] TEUs/day. While the amount of carbon emission density of vessel and equipment activities in kg/TEU is shown in Table 2, with the proportion of carbon emissions of the idle quay crane is about 3% of the carbon emission of quay crane in operating.

4. Model development

Total carbon emissions at container terminals are generated from the arrival of ships and the operation of all cargo handling equipment. In the process of arrival of ships at container terminals, carbon emissions are generated from the transport vessel and loading unloading vessel. Whereas carbon emissions from cargo handling equipment consist of quay crane processing, idle quay cranes, container truck transport and crane processing yards. Vessel arrival rate at berth is uncertain and is limited by berth capacity and the number of vessels still awaiting service on berth, while the vessel rate depends on the number of vessels in berth and container handling rate by quay cranes and the number of containers per vessel. The container handling rate by quay cranes is influenced by the number of containers that must be served and the number of quay cranes assigned to service each vessel. The number of quay cranes assigned to service each ship depends on the quay crane productivity rate and the number of quay cranes available at the container terminal. The number of containers in the storage container yard is affected by the capacity of container trucks and the capacity of yard cranes available. The relationship between variables in system dynamics in container terminals

IOP Conf. Series: Materials Science and Engineering **703** (2019) 012014 doi:10.1088/1757-899X/703/1/012014

is illustrated in the Causal Loop Diagram as shown in Figure 2 and Stock and Flow Diagram in Figure 3.

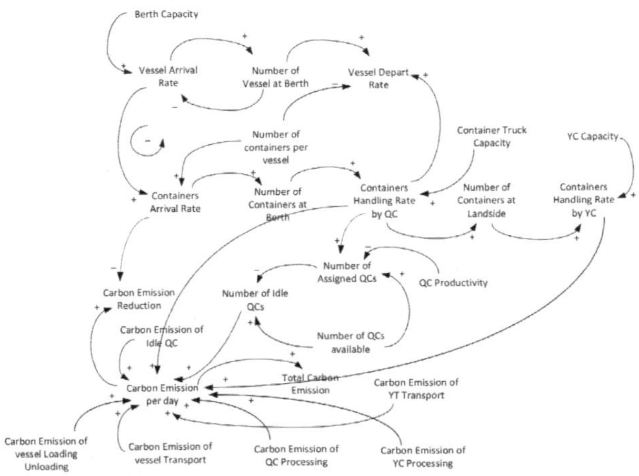

Figure 2. Causal Loop Diagram for system dynamics model of carbon emissions at container terminal.

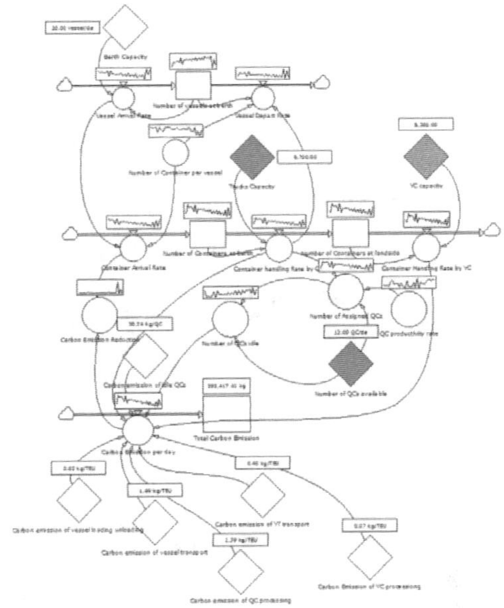

Figure 3. Stock and flow graph for system dynamics model of carbon emissions at container.

List of equations for each variable in the system dynamics model for minimizing the total carbon emissions is shown in Figure 4 below.

Name	Unit	Definition
Berth Capacity	vessel/da	20<<vessel/da>>
Carbon emission of idle QCs	kg/QC	30.24<<kg/QC>>
Carbon emission of QC processing	kg/TEU	1.39<<kg/TEU>>
Carbon emission of vessel loading unloading	kg/TEU	0.02324<<kg/TEU>>
Carbon emission of vessel transport	kg/TEU	1.99<<kg/TEU>>
Carbon Emission of YC processiong	kg/TEU	0.06873<<kg/TEU>>
Carbon emission of YT transport	kg/TEU	0.39645<<kg/TEU>>
Carbon Emission per day	kg/da	('Carbon emission of vessel loading unloading'+'Carbon emission of vessel transport'+'Carbon emission of QC processing')*'Container handling Rate by QC'+('Carbon emission of YT transport'+'Carbon Emission of YC processiong')*'Container Handling Rate by YC'+'Carbon emission of idle QCs'*'Number of QCs idle'
Carbon Emission Reduction	kg/TEU	'Carbon Emission per day'/'Container Arrival Rate'
Container Arrival Rate	TEU/da	'Vessel Arrival Rate'*'Number of Container per vessel'
Container handling Rate by QC	TEU/da	MIN('Trucks Capacity','Number of Containers at berth'/TIMESTEP)
Container Handling Rate by YC	TEU/da	MIN('Number of Containers at landside'/TIMESTEP,'YC capacity')
Number of Assigned QCs	QC/da	CEIL(MIN('Container handling Rate by QC'/'QC productivity rate'/TIMESTEP,'Number of QCs available'),1<<QC/da>>,0<<QC/da>>)
Number of Container per vessel	TEU/vessel	ROUND(NORMAL(500<<TEU/vessel>>,150<<TEU/vessel>>,0.5),1<<TEU/vessel>>)
Number of Containers at berth	TEU	2000
Number of Containers at landside	TEU	3000<<TEU>>
Number of QCs available	QC/da	12<<QC/da>>
Number of QCs idle	QC/da	'Number of QCs available'-'Number of Assigned QCs'
Number of vessels at berth	vessel	3
QC productivity rate	TEU/(da*QC)	ROUND(RANDOM(500<<TEU/da/QC>>,600<<TEU/da/QC>>,0.5),1<<TEU/da/QC>>)
Total Carbon Emission	kg	0<<kg>>
Trucks Capacity	TEU/da	8700<<TEU/da>>
Vessel Arrival Rate	vessel/da	MIN('Berth Capacity'-'Number of vessels at berth'/TIMESTEP,ROUND(RANDOM(8<<vessel/da>>,15<<vessel/da>>,0.5),1<<vessel/da>>))
Vessel Depart Rate	vessel/da	FLOOR(MIN('Number of vessels at berth'/TIMESTEP,'Container handling Rate by QC'/'Number of Container per vessel'),1<<vessel/da>>,0<<vessel/da>>)

Figure 4. List of equations for system dynamics model of carbon emissions at container terminal.

5. Results and discussion

The system dynamics simulation model has run for 30 days to evaluate carbon emissions per day (Figure 5). Based on the available berth capacity, the minimum total carbon emission for 30 days is

393.42 tonnes that obtained by the optimal number of quay cranes is 12 QCs, capacity of container trucks is 8700 TEUs and yard cranes capacity of 8300 TEUs.

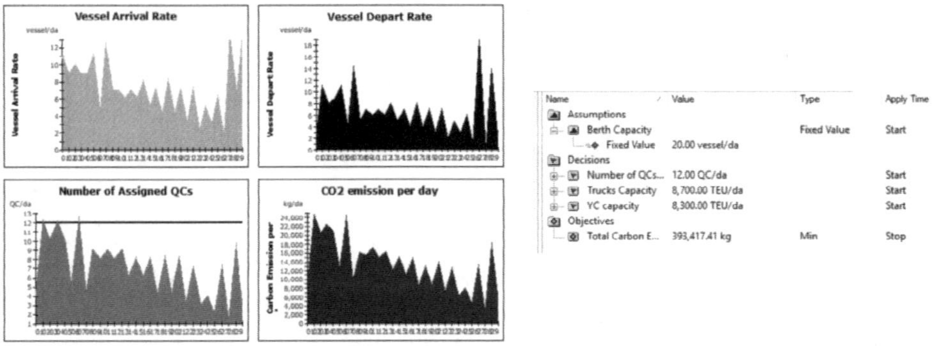

Figure 5. Simulation results of system dynamics model of carbon emissions at container terminal.

6. Conclusions

In this paper, an optimization model has been developed to minimize total carbon emissions in container terminal operations under the uncertainty environment by using a system dynamics simulation model approach. The minimum of total carbon emissions depends on the optimal number of assigned quay cranes for each vessel, the capacity of available container truck and yard cranes in the container terminal. For further research, a total carbon emission model can be developed with taking into account the technological advances in cargo handling equipment.

Acknowledgments

The author would like to thank the two anonymous reviewers for their insightful comments and the editor for helpful suggestions in the revision, which help to improve the presentation of this paper greatly.

References

[1] Wilmsmeier G and Spengler T 2016 Energy consumption and container terminal efficiency *Facilitation of Transport and Trade in Latin America and The Caribbean* **6** 1-10.

[2] Wilmsmeier G 2012 Energy consumption and efficiency : emerging challenges from reefer trade in South American container terminals *FAL Bull.* **329** 1–9.

[3] Yun P, Xiangda L, Wenyuan W, Ke L, and Chuan L 2018 A simulation-based research on carbon emission mitigation strategies for green container terminals *Ocean Eng.* **163** 288–98.

[4] Yang Y C and Lin C L 2013 Performance analysis of electric-rubber tired gantries from a green container terminal perspective *Transp. Res. D.* **23** 9–11.

[5] Acciaro M, Ghiara H, and Cusano M I. 2014 Energy management in seaports: a new role for port authorities *Energy Policy* **71** 4–12.

[6] Geerlings H and van Duin R 2011 A new method for assessing CO2-emissions from container terminals: a promising approach applied in Rotterdam *J. Cleaner Prod.* **19** 657–66.

[7] Sim J 2018 A carbon emission evaluation model for a container terminal *J. Cleaner Prod.* **186** 526–33.

[8] Group R 2006 Specification with guidance at the organization level for quantification and reporting of greenhouse gas emissions and removals ISO 14064-1. 2006.

[9] Eggleston H S, Miwa K, Srivastava N, and Tanabe K 2006 IPCC Guidelines for National Greenhouse Gas Inventories – A primer *Intergov Panel Clim Chang (IPCC)* 1-20.

[10] Liu D and Ge Y 2017 Modeling assignment of quay cranes using queueing theory for minimizing CO 2 emission at a container terminal *Transp. Res. D.* 1–12.

[11] Hu Q, Hu Z, and Du Y 2014 Berth and quay-crane allocation problem considering fuel consumption and emissions from vessels *Comput. Ind. Eng.* **70** 1–10.

[12] Bala B K, Arshad F M, and Noh K M 2017 *System Dynamics: Modelling and Simulation* (Singapore: Springer Science Business Media Singapore).

IOP Conf. Series: Materials Science and Engineering **703** (2019) 012015 doi:10.1088/1757-899X/703/1/012015

Effects of initial concentration, adsorbent mass, pH and temperature to personal care products waste removal with activated carbon as adsorbent

H R Priyantini[1], L Riadi[1], C Effendi[1], F Effendi[1] and A Mitayani[2]

[1]Department of Chemical Engineering, University of Surabaya, Raya Kalirungkut, Surabaya, 60293, Indonesia
[2]Research Center for Electronics and Telecommunication (P2ET), Indonesian Institute of Sciences (LIPI), Sangkuriang, Bandung, 40135, Indonesia

E-mail: rita_priantini@staff.ubaya.ac.id

Abstract. Before being discharged into the water, waste from the personal care products (PCP) industry needs to be treated using the adsorption process. In this study, activated carbon is chosen as adsorbent because of its organophilic property, abundance, and continuous system. The experiment was carried out in batches using 3 variables of COD concentrations in PCP wastes in 250 ppm, 225 ppm, and 200 ppm with the adsorbent mass amount of 0.3 g/l, 0.35 g/l, and 0.4 g/l. In this experiment, the effects of initial COD concentration, adsorbent mass, pH, and temperature on the adsorption process, equilibrium models and mass transfer coefficients of adsorption will be analyzed. Equilibrium models of Distribution Coefficient, Freundlich, and Langmuir Model, were implemented in this experiment. From results of the study, the equilibrium system of PCP waste and activated carbon using Langmuir Equilibrium achieved a removal value of 69.03% at concentration of 200 ppm and the adsorbent mass of 0.4 g/l. To determine the mass transfer coefficient, an external adsorption model was used. The coefficient value of mass transfer coefficient (kc) with the trial method is 3.78x10-8 m/min. Using numerical method with MATLAB, the result is 7.6222x10-8 m/min, while using the analytical approach, the result is 7.413x10-8 m/min.

1. Introduction

The demand of Indonesian market for personal care products (PCP) increases consistently every year at around 10-15% [1]. The higher the PCP products from industry and the community's consumption, the higher the waste will be formed, such as organic waste, pharmaceutical waste and nano-waste namely engineered nanomaterial (ENM). This waste flows to the environment from making process, wrapping, distributing, to personal care product consumption. This waste is also called pharmaceutical and personal care product (PPCP) waste.

There are several methods to reduce PPCP liquid waste level if after secondary treatment, the waste characteristic values still do not meet quality standards and one of them is with tertiary treatment. Quality standard for pharmaceutical liquid waste chemical oxygen demand (COD) level according to East Java Provincial Government Regulation Number 72 Year 2013 [2] is 75 mg/L.

Adsorption is commonly selected for tertiary treatment. Adsorption for liquid waste treatment becomes popular because it is more efficient and more stable compared to biological method that also yields large amount of sludge [3]. Meanwhile, adsorption process is very efficient, easy to design and operate, relatively affordable and not affected by toxicity like biological treatment [4]. Furthermore, removal percentage of organic waste is estimated quite high, around 70-80% [5].

The material often used for conventional adsorption is activated carbon. Activated carbon is commonly used for adsorption because of its large surface area and it does not need adsorbent preparation. However, because of its large surface area, activated carbon regeneration needs a high cost technique. The common regeneration method is thermal regeneration, but it can remove 5-10% carbon because carbon structure is damaged [6].

The research commonly used to investigate the PCP waste adsorption process is generally performed using synthetic waste with synthetic compound concentration parameter. Because of the large amount of component in PCP waste, the parameter used to investigate the PCP waste adsorption process is COD. COD is the amount of oxygen (mg O_2) needed to oxidize organic substances in 1 liter of water sample. So far, there is no research activity that investigates PCP waste adsorption process with COD as the main parameter, so that further research is needed to figure out the PCP waste adsorption characteristic by those two types of adsorbent in order to obtain operation condition and adsorbent type which fits PCP waste treatment the most. The objective of the study is to evaluate the best operation conditions, such as initial concentration of waste, adsorbent mass, pH and temperature for PCP removal using activated carbon as adsorbent.

2. Research method

The system process of this research is described in figure 1. The process consists of 7 steps which is initiated by PPCP waste characterization and ended with adsorption modelling. The modelling is based on finding the most optimum parameter to obtain maximum percentage of removal.

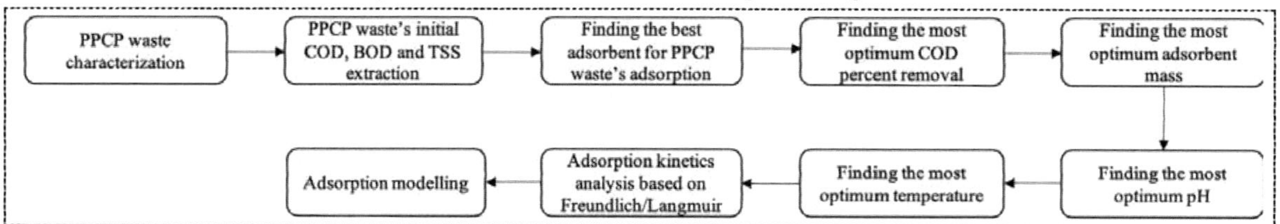

Figure 1. System process diagram of this research.

2.1. PCP waste characterization

Waste characterization included initial total suspended solid (TSS) level test, initial 5-day biological oxygen demand (BOD_5) and initial COD. This procedure determined whether waste needed pretreatment before adsorption process or not,referring to East Java Provincial Government Regulation Number 72 Year 2013 [2] concerning pharmaceutical waste. Initial TSS level test used gravimetry method. Initial BOD_5 level test used reagents: phosphate buffer, $MgSO_4$, $CaCl_2$, $FeCl_3.6H_20$ and KI obtained from BDH. Initial COD test used $K_2Cr_2O_7$ and $AgSO_4$reagents obtained from e-Merck. Activated carbon for this experiment was obtained from Merck.

2.2. Maximum COD percent removal determination at various initial COD concentration of wastewater

The diluting waste at various concentration was carried out in order to figure out maximum COD percentage of removal which could be reached The same adsorbent mass was used for all variables, that is, 0.15 gr. Experiment data were sampled at 0, 1 3, 5, 10, 15, 20, 30, 40, 60, 80, 100 and 120-th minute and waste volume used in this experiment was 500 ml. Adsorption equilibrium data were obtained after 24 hours since the adsorption process started. Variables used in this experiment were: 250 ppm, 225 ppm and 200 ppm.

2.3. Optimum adsorbent mass determination

After obtaining the initial concentration for maximum percentage of removal in previous step, the adsorbent mass for this experiment was varied until the maximum percentage of removal was reached. Experiment data were sampled at 0, 1 3, 5, 10, 15, 20, 30, 40, 60, 80, 100 and 120-th minute and waste volume used in this experiment was 500 ml. Adsorption equilibrium data were obtained after 24 hours since the adsorption process started. Variables used in this experiment were: 0.15 gr, 0.175 gr and 0.2 gr.

2.4. Optimum pH determination

With the previous initial concentration that made up the maximum percentage of removal, initial pH was measured. Then, pH was adjusted until reaching acidic pH (pH=4) and basic pH (pH=11). The adsorption process was performed using previous adsorbent mass which resulted in maximum percent removal. The adsorption results from the maximum percentage of removal in acidic pH, neutral pH, and basic pH were then compared. Experiment data were sampled at 0, 1 3, 5, 10, 15, 20, 30, 40, 60, 80, 100 and 120-th minute and waste volume used in this experiment was 500 ml.

2.5. Optimum temperature determination

With the previous initial concentration, adsorbent mass and pH that made up maximum percentage of removal, temperature was measured and then varied to obtain the optimum temperature that resulted in the maximum removal percentage. Experiment data were sampled at 0, 1 3, 5, 10, 15, 20, 30, 40, 60, 80, 100 and 120-th minute and waste volume used in this experiment was 500 ml. Variables used in this experiment were: 30 °C, 35 °C, 40 °C, 45 °C and 50°C.

3. Results and discussions

This research discusses PCP waste adsorption which is effluent secondary treatment, using activated carbon. Table 1 shows PCP waste characteristic data before entering adsorption process compared to the quality standards from government [2] concerning pharmaceutical waste.

Table 1. PCP waste characteristic

Parameter	Quality Standards [8]	Characteristic Used in Experiment
COD	150 mg/L	860 mg/L
BOD_5	75 mg/L	72 mg/L
TSS	75 mg/L	8 mg/L
pH	6-9	7.9

Activated carbon used in experiment is powder type activated carbon from Merck with surface area of 800 m^2/gr.

3.1. Initial COD concentration effect

Three pollutant concentration variables (250 ppm, 225 ppm and 200 ppm) were used to understand the pollutant concentration effect on the adsorption process in certain adsorbent mass values. This pollutant concentration range was selected based on pollutant concentration for tertiary treatment in industry which has to be the lowest pollutant concentration. The ideal adsorption process was performed in thin solution. Initially, wastewater was diluted with distilled water because PCP waste has COD concentration of 860 ppm. Activated carbon mass used here was 0.15 gr. Experiment data were sampled at 0, 1 3, 5, 10, 15, 20, 30, 40, 60, 80, 100 and 120-th minute. Figure 2 shows COD value change versus time with adsorbent mass of 0.15 gr in various initial concentration.

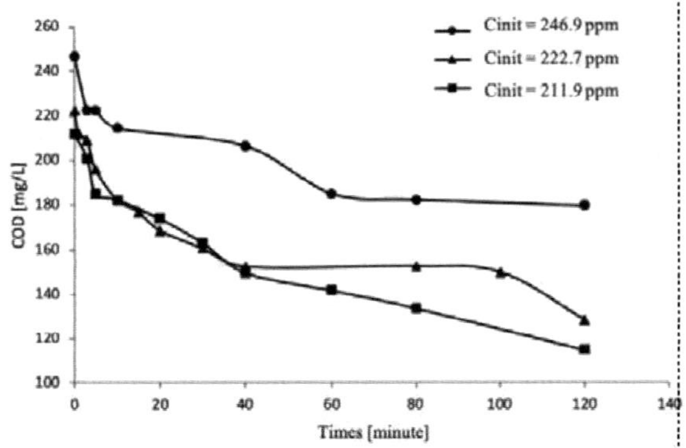

Figure 2. Profile of COD concentration versus time with adsorbent mass = 0.15 gr in various initial concentration.

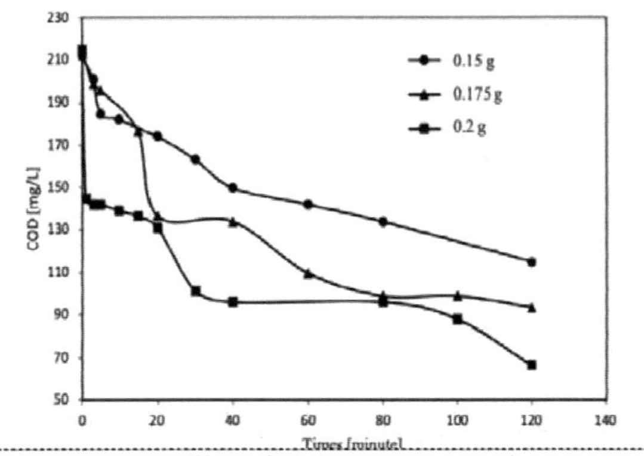

Figure 3. Profile of COD concentration versus time with initial COD concentration = 200 ppm in various initial adsorbent mass.

Table 2. COD percent removal data in various initial COD concentration

COD_0 [mg/L]	246.9	222.7	211.9
$COD_{2\text{-hour}}$	179.559	128.363	114.889
$COD_{,eq}$ [mg/L]	166.087	128.363	104.111
percent removal	32.738%	42.354%	50.867%

Table 3. COD percent removal data in various adsorbent mass

COD_0 [mg/L]	214.5897	214.5897	211.9
Adsorbent mass [gr]	0.2	0.175	0.15
COD_{2hour}	66.3867	93.3327	114.889
$COD_{,eq}$ [mg/L]	66.3867	85.2489	104.111
percent removal	69.0634%	60.2735%	50.867%

Table 2 shows the COD percentage of removal data in various initial concentration (246.9 ppm, 222.7 ppm and 211.9 ppm). Meanwhile, for COD concentration of 250 ppm, after 2 hours of adsorption process, COD concentration becomes 179.559 ppm, where this result does not yet meet the quality standards from government [2].

In the lower concentration, the number of particle bonds among waste particles is lower, hence the adsorption of waste particles to activated carbon becomes more optimum. Furthermore, adsorption process is ideally performed in gas and thin solution, so that if the solution concentration is lower, adsorption process is more optimum [7]. Therefore, for next adsorption process, initial COD concentration of 211.9 ppm is applied.

3.2. Adsorbent mass effect

In varied adsorbent mass experiment, three variables of adsorbent mass were used: 0.15 gr, 0.175 gr and 0.2 gr. Figure 3 shows COD value change versus time in initial COD concentration of 200 ppm with various adsorbent mass.

On Table 3, it can be seen that adsorption process optimum mass is obtained at mass of 0.2 gr, where the greater the adsorbent mass, the greater the active side for adsorbent adherence. Quality standard from government is met for all mass and in the next process, the adsorbent mass of 0.2 gr is used.

3.3. Isotherm and adsorption kinetics

From varied initial COD concentration and adsorbent mass experiment, isotherm for this experiment can be estimated from plotting curves as comparison among Langmuir, Freundlich and distribution coefficient isotherm.

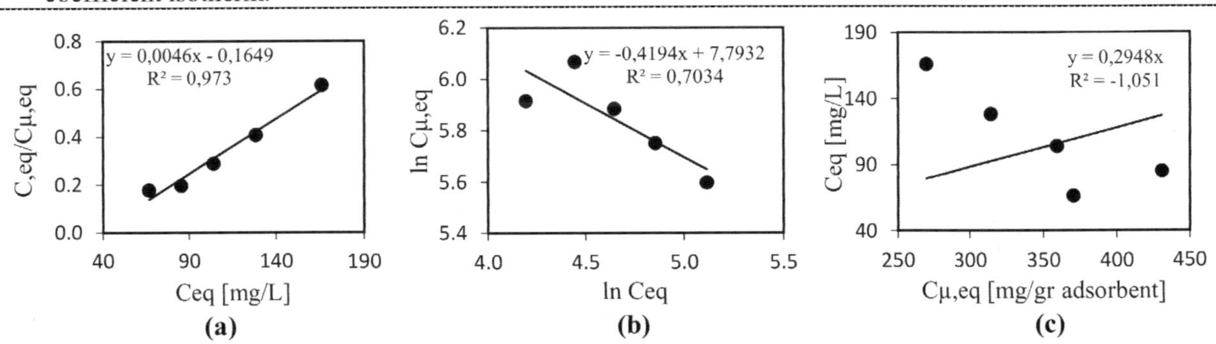

Figure 4. PCP waste adsorption process with activated carbon as adsorbent.**(a)**Langmuir isotherm plot; **(b)**Freundlich isotherm plot;**(c)**Distribution coefficient isotherm plot;

Table 4. PCP waste adsorption isotherm parameter table with activated carbon as adsorbent

ISOTHERM	PARAMETER	R^2
Langmuir	K_L = -0.028 L/mg	0.973
	$C_{\mu,max}$ = 217.391 mg/g	
Freundlich	K_F = 2424.062 mg/g	0.703
	n = -2.384	
Distribution Coefficient	k_d = 0.295 g/L	-1.051

In figure 4, it can be seen that the most suitable equilibrium for this research is Langmuir isotherm. As seen in table 4, Langmuir equilibrium has value of R^2 = 0,973. The value is the closest number to 1 if compared to R^2 values from Freundlich and distribution coefficient equilibrium. This fact shows that PCP waste adsorption process using activated carbon follows monolayer adsorption model as described in equation (1).

$$C_{\mu,eq} = \frac{C_{\mu,max} K_L C_{eq}}{1 + K_L C_{eq}} \tag{1}$$

Analytical approach COD is obtained by using equation (2).

$$\int_{Co}^{C} \frac{dC}{(C - C_{eq})} = -\frac{k_C \times A \times t}{V} \tag{2}$$

Analytical approach is an approach with exponential method resulting in steep declining line then flattening at some point, while experiment data are declining with sloping trend. Therefore, trial method is carried out to find COD value from modelling that approximates experiment value more precisely.COD trial value approximates experiment data better than analytical approach.

In PCP waste adsorption experiment using activated carbon, various k_c(mass transfer coefficient) values ranging from $5.175 x 10^{-8}$to $11.57 x 10^{-8}$with various adsorbent mass and initial concentration of PCP waste were used. Average k_c value obtained with analytical approach is $7.70 x 10^{-8}$. Meanwhile, average k_c from trial method is $3.78 x 10^{-8}$. Because the values are close to one another, it can be concluded that k_c value is not affected by adsorbent mass and initial concentration of PCP waste.

3.4. Effects of pH

In varied pH adsorption process experiment, three variables of various pH were used, that is, pH=4, pH=7.5 and pH=11. pH=4 is used for setting up acidic situation in adsorption process, while pH=11 is used for setting up basic situation in adsorption process. From there, a match between adsorbent and adsorbate in acidic, neutral and basic pH are evaluated. Neutral pH for this waste is 7.5.In this varied pH experiment, fixed initial concentration of 200 ppm and adsorbent mass of 0.2 gr are used. pH value that gives the biggest decrement of COD is 4 (acidic pH) which also means that pH value of 4 results in the biggest percentage of removal of 74.086%. This proves that acidic pH is more suitable for activated carbon. Theoretically, CA possesses point of zero charge (PZC) in pH=5.4. When the pH of solution is below PZC, CA will suffer from protonation, so that positive charge will also decrease. Therefore, CA will adsorb more anionic compounds. The original pH of the solution tends to be basic, that is, is 7.5, thus, to increase the percentage of removal, the pH of the solution should be acidized to less than 5.4.

3.5. Effects of temperature

In this experiment, adsorption process with 5 varied temperatures: 30, 35, 40, 45 and 50^0C was performed. The best temperature for the adsorption process was at 50^0C. It was supported by the value of k_c (equilibrium constant between concentration in liquid and solid state) as can be seen in equation (3). Table 5 shows the effect of temperature and k_c.$C_{\mu,eq}$ denotes adsorbate concentration in adsorbent and C_{eq} denotes adsorbate concentration in solution when those reach equilibrium.

$$k_C = \frac{C_{\mu,eq}}{C_{eq}} \tag{3}$$

Table 5. Effect of temperature for PCP waste adsorption using activated carbon

T [K]	1/T [1/K]	ln k_c
303	0.0033	0.824889
308	0.003247	1.385905
313	0.003195	1.546015
318	0.003145	1.849222
323	0.003096	1.946211

4. Conclusions

In constant activated carbon mass, the biggest percentage of removal value is obtained in the lowest COD initial concentration of 211.9 ppm. In the fixed PCP waste concentration, the biggest percentage of removal value is obtained in the highest varied mass, that is, is 0.2 gr. The highest percentage of removal is reached when COD initial concentration is 200 ppm, adsorbent mass is 0.2 gr, pH of solution is 4 and temperature is 50^0C where percentage of removal value obtained is 73.69%. Mass transfer coefficient value (k_c) with trial method is 3.78×10^{-8} m/min. With computer-aided simulation, it is 7.6222×10^{-8} m/min, while with analytical approach, it is 7.413×10^{-8} m/min.

Reference

[1] U.S. Department of Commerce 2016 Personal Care & Cosmetics Products Cosmetics, (Indonesia: Toiletries and Skincare Market Overviews)
[2] East Java Provincial Government 2013 East Java Provincial Government Regulation Number 72 Year 2013 Concerning Quality Standards of Wastewater for Industry and/or Other Commercial Activities
[3] Bellir K Bouziane I Sadok Boutamine Z Lehocine M Bencheikh and Meniai A H 2012 Sorption Study of a Basic Dye "Genetian Violet" from Aqueous Solutions Using Activated Bentonite (*Energy Procedia* **18**) pp 924 – 933
[4] Ahmaruzzaman Md 2008 Adsorption of Phenolic Compounds on Low-Cost Adsorbents: A Review (*Adv. in Colloid and Interface Sci.* **143**) pp 48-67
[5] Shon H K Phuntsho S Vigneswaran S Kandasamy J 2011 Physico-Chemical Processes for Organic Removal from Wastewater Effluent (*Encyclopedia of Life Support System*: Water and Wastewater Treatment Technologies)
[6] Sufnarski M D 1999 The Regeneration of Granular Activated Carbon Using Hydrothermal Technology (Texas University at Austin, Dept. Of Engineering.)
[7] Fatimah I 2014 Adsorption and Catalysis Using Clay-Based Material (Jakarta: Graha Ilmu)

IOP Conf. Series: Materials Science and Engineering **703** (2019) 012016 doi:10.1088/1757-899X/703/1/012016

The integration of social responsibility into business operation: case study of Indonesian manufacturing industry

E D Rinawiyanti[1,2], C Huang[1], and S As-Saber[1]

[1] School of Management, RMIT University, Melbourne, Australia

[2] Industrial Engineering Department, University of Surabaya, Raya Kalirungkut, Surabaya 60293, Indonesia

E-mail: estidwi@staff.ubaya.ac.id

Abstract. Prior studies argue that CSR should be integrated with business operations to get its benefits. However, the empirical research on how CSR and business operations can be incorporated is still limited. This study aims to investigate how to integrate CSR and business operations and to identify the impact of such integration on company performance. Using a sample of 342 manufacturing companies in Java, Indonesia, PLS-SEM was employed to create a path model depicting the relationships between functional integration and company performance. The results showed that Cost, Quality, Supplier, and Employee, have significant positive relationships on company performance, while Innovation and Customers have no impact on company performance.

1. Introduction

Corporate Social Responsibility (CSR) can be defined as "... a concept where-by companies integrate social and environmental concerns in their business operations and their interaction with their stakeholders on a voluntary basis" [1]. CSR is the most frequently used term to point out the correlation between companies and society [2] and has become a concern for all businesses [3]. Several prior studies emphasize the incorporation of CSR into business operation to improve organizational performance financially and non-financially. But, there are not many research works on how companies attempt to integrate CSR activities into their business [4]. There are still shortcomings of particular concepts, mechanisms for integration activities, and study on specific objectives in the field of social responsibility [5]. Empirical studies on the integration of corporate sustainability into strategic management are still needed [6] because it is still unclear how CSR and business operation can be integrated as well as how such integration between them can affect company performance. Therefore, this study aims to investigate how CSR and business operations are integrated and to examine the impacts of such integration on the organisational performance.

2. Theoretical background

The development of CSR practices depends on how they are integrated into current business practices [7, 8] which enables the companies to achieve not only significant social advantages but also meaningful business-related benefits [9]. CSR implementation can involve a 'built-in' and 'bolt-on' approach. The former is strategic through incorporating socially responsible behaviours into companies' operations, processes, and decision-making. The latter is more potential by embracing social activities that extend beyond current business operations [10]. The former should involve mainstream functions, e.g., production, logistics, and quality control [11].

Some researchers highlight the necessity of incorporating CSR actions into the core activity of the value chain [12, 13, 14]. Companies should consider the social issues connected to the company's core activities, both the primary and supporting activities as a part of the value chain which consists of suppliers, customers, and specific tools [14]. Thus, they can evaluate which need to be enhanced to broaden the social agreement [12]. Prior studies address on four competitive priorities among manufacturing companies: low cost, quality, delivery performance (speed and reliability), and manufacturing flexibility [15, 16]. Another essential dimension is innovation [17] as one of the main drivers in the strategic orientation of a company [18]. Human Resource Management is also crucial in the integration process [19] by providing employees with the willingness, training, and motivation, necessary to apply CSR actions and initiatives [20].

Stakeholder theory claims that organisations have obligations not only to shareholders but also to stakeholders as groups and individuals who can affect, or are affected by, the achievement of a company's mission [21]. As the integration of CSR into business operation need to count in the stakeholders' concerns and objectives in the long-term perspective [22, 23], this study uses stakeholder theory to examine stakeholders' relationship in the functional integration. Through functional integration, CSR activities are incorporated into the business operation so that the companies ensure that their actions benefit them economically and socially [24, 25]. Therefore, we suggest the hypothesis that the functional integration of CSR and business operation has a positive impact on company performance. Partial Least Square Structural Equation Modelling (PLS-SEM) was employed to predict and explain measured constructs [26] to verify the hypothesis.

3. Research methodology

3.1 Sample and sampling process

This study was conducted in the manufacturing industry in Java, Indonesia, since approximately 2.8 million manufacturers (64.29%) are located in Java. Furthermore, Java contributes over 70% of Indonesian national GDP [27]. The samples were selected from eight industrial estates in Java: (1) Surabaya Industrial Estate Rungkut, (2) Sidoarjo Industrial Estate Berbek, (3) Kawasan Industri Gresik, (4) Pasuruan Industrial Estate Rembang, (5) Ngoro Industrial Park (Mojokerto), (6) Kawasan Industri Wijayakusuma (Semarang), (7) Jakarta Industrial Estate Pulogadung, and (8) Kawasan Industri Jababeka (West Java). This study employed the purposive sampling (non-probability sampling) as the samples were selected from the Manufacturing Industrial Directory 2017 and tenant list of the relevant industrial estates based on the conformity with the sample criteria [28].

3.2 Data collection

A survey using questionnaires was employed from June to September 2018 and administered both mail, e-mail, and online, as self-completion methods [29]. The questionnaires were addressed to senior managers because of their comprehensive understanding of the company strategy and practices related to CSR, and, therefore, would be competent to fill in the questionnaire. From 1,055 questionnaires distributed, 505 questionnaires were returned. After data screening, 342 cases with the response rate of 32.42% fitted the sample criteria and were used for further analysis.

4. Findings and discussion

4.1. The profile of respondents

Details on company characteristics are provided as in Table 1, signifying that the sample represents a variety of manufacturing companies. Most companies are manufacturers of food and beverage (21.1%). The majority of companies (84.8%) have more than 100 employees, and half of the companies have been set up and operating in the range of 21 and 50 years. A three-quarter (73.7%) of the companies have private ownership, and most of them (67.5%) are in East Java.

Table 1. Company profile.

Variable	Frequency (n=342)	Percentage (%)	Variable	Frequency (n=342)	Percentage (%)
Main product			Number of employees		
Food and beverage	72	21.1	20 - 99	52	15.2
Tobacco	5	1.5	>100	290	84.8
Textile	11	3.2	Company's age (years)		
Leather and footwear	6	1.8	< 5	16	4.7
Goods from wood, handicraft	4	1.2	5-10	32	9.4
Paper	13	3.8	11-20	56	16.4
Coke and refined petroleum products	3	0.9	21-50	184	53.8
Chemicals and chemical products	44	12.9	> 50	54	15.8
Pharmaceuticals and medicinal chemical	7	2.0	Company's ownership		
Rubber and plastic products	34	9.9	Multinational	81	23.7
Basic metals	6	1.8	State-ownership	9	2.6
Non-metallic mineral products	23	6.7	Private	252	73.7
Fabricated metal products	39	11.4	Company's location		
Automotive	23	6.7	East Java	231	67.5
Computers, electronic and optical products	10	2.9	Centre Java	18	5.3
Repair and installation of machinery and equipment	5	1.5	West Java & Jakarta	93	27.2
Furniture	11	3.2			
Machinery and electrical equipment	20	5.8			
Other manufacturing	6	1.8			

4.2. Structural Equation Modelling (SEM) Analysis

Figure 1 presents the model of functional integration and company performance involving six constructs of functional integration, i.e., Cost, Innovation, Quality, Supplier, Customer, and Employee. To provide an understanding of CSR impacts on the company performance [30], organisational performances are measured through four constructs: financial performance (CFP) in strategic level, operational performance (COP) and social performance in tactical and operational level, including customer performance (CCP) and employee performance (CEP) [31].

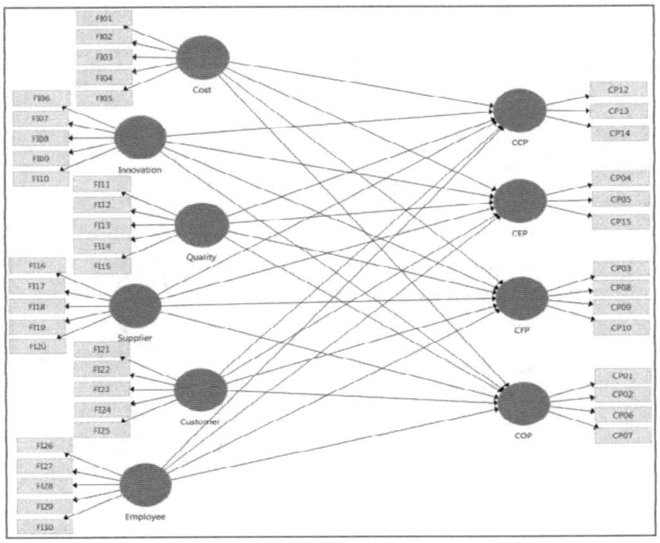

Figure 1. Model of functional integration and company performance.

Table 2. Indicators measurement.

Indicator	Loading	Indicator	Loading
FI01	0.771	FI23	0.830
FI02	0.766	FI24	0.731
FI03	0.804	FI25	0.818
FI04	0.805	FI26	0.811
FI05	0.781	FI27	0.860
FI06	0.721	FI28	0.787
FI07	0.851	FI29	0.877
FI08	0.848	FI30	0.818
FI09	0.789	CP12	0.873
FI10	0.776	CP13	0.860
FI11	0.810	CP14	0.842
FI12	0.832	CP04	0.848
FI13	0.845	CP05	0.886
FI14	0.814	CP15	0.835
FI15	0.834	CP03	0.786
FI16	0.756	CP08	0.855
FI17	0.754	CP09	0.891
FI18	0.844	CP10	0.773
FI19	0.767	CP01	0.759
FI20	0.791	CP02	0.821
FI21	0.810	CP06	0.812
FI22	0.795	CP07	0.827

4.2.1 Assessment of the measurement model. Table 2 displays the indicators measured using the 5-point scale from 1='strongly disagree' to 5='strongly agree'. In terms of company performance, subjective measures were used [32] by asking respondents to rate their company's performance relative to their competitors [33] over the most recent 3-year period using the 5-point scale from 1='much longer/much worse/much lower' to 5='much shorter/much better/much higher'. All of 44 indicators have loadings above 0.7 after eliminating two indicators of CCP and CEP with factor loadings below 0.7 [26]. As summarised in Table 3, all of ten constructs have Cronbach's alpha more than 0.8, composite reliability for all constructs is in the range of 0.89 and 0.92, exceeding the threshold value of 0.7 [28, 34], and AVE is greater than the threshold of 0.50. Results show that the heterotrait-monotrait ratio of correlations (HTMT) used to assess discriminant validity were significantly different from one [35] and below 0.90 [36]. Accordingly, internal consistency reliability, convergent validity, and discriminant validity have been established for the model.

Table 3. Constructs measurement

Construct	AVE	Cronbach's Alpha	Composite Reliability	R^2 value	Q^2 value
Cost	0.617	0.845	0.889		
Innovation	0.638	0.857	0.898		
Quality	0.684	0.885	0.915		
Supplier	0.613	0.842	0.888		
Customer	0.636	0.856	0.897		
Employee	0.691	0.888	0.918		
Customer Performance (CCP)	0.737	0.822	0.894	0.361	0.242
Employee Performance (CEP)	0.734	0.818	0.892	0.434	0.291
Financial Performance (CFP)	0.685	0.845	0.896	0.370	0.232
Operating Performance (COP)	0.648	0.819	0.880	0.389	0.230

4.2.2 Assessment of the structural model. The structural model assessment includes the collinearity, the significance, and relevance of the structural model relationship, the coefficient of determination (R^2), the effect size (f^2), and the predictive relevance (Q^2) [26]. The result shows that VIF values for all indicators are below 5, indicating no significant levels of collinearity detected among the indicators and the constructs [26]. Table 4 displays the path coefficient resulted from bootstrapping procedure with 5,000 boot-strap samples [26] at 500 observations in the original data with no sign changes option [37] with a significance level of 5% [34]. As shown in Table 4, 21 of 24 paths have a positive direct effect, and three of them show the negative direct effect. Eight positive direct effects are significant because their t-value are higher than the critical value at p-value 5%, this is, Cost → CCP, Cost → COP, Employee → CEP, Quality → CCP, Quality → CFP, Supplier → CCP, Supplier → CEP, and Supplier → CFP. Other three direct effects are significant at p-value 10%: Cost → CFP, Employee → CFP, and Quality → COP. Among the positive and significant paths, the most substantial direct effect is on Supplier → CEP (0.326), then Employee → CEP (0.291), and finally Cost → COP (0.269) as well as Quality → CCP (0.269). Based on those results, the hypothesis were accepted that the functional integration of CSR and business operation has positive and significant impacts on company performance. Table 3 exhibits that four constructs of company performance are weak to medium predictors with R^2 value from 0.361 to 0.434. Table 4 shows that six predictors have weak effect size (f^2) on company performance. As presented in Table 3, the Q^2 values are above 0, signifying that the exogenous constructs have excellent predictive relevance for all endogenous constructs [26].

Table 4. Path coefficient, T value, P value, and f^2 value.

Path	Path coefficient	T value	P value	f^2 value	Path	Path coefficient	T value	P value	f^2 value
Cost -> CCP	0.138	2.307	0.021	0.017	Innovation -> CCP	0.062	0.836	0.403	0.002
Cost -> CEP	0.093	1.422	0.155	0.009	Innovation -> CEP	-0.030	0.440	0.660	0.001
Cost -> CFP	0.121	1.831	0.067	0.014	Innovation -> CFP	0.102	1.438	0.150	0.006
Cost -> COP	0.269	4.225	0.000	0.069	Innovation -> COP	0.028	0.371	0.711	0.000
Customer -> CCP	0.030	0.390	0.697	0.000	Quality -> CCP	0.269	3.650	0.000	0.039
Customer -> CEP	-0.058	0.813	0.416	0.002	Quality -> CEP	0.134	1.552	0.121	0.011
Customer -> CFP	-0.073	0.923	0.356	0.003	Quality -> CFP	0.244	3.155	0.002	0.033
Customer -> COP	0.083	0.973	0.331	0.004	Quality -> COP	0.146	1.761	0.078	0.012
Employee -> CCP	0.033	0.374	0.709	0.001	Supplier -> CCP	0.179	2.151	0.032	0.018
Employee -> CEP	0.291	4.177	0.000	0.053	Supplier -> CEP	0.326	4.096	0.000	0.069
Employee -> CFP	0.155	1.945	0.052	0.013	Supplier -> CFP	0.169	2.165	0.030	0.017
Employee -> COP	0.103	1.231	0.218	0.006	Supplier -> COP	0.125	1.641	0.101	0.009

5. Conclusion

This study is intended to investigate how CSR and business operations are integrated and to examine the impacts of such integration on the company performance. The structural model assessment revealed that functional integration has significant positive results on company performance. Specifically, Cost has substantial impacts on the customer, operating, and financial performance, while Employee has effects on the employees and financial performance. Quality relates to the customer, operating, and financial performance significantly, and Supplier gives effects on the customer, employee, and financial performance. The findings from this study contribute to additional empirical evidence of the CSR integration and its impacts on company performance, particularly in the manufacturing industry.

Acknowledgments

Authors acknowledge support from the Indonesian Endowment Fund for Education (LPDP) for its grant to conduct this study.

References

[1] Commission E 2011 *Communication from The Commission to The European Parliament, The Council, The European Economic and Social Committee and The Committee of The Regions: A Renewed EU Strategy 2011-14 for Corporate Social Responsibility*
[2] Branco M C and Rodrigues L L 2006 *J. Bus. Ethics.* **69** 111-12
[3] Martinuzzi A and Krumay B 2013 *Journal of Change Management* **13** 424-43
[4] Yuan W, Bao Y and Verbeke A 2011 *J. Bus. Ethics.* **101** 75-92
[5] Tonysheva L L, DVC 2016 *Asian Social Science* **12** 115-23
[6] Engert S, Rauter R and Baumgartner R J 2016 Exploring the integration of corporate sustainability into strategic management: a literature review *J. Clean. Prod.* **112** 2833-50
[7] Marín L, Rubio A and de Maya S R 2012 Competitiveness as a strategic outcome of corporate social responsibility *Corporate Social Responsibility and Environmental Management* **19** 364-76
[8] Marques-Mendes A and Santos M J 2016 Strategic CSR: an integrative model for analysis *Soc. Responsib. J.* **12** 363-81
[9] Bhattacharyya S S 2010 Exploring the concept of strategic corporate social responsibility for an integrated perspective *European Business Review* **22** 82-101
[10] Insight E 2016 *Strategic Direction* **32** 27-9
[11] Busaya V, Kalayanee K and Gary N M 2009 *Social Responsibility Journal* **5** 178-99
[12] Filippo V, Michele R and Antonello G 2016 *J. Manag. Dev.* **35** 1323-43
[13] Rangan K, Chase L A and Karim S 2012 *Why Every Company Needs a CSR Strategy and How to Build It*, Harvard Business School, April 5, 2012, Working Paper
[14] Witek-Hajduk M K and Zaborek P 2016 *Sustainability* **8** 93
[15] Chi T 2015 Business contingency, strategy formation, and firm performance: an empirical study of chinese apparel SMEs *Administrative Sciences* **5** 27-45
[16] Ward P T, Duray R, Keong Leong G and Sum C C 1995 *J. Oper. Manag.* **13** 99-115
[17] Theodorou P and Florou G 2008 Manufacturing strategies and financial performance—the effect of advanced information technology: CAD/CAM systems *Omega* **36** 107-21
[18] Baumgartner R J 2014 *Corporate Social Responsibility & Environmental Management* **21** 258-71
[19] Guadamillas-Gómez F, Donate-Manzanares M J and Škerlavaj M 2010 The integration of corporate social responsibility into the strategy of technology-intensive firms: a case study *Zbornik radova Ekonomskog fakulteta u Rijeci: časopis za ekonomsku teoriju i praksu* **28** 9-34
[20] Waring P and Lewer J 2004 *J Bus Ethics* **52** 99-108

[21] Freeman, RE 1984 *Strategic Management: A Stakeholder Approach* (Cambridge University Press)
[22] Gazzola P and Colombo G 2014 *Cross-Cultural Manag. J.* **16** 331-8
[23] Maon F, Lindgreen A & Swaen V 2009 *J. Bus. Ethics.* **87** 71-89
[24] Porter M and Kramer M 2006 Strategy and society: the link between competitive advantage and corporate social responsibility *Harv. Bus. Rev.* **84** 78
[25] Werre M 2003 *J. Bus. Ethics.* **44** 247-60
[26] Hair J, Hult G, Ringle C M and Sarstedt M 2017 *A primer on partial least squares structural equation modeling (PLS-SEM)* Second ed (Los Angeles Sage)
[27] BPS 2017 *Analisis Hasil Listing: Aglomerasi Industri Manufaktur di Indonesia* (Indonesia Badan Pusat Statistik) 978-602-438-178-3
[28] Zikmund W, D'Alessandro S, Winzar H and Babin B 2017 *Marketing Research* (Australia Cengage Learning)
[29] Hair J, Celsi M, Money A H, Samouel P and Page M J 2011 *Essentials of Business Research Methods*, Second ed (New York ME Sharpe Inc)
[30] Waagstein P R 2011 *J. Bus. Ethics.* **98** 455-66
[31] Gyusun Hwang S H, Jun S B and Park J W 2014 *International Journal of Innovation, Management and Technology* **5** 50-5
[32] Kim E, Nam D-i and Stimpert J L 2004 *Journal of Business Strategies* **21** 19-45
[33] Wall T D, Michie J, Patterson M and Wood S J 2004 *Personnel Psychology* **57** 95-118
[34] Hair J, Ringle C and Sarstedt M 2011 *The Journal of Marketing Theory and Practice* **19** 139-52
[35] Sarstedt M, Ringle C M and Hair J F 2017 *Handbook of Market Research* (Springer International Publishing AG) 1-41
[36] SmartPLS 2014 *Discriminant Validity Assessment* SmartPLS GmbH, viewed 7th January 2019, <https://www.smartplscom/documentation/algorithms-and-techniques/discriminant-validity-assessment>
[37] Henseler J, Ringle C M and Sinkovics R R 2009 *The use of partial least squares path modeling in international marketing* (Emerald Group Publishing Limited)

IOP Conf. Series: Materials Science and Engineering **703** (2019) 012017 doi:10.1088/1757-899X/703/1/012017

A kinetic study of oil-in-water emulsion formation stabilized by rice husk ash and lecithin

L Sapei, S W Kurniawan and A P Siantoro

Department of Chemical Engineering, Faculty of Engineering, University of Surabaya, Raya Kalirungkut, Surabaya 60293 East Java, Indonesia

Email: lanny.sapei@staff.ubaya.ac.id

Abstract. Many food products are in the forms of emulsions whereby at least one phase is dispersed in other phases. Since emulsions consist of different immiscible phases, they tend to be unstable with time despite the incorporation of emulsifiers. In this study, the stabilities of oil-in-water (O/W) emulsions prepared using emulsifier mixtures of rice husk ash (RHA) and lecithin were monitored with time and destabilization kinetics of emulsion were studied. Rice husk ash of 2% was combined with various lecithin concentrations ranged from 0.5% to 3%, and they were added into the oil phase. The emulsification was carried out using a rotor stator homogenizer. The results showed that the rate of emulsion destabilization followed the first order kinetic model. The lowest value k_1 of 0.4 $x10^{-3}$ $minute^{-1}$ was obtained from the combination of 2% RHA and 2% lecithin. This implied the synergism action of RHA with lecithin in stabilizing the emulsion. The mixtures of lecithin and RHA seemed to be quite promising in stabilizing the O/W emulsion system for designing healthy O/W emulsion food-based products.

1. Introduction

Food is an essential basic need of human being. Therefore, the overall quality of food products such as appearance, texture, flavour, nutrient, and stability has to be considered. Stabilizers are usually added into food formulations in order to improve the texture and appearance of food products. Emulsifiers are amongst the stabilizers which are commonly added into food formulations containing oil and water phases. Emulsifiers generally consist of one or more hydrophobic groups and one or more hydrophilic groups. Due to its amphiphilic nature, emulsifiers are adsorbed on the oil-water interfaces while reducing the surface tension arising between them and help dispersion of the dispersed phased in the continuous phase. There are increasing attempts to reduce the use of polymeric emulsifiers since they may cause some adverse effects, such as air entrapment, foaming, irritancy, and detrimental interaction with living matter [1] and also obesity [2].

Emulsion stabilized by particles which is so called Pickering emulsion has recently been gaining a great interest since it may impart an extreme stability to emulsion by hindering droplets coalescence, may reduce toxicity thus safer to use *in vivo*, and may bestow other preferred traits such as porosity, responsiveness, etc [3]. Solid particles such as carbon, silica, clays, iron oxides, hydroxides, metal sulfates, and alumina have been used as stabilizers of either O/W or W/O emulsions [4-7]. Recently, some other solids such as hydroxyapatite, magnetic nanoparticles, chitosan, cyclodextrin, nanotube, and some food-grade stabilizers and organic particles, including Janus colloidal particles, microsphere, and microcapsule can effectively serve as Pickering emulsifiers [3, 8-9]. Therefore, the applications of Pickering emulsions have become widespread for examples as catalysts, delivery vehicles, porous scaffolds, stimuli-responsive materials, and so on [3].

Silica as an abundant material found in the earth crust has also been richly accumulated in several parts of plants such as horsetail outer epidermal layer [10-12] and rice husk. Silica accumulated in the living tissues is generally known as biosilica [10-11], which is considered safe, and there is No Observed Adverse Effects Level (NOAEL) of 50,000 ppm (mg/L) for dietary silica [13]. Silica is also amongst the additive for foods (E 551) and is typically used as an anti-caking agent. There have been many evidences supporting the potential health benefits of silica such as its contribution to collagen formation and bone health, as well as prevention of toxicity to the brain and maintenance of blood vessel integrity [13]. Therefore, biosilica has a great potential to be incorporated in food systems not only due to its function as emulsifier but also due to its inherent health benefit.

There have been few investigations on the use of biosilica as emulsifiers. Sapei *et al.* investigated the use rice husk biosilica combined with lecithin [14] and Tween 20 [15] for the stabilization of O/W emulsion and the use of rice husk silica/ Tween-20 for stabilizing the double W/O/W emulsion [16]. It turned out that the incorporation of biosilica together with emulsifiers enhanced the emulsion stability. The destabilization of emulsion with time followed the first kinetic model implying the predominant coalescence phenomena during the emulsion instability process. The lowest destabilization constants obtained were about $2.3x10^{-3}$ min^{-1} and $1.5x10^{-3}$ $min^{-1}$for the O/W emulsions stabilized with rice husk silica combined with lecithin and Tween-20, respectively. Some other researchers used colloidal hydrophilic silica particles (Aerosil) combined with other emulsifiers such as monoolein, lecithin, Tween-60, sodium caseinate) in order to develop "food grade" O/W emulsions [17-19]. It has been observed that use of mixed silica particles and emulsifiers produced O/W emulsions with higher stability due to the formation of smaller droplet sizes compared to those stabilized by either emulsifiers or silica particles. There has been no investigation on the use of rice husk ash for stabilizing O/W emulsion yet. In fact, rice husk ash consists of about ~ 94% silica and other compounds containing metals such as K, Mg, Ca, Fe, etc [20]. This research aimed to investigate the potential use of rice husk ash combined with lecithin to enhance the stability of O/W emulsion based on the emulsion destabilization kinetic study.

2. Materials and Methods

2.1 Materials
Lecithin (PT. Brataco, Surabaya, Indonesia); rice husk was obtained from paddy milling in Mojokerto, East Java, Indonesia; palm cooking oil (Sunco, PT. Milkie Oleo Nabati Industri, Bekasi, Indonesia); demineralized water.

2.2 Preparation of rice husk ash
Rice husks were separated from the dust and debris and then washed with water and finally rinsed with demineralized water prior to drying in an oven (Memmert, Germany) at 105°C for 2 hours. Dried rice husks were burnt in the furnace (Ney VULCAN D-550, Dentsply Ceramco, USA) to remove the organic compounds at 750°C for 5 hours. Rice husk ash (RHA) which was gray in color consisted of silica of about 94% [20]. The rice husk ash was subjected to milling followed by screening with 200 mesh screen prior to further use as the emulsifier.

2.3 Preparation of O/W emulsion stabilized with RHA and lecithin mixtures
Oil phase with the fraction of 0.3 was heated up to 50°C followed by the gradual addition of varying lecithin concentrations of 0.5 – 3% under stirring with a magnetic stirrer. RHA of 2% was subsequently added little by little into the oil. Mixing was continued to 15 minutes until lecithin was well solubilised in the oil. Aqueous phase consisting of demineralized water was heated to 50°C prior

IOP Conf. Series: Materials Science and Engineering **703** (2019) 012017 doi:10.1088/1757-899X/703/1/012017

to mixing with the oil phase. The emulsification was carried out using a rotor-stator homogenizer (IKA T25 digital ULTRA TURRAX, *Germany*) at 20,000 rpm for 5 minutes after aqueous phase was poured into the oil phase. The emulsion was kept in a transparent 40 ml glass vial (ID= 25 mm, height= 95 mm) until the height of ~ 5.65 cm from the bottom was attained and then stored at a room temperature of ~ 28°C. The percentages used belonged to weight %.

2.4 Determination of emulsion stability

The emulsion stability was simply determined using equation (1). The height of emulsion was measured at a certain time interval, i.e. every 1 minute for the first 30 minutes, followed by the measurement in every 5 minutes until 60 minutes, and then the measurement was taken in every 10 minutes until 120 minutes and each day up to 7 days. The stable emulsion layer was indicated by a milky appearance with the formation of neither cream nor silica particles sediment.

$$\%S = \frac{h_t}{h_0} x 100\% \quad (1)$$

whereas h_t is emulsion height at a certain time and h_0 is initial emulsion height.

2.5 Determination of kinetic models of destabilization rate of o/w emulsion.

The data used for the kinetic models determination was within 0-120 minutes since the emulsion stability remained more or less constant after 120 minutes. The destabilization rate of o/w emulsion was fitted using the zero order and first order kinetic models. The most appropriate kinetic order would be chosen based on the best obtained correlation coefficient (R^2) calculated using the least square procedure. The emulsion destabilization rate could be written in an equation as follows [21]:

$$-r = -\frac{dS}{dt} = k\ S^a \quad (2)$$

$$S = S_0 - k_0 t \quad (3)$$

$$ln(S) = ln(S_0) - k_1 t \quad (4)$$

whereas $-\frac{dS}{dt}$ the rate of emulsion destabilization, a the order of emulsion destabilization rate, k_0 and k_1 the emulsion destabilization rate constants for the zero order (% stability/ minute) and for the first order (per minute), respectively, t the storage time (minute), S the percentage of emulsion stability after time t and S_0 the initial emulsion stability percentage.

3. Results and Discussion

An emulsion is a mixture of at least two immiscible liquids whereas one phase is dispersed in another phase. An emulsifier is required to stabilize the dispersion of the dispersed phase in the continuous phase by lowering the interfacial tension between the two phases thus droplets coalescences or flocculation could be hindered. However, O/W emulsion will definitely undergo instability with time indicated by creaming due to the inherent thermodynamic instability of the emulsion. The macroscopic observation of O/W emulsion stabilized by 2% RHA and 2% lecithin could be seen in Figure 1. It turned out that the emulsion was not stable right after the emulsification (t= 1 min). Creaming and oil layer formation was seen as the results of oil droplets coalescences. RHA seemed to be suspended in the cream layer rather than being adsorbed on the oil-water interfaces. Some RHA was deposited at the bottom of the vials due to sedimentation. Possibly, RHA which was mainly composed of hydrophilic silica did not well intermixed with lecithin in the oil phase, thus lowering its emulsifying capacity on the interfacial layer. It could have been dispersed in the aqueous phase to improve its adsorption in the oil-water interfaces. Previous investigation used oil phase as the loading phase of hydrophilic silica, however sonication was applied to improve the silica dispersity [22]. The stability of the emulsions was slightly decreasing with time within the range of 50-60% as depicted in Table 1 which was hardly seen in Figure 1. After 7 days, the cream layer seemed thinner since the oil was out from the cream leading to the increase of oil layer. The RHA tended to be sedimented which was not clearly seen in Figure 1. The emulsion instability was mainly due to the coalescences of oil droplets leading to the enlargement of dispersed oil droplets as could be seen in Figure 2.

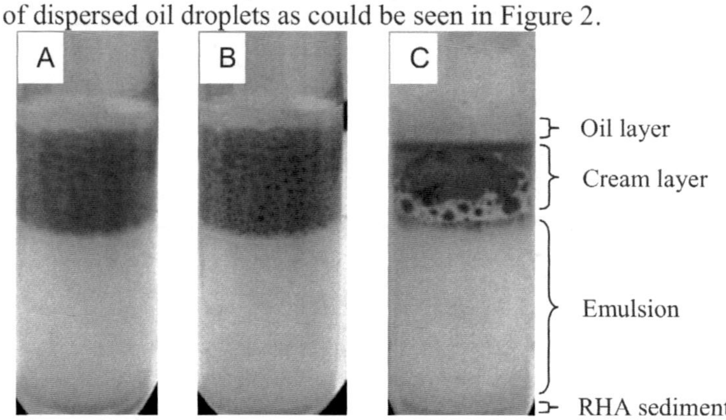

Figure 1. Stability of O/W emulsions stabilized with 2% RHA and 2% lecithin after being stored at a certain time. A) 1 min; B) 120 min; C) 7 days.

Table 1. O/W emulsion stability stabilized using RHA/ lecithin.

Emulsifiers	Emulsion Stability, S (%)		
	t = 1 min	t= 60 min	t= 120 min
2% RHA	60.18	56.64	54.87
1.5% lecithin	61.95	53.10	53.10
0.5% lecithin + 2% RHA	61.95	53.10	51.33
1% lecithin + 2% RHA	61.95	56.64	53.10
1.5% lecithin + 2% RHA	60.18	58.41	56.64
2% lecithin + 2% RHA	61.95	60.18	58.41
2.5% lecithin + 2% RHA	60.18	58.41	56.64
3% lecithin + 2% RHA	60.18	58.41	56.64

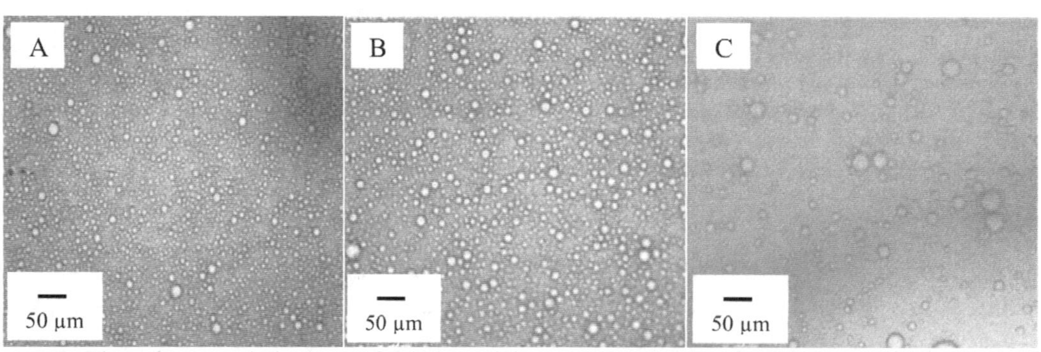

Figure 2. The microscopic images of O/W emulsion stabilized with 2% RHA and 2% lecithin after being stored at a certain time. A) 1 min; B) 120 min; C) 7 days.

It was obvious from table 1 and figure 3 that the emulsion stability after 60 or 120 min was improved when lecithin concentration was increased up to 2% and then decreased. Furthermore, the emulsion stability was much improved when combined RHA/lecithin was used instead of single emulsifier of RHA or lecithin only. The kinetics study was conducted to confirm the effect of lecithin addition on RHA stabilized O/W emulsions as depicted in table 2. The calculated k constant represented the rate of emulsion stability reduction as time elapsed.

The destabilization process of emulsion followed the first order rather than zero order kinetic models based on the best obtained correlation coefficient (R^2). This was in line with the previous study [14-15] implying the predominant coalescences of dispersed oil droplets as the main factor of emulsion instability. The lowest destabilizing rate constant k_1 of 0.4 x 10^{-3} min^{-1} was obtained for the emulsion stabilized with 2% RHA combined with 2% lecithin. Higher lecithin concentration reduced the emulsion stability indicated by the higher k_1. Lecithin was effective in stabilizing W/O emulsion due to its hydrophobicity therefore the excess of lecithin could lower O/W emulsion stability. Lecithin concentration seemed to influence the stability of emulsion stabilized with RHA in contrast to emulsions stabilized with rice husk silica/ lecithin [14]. Furthermore, the obtained k_1 was much lower with the order of 1 in comparison to those obtained from rice husk silica/ lecithin [14] and rice husk silica/ Tween-20 [15]. This may indicate a higher emulsifying capacity of RHA compared to rice husk silica. Probably this was due to the presence of salts containing metals such as Na, K, Ca, Mg retained in RHA which helped reduce the interfacial tension between oil and water.

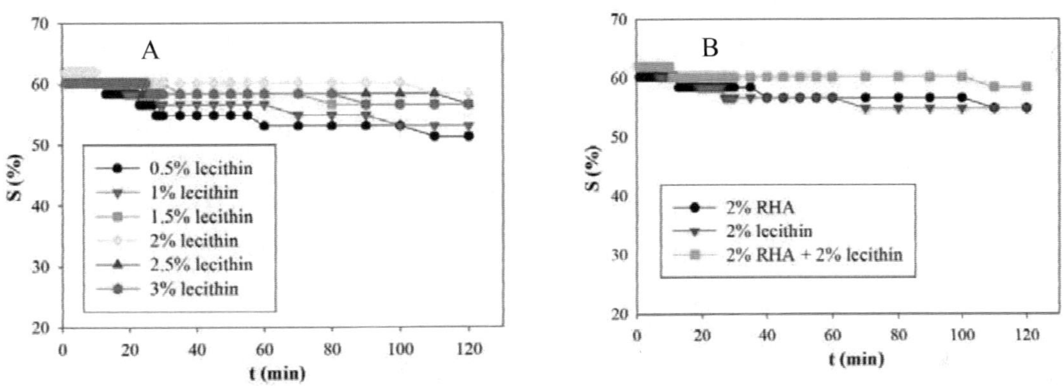

Figure 3. Emulsion stability *vs* time. A) Comparison of stability of emulsions prepared using 2% RHA with varying lecithin concentrations; B) Comparison of stability of emulsions prepared using combination RHA/lecithin and single emulsifier.

Table 2. Kinetic emulsion destabilizing rates constants and R^2 values according to zero-order and first order kinetic models of O/W emulsions stabilized using 2% RHA and various lecithin concentrations.

Emulsifiers	Orde 0		Orde 1	
	k_0 (% stability/min)	R^2	k_1 x 10^3 (min^{-1})	R^2
2% RHA	0.0438	0.8768	0.8	0.8856
2% lecithin	0.0943	0.8420	1.6	0.8532
0.5% lecithin + 2% RHA	0.0894	0.8085	1.6	0.8275
1% lecithin + 2% RHA	0.0812	0.8982	1.4	0.9104
1.5% lecithin + 2% RHA	0.0445	0.9936	0.8	0.9949
2% lecithin + 2% RHA	0.0266	0.8182	0.4	0.8247
2.5% lecithin + 2% RHA	0.0264	0.8885	0.5	0.8940
3% lecithin + 2% RHA	0.0374	0.9398	0.6	0.9439

The synergism between silica particles with emulsifiers has been reported in many publications [14-15, 17-18, 22]. The presence of emulsifiers modified the wettability of solid particles and most likely enhanced the inter-particle interactions [23]. The adsorption of small amount of lecithin in the interfacial layer helped reduce the interfacial tension between the two liquid phases followed by the adsorption of RHA which provided a steric hindrance surrounds the oil droplet. A rigid interfacial film consisting of RHA particles and lecithin could also contribute to the improved emulsion stability.

4. Conclusions

The use of mixed rice husk ash (RHA) with lecithin for stabilizing O/W emulsions was studied and supported by the investigation of emulsion destabilization kinetics. Both RHA and lecithin were loaded into the oil phase prior to mixing with the aqueous phase. It turned out that the increase of lecithin concentrations up to 2% improved the stability of RHA stabilized emulsions. The excess of lecithin concentration decreased the emulsion stability. Lecithin adsorbed on the interfacial layer could help decrease the interfacial tension and together with adsorbed RHA particles could form a rigid film which provided a steric hindrance against droplet-droplet coalescence. The emulsion destabilization kinetics followed the first order model. The combination of 2% RHA and 2% lecithin resulted in the lowest k_1 value of 0.4 $x10^{-3}$ min^{-1}, which was 4 times smaller and 2 times smaller than k_1 obtained from lecithin only and RHA only, respectively. RHA/ lecithin stabilizing emulsions seemed promising for developing healthier food emulsion products.

Acknowledgment

The research was partially funded by Ministry of Research, Technology and Higher Education of the Republic of Indonesia under the research grant scheme of "Fundamental Research" 2019 under the contract number: 004/SP2H/LT/MULTI/L7/2019.

References

[1] Frelichowska J, Bolzinger M and Chevalier Y 2010 *J. Colloid Interface Sci.* **351** 348–56
[2] Simmons A L, Schlezinger J J and Corkey B E 2014 *Curr. Obesity Rep.***3** 273–85
[3] Yang Y, Fang Z, Chen X, Zhang W, Xie Y, Chen Y and Yuan W 2017 *Frontiers in Pharmacol.* **8** 287
[4] Binks B P 2002 *Curr. Opin. Colloid Interface Sci.* **7** 21–41
[5] Wang W, Zhou Z, Nandakumar K, Xu Z and Masliyah J H 2004 *J. Colloid Interface Sci.* **274** 625–30
[6] Dyab A 2012 *Colloids Surf. A* **402** 2–12
[7] Xie X, Wang Y, Li X, Wei X and Yang S (2018) *Colloids Surf. A* **558** 65–72
[8] Dai L, Sun C, Wei Y, Mao L and Gao Y 2018 *Food Hydrocolloids* **74** 239–48
[9] Dai L, Zhan X, Wei Y, Sun C, Mao L, McClements D J and Gao Y (2018) *Food Hydrocolloids* **85** 281–90
[10] Sapei L, Gierlinger N, Hartmann J, Nöske R, Strauch P and Paris O 2007 *Anal. Bioanal. Chem.* **389** 1249–57
[11] Sapei L, Nöske R, Strauch P and Paris O 2008 *Chem. Mater.* **20** 2020–25
[12] Gierlinger N, Sapei L, and Paris O *Planta* 2008 **227** 969–80
[13] Martin K R 2007 *J. Nutr. Health Aging* **11** 94–8
[14] Sapei L, Damayanti O and Liliana L 2017 *ASEAN J. Chem. Eng.* **17** 8–21
[15] Sapei L, Sandy I, Suputra I and Ray M 2017 *IOP Conf. Ser. Mater. Sci. Eng.* **273** 012023
[16] Sapei L, Adiarto T, Handomo R and Chandra S H 2018 *MATEC Web Conf.* **215** 01028
[17] Pichot R, Spyropoulus F and Norton I T 2009 *J. Colloid Interface Sci.* **329** 284–91
[18] Pichot R, Spyropoulus F and Norton I T 2010 *J. Colloid Interface Sci.* **352** 128–35
[19] Pichot R, Spyropoulos F and Norton I T 2012 *J. Colloid Interface Sci.* **377** 396–405
[20] Sapei L and Suseno N 2018 *Indonesian Patent (registered)* P00201802837
[21] Levenspiel O 1999 *Chemical Reaction Engineering* 3rd *ed* (New York: John Wiley & Sons, Inc) pp 47–51
[22] Eskandar N, Simovic S and Prestidge C 2011 *J. Colloid Interface Sci.* **358** 217–25
[23] Tambe D E and Sharma M M 1993 *J. Colloid Interface Sci.* **157** 244–53

IOP Conf. Series: Materials Science and Engineering **703** (2019) 012018 doi:10.1088/1757-899X/703/1/012018

A systematic literature review for developing sustainability assessment tool: formulating the state of the art and future direction

Y Sari[1,2], **A Hidayatno**[1], **A Suzianti**[1] **and M Hartono**[2]

[1] Department of Industrial Engineering, Faculty of Engineering, Universitas Indonesia, Jalan Kampus UI, Depok, West Java, Indonesia
[2]Department of Industrial Engineering, Faculty of Engineering, University of Surabaya, Jalan Raya Kalirungkut, Surabaya, East Java, Indonesia

E-mail: akhmad@eng.ui.ac.id, ysari@staff.ubaya.ac.id

Abstract. Sustainability, with the concept of "triple bottom line", has been growing as a "*trending topic*" recently. It is no longer a paradox; it is now about how to turn this challenge or opportunity into day-to-day management decisions to achieve advantage in both financial performance and corporate sustainability. In conjunction with it, there is a growing need for a better way to develop, assess and measure organization's sustainability performance. This paper discussed a three-stage systematic literature review that is being deployed as the research framework, to scope how sustainability assessments have been developed recently and to formulate the State of the Art for future research. It reviewed 63 selected articles out of 875 papers from Science Direct database. Based on our review, as the novelty proposed by this publication, there were two major approaches used in developing sustainability assessment i.e. criteria-based and model-based approaches. Both approaches had similar popularity in publication: 57% for criteria-based and 43% for model-based. Each had its strengths and weaknesses, but the model-based approach was chosen for future development of the conceptual framework. Furthermore, the review indicated that several main concepts were being used for sustainability assessment and guided the future development into the concept of Business Process Management and Corporate Sustainability.

1. Introduction

Organizations are facing more challenges in the globalized world because society now demands them to be also responsible on creating a more sustainable world. They are forced to think about the impact of their actions in a more diverse way to include social and environment responsibility. The application of sustainability concept has been interpreted differently by industry and academia in Indonesia. For example, Unilever, a well-known consumer goods manufacturer in Indonesia, helped to develop the welfare of farmers and suppliers of raw materials or to reduce the use of chemicals that had negative impacts to the environment through its "*Sustainable Living Plan*" program. Djarum Foundation, a prominent foundation in Indonesia, built building and laboratory facilities for universities through its corporate social responsibility program. Some related studies ([1], [2], [3], [4]) revealed that the sustainability concept has been applied in many aspects such as performance measurement, economic, educational and social life.

In order to achieve corporate sustainability as a competitive advantage, organizations need to manage their business process especially at the strategic level. Business processes are divided into: (a) core processes, (b) support processes, (c) managerial processes ([5], [6]). The role of business process was initially mentioned by [7] and later developed by [8] who described that managing the core process would create a competitive advantage of an organization, controlling the support process would enable competitive advantage, meanwhile the managerial process would ensure sustainable competitive advantage.

In terms of managing the business processes, a methodology for Business Process Improvement which was formed by [9] described that the initial phase of seven-step Business Process Improvement is the need of readiness assessment. To measure the readiness level of an organization before any business process improvements are made, an assessment tool that is designed appropriately is obviously needed. The readiness of an organization can be measured through a well-designed assessment tool so that the result of its assessment provides the foundation for the analysis or improvement.

For that same purpose, the development of sustainability assessment tool becomes necessary; it will be a set of methods, forms or maps that organizations can use to assess their readiness in initiating sustainability efforts or to chart their progression in achieving their target of sustainability. It now turned out as some research questions, namely *"What sustainability indicators will be used? How should the indicators be selected? How should the measurement process be done? Should there be any framework constructed to carry out the assessment process?"*. Therefore, this paper review aimed to conduct a systematic literature review, to scope how sustainability assessments had been developed recently and to provide viewpoint and foundation for the development of sustainability assessment.

2. Research Method

A systematic literature review was introduced in this article as an effective method to review multiple research papers or publications, in which the process was conducted through three stages [10]:(1) stage of planning the review; (2) stage of conducting the review; (3) stage of reporting and dissemination.

Table 1. Three stages of Systematic Literature Review (SLR) as research method

Stage	**Activities**
Stage1	***Planning the review*** Review basic concepts of sustainability and perform meta-analysis on sample of 15 related published articles, which aimed to: a. Gain comprehensive understanding b. Identify the terms for searching articles and publication year • Terms being used to search for the paper in title, abstract, and keywords are: *(sustainability indicator OR sustainability metric) AND (sustainability assessment OR sustainability performance OR performance measurement OR sustainability performance measurement) AND (corporate sustainability OR corporate sustainability model OR strategic management)* • Publication year: *2010 - 2016* c. Determine targeted publishers • Targeted publishers: Science Direct, Emerald, Springer, Taylor & Francis, Wiley • In this review paper, the review process was initially attempted to Science Direct d. Define criteria for classifying the articles to support the analysis
Stage2	***Conducting the review*** Select articles to be reviewed, based on the terms and criteria defined in previous stage: a. From title, abstract and keywords: filtering the papers with the defined terms. Total search: 875 articles. b. From full paper, skimming reading on introduction, research methods and conclusions, to select valid articles for the review, classifying articles according to the classification criteria. Valid: 63 articles (7.2%) c. From selected full paper: focused reading, deep understanding and analysis about sustainability indicator and corporate sustainability
Stage3	***Reporting & Dissemination*** a. After selecting articles (2nd round of previous stage) from electronic database Science Direct and after 30% reading (3rd round on previous stage),

Stage	Activities
	b. Prepare initial draft of the review result to be published in the international conference. c. Conclude main aspects of SLR and future directions for conceptual framework of SAT to assess sustainability initiatives in corporate level

As recommended by [12] and deployed by [11], these stages of systematic literature review were also performed in this review paper, the detailed was shown in Table 1. In the first stage, the literature review was planned by reading the references (books) that contained basic concepts of sustainability as well as performing meta-analysis. The planning stage was aimed to: (a) gain the comprehensive understanding about the research topic, (b) identify proper terms for searching the paper in the next stage of review, (c) determine the targeted journal publishers or electronic databases and (d) define the criteria to classify the articles for data analysis.

3. Results & Discussions

After the planning stage of systematic literature review, the review process continued by searching the articles from scientific database Science Direct, giving the result of 875 articles whose titles, abstracts or keywords fitted the defined terms and publication years. Then, by skimming reading to introduction, research methods and conclusions, there were 63 (or 7.25% of 875) articles valid as selected papers for further analysis. For the 63 selected articles, the meta-analysis was performed to collect information about: authors, publication year, publisher name, industry sector, underlying concept to formulate the indicators/models, and outcoming model/framework. The following results and discussions were made based on the meta-analysis.

Based on the review, to assess sustainability performance of an organization, there are two primary approaches in developing sustainability assessment:

A. ***Criteria-based approach.*** Many publications showed that the formulation of indicators in the sustainability assessment often used the concept of sustainability or Triple Bottom Line which comprised of three aspects: people, profit and planet. The sustainability indicators could be derived from literature review ([13], [14]) or selected based on some criteria ([15], [16]). It designed a general set of sustainability indicators that were suitable to be applied in all sectors but it was limited to a specific industry [15] meanwhile [16] derived sustainability indicators to assess an eco-industrial park, and those specific indicators were limited to be used for any other industries.

B. ***Model-based approach.*** Some sustainability assessments were also done through the development of a model. For example, [17] proposed the sustainability aspect into the Balanced Scorecard method [18], known as the Sustainability Balanced Scorecard (SBSC).

Table 2. Two primary approaches in formulating indicator, with their strengths & weaknesses

Approach	Strengths	Weaknesses
Criteria based selection	• It provides a general set of common sustainability indicators to easily be chosen • These common indicators are also available for specific industry • It can be easily integrated with the current organization strategy as new key performance indicators	• Indicators in the criteria can be difficult to choose due to its similarities • Selection of the indicators can be subjective without measuring the appropriateness and quality of the indicators used. A right way of filtering the indicators is necessary. • Organizations might have difficulties in identifying whether they have achieved sustainability or not
Model based selection	• It creates a clear ideal model on how a sustainable organizations operates and functions. • The gap between ideal and current conditions will make it easier to plan and develop process and activities • It creates a more robust sustainability indicators because it is based on a proven	• It debates on how ideal model is actually ideal • Some models are industry specific, that might not be easy to generalize • Some models might require major changes in the current strategy of the organizations

	ideal model • With the ideal model in sight, organizations can develop stages to measure their closeness on achieving their ideal conditions	

As mentioned previously, it was found that there were two primary approaches in formulating and selecting sustainability indicators, and based on the review, both approaches have their strengths and weaknesses as described in Table 2. Then, both approaches were set as the criterion for classification. For criteria-based approach, it was found that some research determined the indicators directly (from literature review and underlying concepts then the indicators were formulated) or indirectly (a particular mechanism was applied to filter a set of indicators that being derived from literature review and underlying concept). The review showed that, from these 63 selected articles, there were 27 selected articles (43% out of 63) in model-based approach and 36 articles (57% out of 63) articles in criteria-based approach. For the criteria-based approach, it can still be divided into: 43% were not using any filtration methods (directly) to formulate the indicators and 14% were using certain filtration methods (indirectly). Descriptive statistics based on selected papers were conducted. Figure 1 showed that articles within these topics of interest (i.e. sustainability indicator or corporate sustainability model) have been increasing for the last five years, from 5 articles in 2012 to 23 articles in 2016. The same trend also occurred to the articles which used criteria-based approach with no filtration (CNF) and model-based approach (M), there were 2 articles of CNF in 2012 increasing to 11 articles in 2016 and 1 articles of M in 2012 increasing to 9 articles in 2016. Given the number of articles, 1 article of criteria-based model with filtration (CF) in 2012, and there was no related publication in 2011, 2012 and 2013, and 3 articles in 2016; it indicated articles of criteria-based approach with filtration were rare and had no trend in quantity year to year.

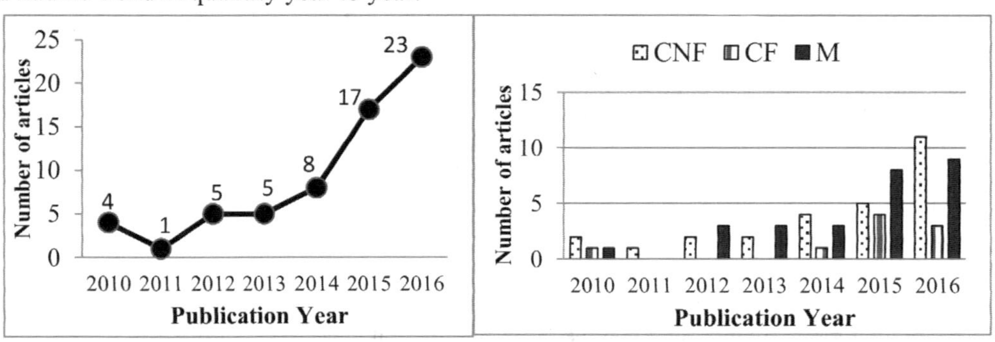

Figure 1. Distribution of number of publications per year
(left) for topics of sustainability indicators or models, (right) for specific classification
Notes: CNF: criteria-based approach no filtration, CF: criteria-based with filtrations, M: model-based)

Table 3 indicated that about 33% of the selected articles were discussing about sustainability assessment for general corporations, meanwhile 67% of them were discussing the same issue for specific industry sectors which majority covered the sectors of supply chain & logistic, public service and information technology. Among 36 selected articles using criteria-based approach, environmental dimension was the highest percentage (89%) being used to assessment sustainability. However, the development of indicators for other dimensions, namely social and economics (81% and 64%) were also increased recently. Table 4 also summarized that qualitative and quantitative indicators have almost equal percentage of usage.

Table 3. Distribution of Industrial Sectors

Industry Sector	Percentage	# of articles
General corporation	33	21
Specific industry	67	42
For 42 specific industry sector, there were:		**# of articles**

Supply Chain & Logistics (e.g. retailer, port, logistics service provider)	6
Public Service (e.g. urban design, water service, park, electricity)	5
Information Technology (e.g. green IT, service centre, telecommunication service)	3
Manufacturing, Tourism, University,Micro Small Medium Enterprise	3 foreach industry
Automotive, Energy, Forestry, Mining, Textile	2 foreach industry
Fashion, Fishing, Forensic, Hospital,Local Government, Project Management	1 foreach industry

Table 4. Criteria-based approach: dimensions and type of sustainability indicators

Criteria-based approach	**# of Articles**	Dimensions of Sustainability				Type of Indicators	
		Environment	**Social**	**Economic**	**Additional**	**Qualitative**	**Quantitative**
No Filtration	27	23	20	16	12	20	16
Filtration	9	9	9	7	7	5	6
Total	*36*	*32*	*29*	*23*	*19*	*25*	*22*
Percentage		**89%**	**81%**	**64%**	**53%**	**69%**	**61%**

For model-based approach, there were 27 articles being selected because the researches discussed the development of sustainability models using certain basic concepts. Table 5 listed the underlying concepts that used to develop the model. Given the percentage of usage, Business Process Management was placed as the most-frequent-used underlying concept; with the percentage of usage was 37%. It was followed by Corporate Sustainability and Supply Chain Management as the second and third highest percentages, with the percentages of usage were 33% and 22% respectively.

Table 5. Model-based approach: distribution of underlying concepts and developed models

Underlying Concepts	**# of articles (percentage)**	**Developed Models**
Business Process Management (BPM)	10 (37%)	Process Analysis Method, Green Performance Indicators, Sustainable Management System, Systemic Navigation Framework, Sustainability exploitation & exploration, Corporate Sustainability Activity framework, Triple-layered Business Model Canvas, Sustainability Business Model, Campus Sustainability Assessment, Sustainable Strategic Management
Corporate Sustainability (CS)	9 (33%)	Process Analysis Method, Green Performance Indicators, Sustainable Management System, Systemic Navigation Framework, Sustainability exploitation & exploration, Corporate Sustainability Activity framework, Sustainability Business Model, Campus Sustainability Assessment, Sustainable Strategic Management
Supply Chain Management (SCM)	6 (22%)	Sustainability SCM, Green SCM, Multi-methodological approach, Risk Management Approach, Sustainability Supply Chain Performance, Assessment of Supply Chain Sustainability Framework
Triple Bottom Line (TBL)	5 (19%)	Urban Water Service Sustainability, Economic and Socio-economic assessment, Efficiency Model, Triple-layered Business Model Canvas, Product life cycle phases
Engineering Management (EM)	4 (15%)	Energy Management Maturity Model, Energy Efficiency Model, Sustainable Systems Integration Model-Metrics Framework, Energy Efficiency Model
Balanced Scorecard (BSC)	3 (11%)	BSC Strategic Framework, BSC Model, Sustainability BSC (SBSC)
Product Life Cycle	1 (4%)	Product life cycle phases

4. Conclusion

This paper described on how the three-stage systematic literature review gave a scientific contribution in providing, for the best of authors' understanding, a helicopter-view about the expansion of the publications related to the interest topic. It started from (i) planning the literature review by defining specific terms to search the articles, determining targeted publisher, publication year and classification criteria, (ii) conducting the review by skimming the title, keywords, abstract and reading the introduction, methodology and conclusion of selected articles, then (iii) reporting the result by concluding main issues and providing guidance for future directions.

Regarding to the topic of interest i.e. the development of sustainability assessment tool, the systematic literature review filtered only 63 articles or 7.2% out of 875 articles from Science Direct electronic database; however, the distribution of number of publications has shown an uptrend within last five years. It identified two primary approaches that were used in designing the sustainability assessment i.e. criteria-based approach (assessment indicators were derived directly or indirectly from literature review) or model-based approach (the development of a certain model was done prior to the formulation of the assessment indicators).

In general, there were two major issues to be concluded from the result of the literature review. Firstly, both approaches had similar popularities in publications; there were 27 articles (or 43%) of 63 selected articles that using model-based approach, another 43% of them were criteria-based approach with no filtration and the rest 14% were criteria-based using a specific filtration method to select the indicators. Furthermore, the review indicated that the approaches were balance in strengths and weaknesses, for instance, it was easy for criteria-based approach to formulate a set of indicators, but it might be difficult because of the similarities among indicators and the subjectivity in selection. For future direction, with model-based approach, it could create a more robust indicator based on a proven ideal model but it might be difficult to find how ideal a model was. However, a model-based approach appears more appropriate because a model can be an identifier to distinguish an assessment tool from others; a model can describe the constructs and their relationships. Business Process Management and Corporate Sustainability Model will be studied further to determine the constructs of the model. The model will be developed for general corporations and will be validated by applying it to some case studies.

References

[1] Said A M, Ahmadun F, Paim L and Masud J 2003 Environmental concerns, knowledge and practices gap among Malaysian teachers *Int. J. Sustainablity in Higher Educ.* **4** pp 305–13
[2] Thogersen J 2005 How may consumer policy empower consumers for sustainable lifestyles? *J. Consum. Policy* pp 143–78
[3] Adams C, Muir S and Hoque Z 2014 Measurement of sustainablity in the public sector sustainability accounting *Manage. Policy J.* **5** pp 46–63
[4] Sari Y, Hadiyat M A and Loa J L 2015 The modelling of sustainable lifestyle towards the readiness of ASEAN Economic Community by Structural Equation Modeling (case study: city of Surabaya) *Proc. Conf. Prod. Syst. XI and Conf. Manage. Qual. Eng.*
[5] AMICE 1989 *CIMOSA: open systems Architecture for CIM: Springer-Verlag/ESPRIT Consortium* (Berlin: AMICE)
[6] Childe S J, Maull R S and Bennett J 1994 Frameworks for understanding business process re-engineering *Int. J. Oper. Prod. Manage.* **14** pp 23–34
[7] Bititci U and McCallum N 2004 Understanding and managing the manage processes *Biennial Conf. Perform. Meas. Assoc.* pp 1–21
[8] Mackay D, Bititci U, Maguire C and Ates A 2008 Delivering sustained performance through a structured business process approach to management *Measuring Bus. Excellence* **12** pp 22–37
[9] Adesola S and Baines T 2005 Developing and evaluating a methodology for business process improvement *Bus. Process Manage. J.* **11** pp 37–46
[10] Tranfield D, Denyer D and Smart P 2003 *Br. J. Manage.* **14** pp 207–22
[11] Morioka S N and de Carvalho M M 2016 *J. Cleaner Prod.* **136** pp 134–46
[12] Littell J H, Corcoran J and Pillai V 2008 *Systematic reviews and meta-analysis* (Oxford: Oxford University Press)
[13] Kang Y, Ryu M H and Kim S 2010 *J. World Bus.* **45** pp 415–21
[14] Antolin-Lopez R, Delgado-Ceballos J and Montiel I 2016 Deconstructing corporate sustainability: a comparison of different stakeholder metrics *J. Cleaner Prod.* **136** pp 5–17
[15] Rahdari A H and Rostamy A A A 2015 *J. Cleaner Prod.* **108** pp 757–71

[16] Valenzuela-Venegas G, Salgado J C and Díaz-Alvarado F A 2016 *J. Cleaner Prod.* **133** pp 99–116
[17] Figge F, Hahn T, Schaltegger S and Wagner M 2002 The sustainability balanced scorecard–linking sustainability management to business strategy *Bus. Strategy Environ.* **11** pp 269–84
[18] Kaplan R S and Norton D P 1996 *Using the balanced scorecard as a strategic management system* (Harvard: Havard Business Review)

IOP Conf. Series: Materials Science and Engineering **703** (2019) 012019 doi:10.1088/1757-899X/703/1/012019

Controlled release fertilizer based on starch chitosan encapsulation

E Savitri, E Purwanto, A N Kodrat and E Yonathan

Department of Chemical Engineering, University of Surabaya, TG Building 5th floor, Raya Kalirungkut, Surabaya, 60293, Indonesia

E-mail: savitri_ma@staff.ubaya.ac.id

Abstract. Phosphor is the main component in a triple super phosphate (TSP) fertilizer which is needed by plants. However, phosphorus is very soluble in the water and makes it easily removed from the soil flowing into the river, and causing algae bloom problems. The solution for the problem is by using the fertilizer coated with starch-chitosan based material. The coated hydrogel bead fertilizer was expected to have a phosphorus controlled-released behaviour and retain the phosphor to be able to stay for a longer time in the soil. This research dealt with studying phosphorus releasing behaviour of starch-chitosan based fertilizer, water uptaking characteristic (swelling), and the morphology of the materials. The commercial granular fertilizer TSP was coated using starch-chitosan suspension. The coatings were conducted by spraying technique. The chitosan-starch mass ratio used were 100:0, 30:70, and 70:30, otherwise the starch percentages were made of 1%, 2%, and 3%. After coated by starch-chitosan suspension, the fertilizers were dried and used on the soil media. The soil which contained coated fertilizer was regularly watered and the phosphorus content was analyzed.The study showed that the water uptaking of the coated fertilizer was higher than that of the conventional uncoated fertilizer. The optimum condition of the chitosan-starch based coating was 30:70 and 2% of starch percentage. Furthermore, the higher chitosan content yielded the higher water uptaking capacity. This phenomenon is supported by the Scanning Electron Microscope (SEM) result. The coating fertilizer with higher chitosan content (ratio of 70:30, 2% starch) had more pores than that of lower chitosan content (ratio of 30:70, 1% starch).

1. Introduction

Phosphorus is the important component in triple super phosphate (TSP) fertilizer. The component is easily dissolved and carried away by water which causes eutrophication problems [1]. According to [2], the phosphorus absorbed by plants is only 10-15% while the rest is released into the waters. This means that fertilizer efficiency is low because a lot of phosphorus is wasted.

Fertilizers can be modified to produce the fertilizer which has slower release behaviour than that of the conventional fertilizer. The purposes of modification are to control the release of nutrients into soil and to prolong the availability of the nutrients in the soil [3].

The modification can be conducted by coating the fertilizer using polymers such as chitosan and starch. Chitosan is a biopolymer which has an amine group produced from deacetylation of chitin and can be extracted from shrimp waste [4]. Besides having controlled release behavior, both chitosan and starch have the ability to swell (save water). Those characteristics make them have a dual function on fertilizer, as a controlled release system and also as a water storage system that maintains the availability of water in the soil, especially in the dry season.

This study aims to synthesize and characterize chitosan-starch-coated fertilizers. The characterization includes the morphology and swelling properties of coated fertilizers. Then, the performance of coated TSP fertilizers is analyzed by the release rate of the component. The effect of the chitosan : starch mass ratio as coating material is also analyzed.

2. Experimental

The research was carried out in three steps of processes. The steps consist of the synthesis of chitosan-starch coated fertilizers, the characterization of the coated fertilizer, and the analysis and modelling of phosphorus release rate.

2.1. Synthesis of hydrogel bead fertilizer

Chitosan solution 1% (w/w) was prepared by dissolving chitosan powder in 1% (v/v) acetic acid solution. The starch gel was made by heating 1, 2, and 3% (w/v) of tapioca flour in distilled water at 76 °C. Chitosan solution and starch gel with mass ratio (100:0, 70:30, 30:70) were stirred for 30 minutes. Then, the mixture was added 0.8 %w/w citric acid solution as a crosslinking agent and stirred for 2 hours. The coating process was conducted by spraying the coating mixture onto the entire surface of the fertilizer. After that, the coated fertilizer was dried at 60 °C for 12 hours. The fertilizer coating process was repeated for 5 layers.

2.2. Characterization of hydrogel bead fertilizer

The characterization of coated fertilizers was analysed by studying the morphology and the swelling behaviour of the coating system. Morphological analysis was carried out using Scanning Electron Microscope (SEM). The morphology of the coated fertilizer surfaces was observed using an SEM FEI INSPECT S-50, Netherlands. The surface of the fracture was attached onto double-sided carbon tape and subsequently placed in a sample holder.

The swelling analysis was carried out by immersing dry-coated fertilizer samples in water for definite time and then weighing them. The procedure was repeated until the hydrogel bead fertilizers achieved the equilibrium. It was assigned by constant weight of the samples. Furthermore, the swelling parameters were analysed by comparing the changes in the coated fertilizer mass at a certain time to the dry hydrogel mass of coated fertilizer.

2.3. Controlled release of hydrogel bead fertilizer

A certain amount of coated fertilizer and topsoil were placed in a polybag that has been perforated. The arrangement on polybag was placed of soil-coated fertilizer-soil in series. Water reservoir was placed below the polybag to keep the water drain. Furthermore, the analysis was conducted by watering each system with some amount of water for 7 days. After the watering process, the soil sample was analysed using spectrophotometer to find out how much phosphorus was released in the soil. The phosphorus in the sample was quantitatively determined via colorimetry by reacting orthophosphate with ammonium molibdovanadat to form a yellow complex molibdovanadat phosphoric acid compound. Yellow liquid containing phosphorus was analysed by spectrophotometer.

3. Results and Discussions

The commercial TSP fertilizer which has 36 %w/w phosphorus content was increased in its efficiency by modifying the fertilizer. Modification was conducted by coating the fertilizer surface with chitosan-starch mixture. The variation systems consist of 1% chitosan : starch with ratio 100: 0 (A), 70:30 (B), 30:70 (C)- 1% starch and 1% chitosan : starch 70 : 30 with 1% starch(D), 2% starch (E), 3% starch (F) and non-coating fertilizer (G). The phosphorus released in the soil was analysed on the same treatments for each variations. The optimum results of the variations were analysed on the morphology, swelling ability, and the release rate of phosphorus content.

3.1. Morphology of chitosan-starch coating surface

The morphology of chitosan-starch coating on samples C and E was analyzed by Scanning Electron Microscope (SEM). The samples C and E have the optimum performance on phosphorus release. The C sample was coated fertilizer which has chitosan: starch ratio 30:70 ratio and 1% starch concentration, while sample E was coated fertilizer which has chitosan : starch ratio 70:30 and 2% starch concentration.

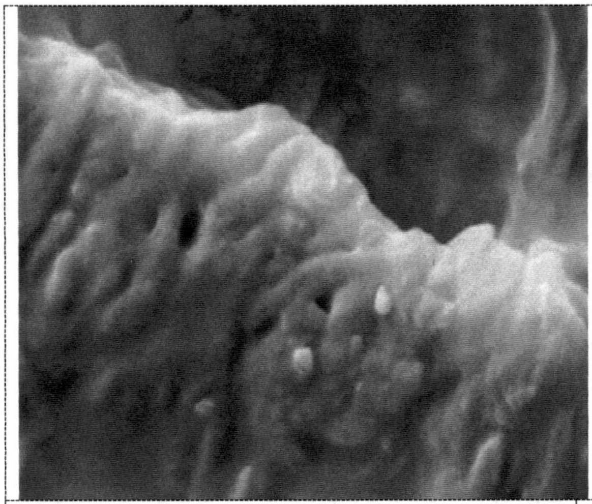

Figure 1. The morphology of coated fertilizer for chitosan : starch ratio 30 : 70 and 1 % starch solution (sample C)

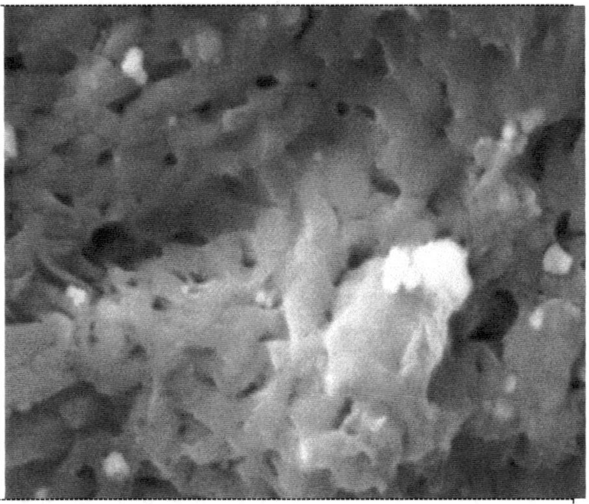

Figure 2. The morphology of coated fertilizer for chitosan : starch ratio 70 : 30 and 2 % starch solution (sample E)

Figure 1 and 2 show that the film coating consists of fiber matrix and some granulars distributed on the matrix. The fiber matrix is chitosan structure while the granular is starch component. The arrangement of chitosan fiber created pores on the film structure. The starch granular was attached on the chitosan matrix by crosslink linkage. The crosslink linkage occured because the hydroxyl (OH) group in starch was replaced by an amine group (NH_2) from chitosan [5]. The matrix structure with granular became the surface to barrier the phosphorus content of TSP fertilizer release. The sampleC (Figure 1) was more dense than the sample E (Figure 2). Sample C had more starch content, whereas the sample E had more chitosan content. An increase in starch content caused the coating pores to become more dense than that of higher chitosan content. The pore size in sample C was 200-300 nm while in sample E was 200-600 nm.

Besides the size of pores on the matrix coating, there were different number of pores for both of the samples. The sample E showed more number of pores than that of sample C. It indicates that sample E is easier to pass phosphorus content than that of sample C. This is proven by the amount of phosphorus on the soil for period of release as shown in Figure 3.

From Figure 3, it can be seen that on the first three days, both sample C and E had the same phosphorus release behaviour, but on the next day the amount of phosphorus in the soil in sample E was lower than that of sample C. Sample C was able to maintain more amount of phosphorus in the soil than that of sample E. According to the results , it can be stated that the slow release behaviour of coated fertilizer can be obtained by creating the dense surface with less pore on the matrix structure.

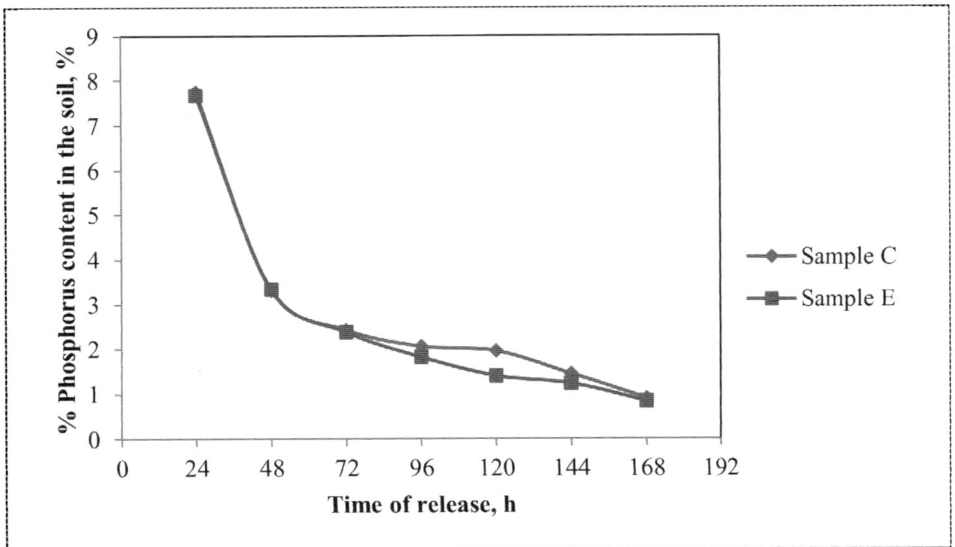

Figure 3. The percentage of phosphorus content on the soil after period of release because of watering

3.2. Swelling Analysis

Swelling analysis was done to determine the water absorption capacity of the coating system. Table 1 shows the results of swelling analysis for each variation of fertilizer coating.

Table 1. The swelling behaviour for different coating systems on fertilizer

Type of fertilizer	Starch content (%)	Chitosan : Starch ratio	Swelling (%)
A	1	100 : 0	24.658
B	1	70 : 30	24.735
C	1	30 : 70	18.422
D	1	70 : 30	25.099
E	2	70 : 30	18.917
F	3	70 : 30	20.282
G	-	-	7.828

When compared with non-coating fertilizer (sample G), it can be seen that the presence of coatings (sample A - F) increased the percentage of swelling. Increasing the swelling percentage affected the amount of water that can be absorbed by the coating film was also an increase. Table 1 shows that the presence of chitosan in coating system increased the percentage of swelling, where in samples A, B and D there was relatively more chitosan than starch. During the starch gelatinization process, the hydrophobic hydroxyl groups in the starch chain would lead to place on the outer side [6].As a result of the condition, the more amount of starch content in the coating systems, the more hydrophobic the coating systems tend to be and the smaller the swelling percentage is.

The presence of starch in the coating systems was proven to have an effect on the reduction on swelling, but the physical strength of the coating was improved by increasing the chitosan content. The result is supported by the research of Xu, et al. [7], that films made from chitosan have poor water vapor barrier characteristics or high Water Vapor Transmission Rate (WVTR) compared to films that have been modified with starch. Water vapor barrier or Water Vapor Transmission Rate (WVTR) is a parameter that indicates whether or not a layer is penetrated by water. A high WVTR or poor water vapor barrier indicates that chitosan films are easily passed by water vapor. In conclusion, the addition of starch will slow the release of phosphorus content because of reducing the ability of the coating to absorb water.

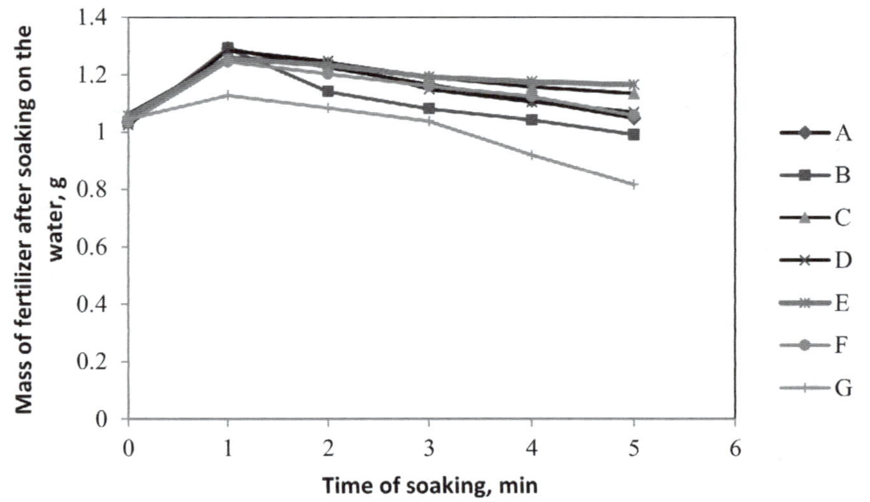

Figure 4. Mass of fertilizer after soaking on the water for period of time

Figure 4 shows that the increase in fertilizer mass after soaking on the water had the maximum value in the first minute. It can be stated that the process of water absorption or swelling process of coated fertilizers occured quickly. The phenomenon can be related that the phosphorus diffusion process out of fertilizer occured slower and the phosphorus release process followed a diffusion controlled transport mechanism. If the swelling rate is much faster than that of the diffusion rate, the controlling mass transfer is the diffusion rate.

3.3. The Performance of Phosphorus Released by Coated Fertilizers

The release behaviour performance of coated fertilizer can be conducted by measuring the amount of phosphorus release in to the soil with a spectrophotometer.

Table 2 shows that the coated fertilizers had slower release behavior than that of non-coated fertilizers. The increasing amount of chitosan could produce the strong matrix of layer but gave more pores, otherwise the increasing amount of starch could enhance the hydrophobicity of coating system. The sample C gave the best performence as slow release fertilizer because it had the best physical properties as surface barrier to retain the phosphorus on the fertilizer. It can also be stated that the effect of the chitosan : starch ratio is greater to release behaviour than that of the starch content on coating system.

Table 2. The Amount of Phosphorus Release on to soil at different sample for period of release

Type of Fertilizer	Percentage of Phosphorus Content (%)						
	24 h	48 h	72 h	96 h	120 h	144 h	168 h
A	7.775	3.537	1.611	1.617	1.174	1.131	0.499
B	7.651	3.334	2.236	1.669	1.062	1.104	0.499
C	7.745	3.316	2.430	2.059	1.956	1.448	0.904
D	7.671	3.448	2.317	1.422	1.230	1.091	0.585
E	7.668	3.337	2.377	1.820	1.402	1.231	0.835
F	8.378	3.613	2.156	1.712	0.954	0.734	0.838
G	8.449	3.043	2.236	1.679	1.173	0.987	0.582

The results of this study can be compared with the research of Perez, J. and Francois, N. [8] who examined the behavior of nitrogen release in chitosan-starch coated fertilizers. The results of the study indicated that the higher amount of chitosan in the coating system, the more nitrogen can be released from fertilizer. At various chitosan : starch ratio 100:0; 30:70; and 20:80 nitrogen releases reached 95%, 73% and 80%, respectively. From this study it can be stated that the starch content for optimum performance is in the coating system which has composition chitosan:starch ratio as high as 30:70. The optimum coating is a coating which has the best physical properties so that the release of phosphorus is minimum.

4. Conclusions

The presence of chitosan on the coating system is responsible for producing a porous matrix surface while the availability of starch tends to reduce the number of pores in the surface. The swelling analysis shows that the presence of chitosan-starch coatings increases the water absorption ability. The greater the content of chitosan, the higher the percentage of swelling. The release behaviour analysis shows that the optimum of coated fertilizer is the fertilizer which has a mass ratio of chitosan:starch by 30:70, and starch concentration of 1%.

References

[1] Zhang M, Gao B, Chen J, Li Y, Creamer A E and Chen H 2014 *Chem. Eng. J.* **255** 107
[2] Barber S A 1976 *Agron. Res. For FoodAsa Spec. Pub. Bull* 26
[3] Yamamoto C F, Peirera E I, Matoso L H, Matsunaka T and Ribeiro C 2016 *Chem. Eng. J.* **287** 390
[4] Zang J J, Jia S Y, Liu Y, Wu S H and Zhang Y B 2012 *Catal. Commun.* **27** 73
[5] Singh V and Kumari K 2014 *Int. J. Mater. and Biomater App.* **4** 7
[6] Wiącek A E 2015 *J. Food Hydrocoll.* **48** 228
[7] Xu Y X, Kim K M, Hanna M A and Nag D 2005 *J. Ind. Crops and Prod.* **21**185
[8] Perez J and Francois N 2016 *Carbohyd. Polym.***148** 134

IOP Conf. Series: Materials Science and Engineering **703** (2019) 012020 doi:10.1088/1757-899X/703/1/012020

Price and inventory policy strategy model in a price sensitive dual channel supply chain structure considering product substitution

R Y H Silitonga and N Christina

Department of Industrial Engineering, Institut Teknologi Harapan Bangsa, Dipati Ukur No. 80, Bandung, 40132, Indonesia

E-mail: roland@ithb.ac.id

Abstract. Dual channel supply chain structure, which utilizes both online and traditional offline channel, is widely adopted by a lot of companies. The structure offers better market coverage and profitability. Despite the advantages, it also brings some challenges to coordinate the strategy of the channel. This paper proposes a pricing and inventory policy model to achieve better performance of the structure, where a product substitution was introduced in the offline channel. A price sensitive linear demand function was also incorporated in this EOQ model. The model then evaluated by analysing the profit in two scenarios, the Stackelberg Leadership scenario and Total Profit scenario. The results show that the total profit of this structure is higher under Total Profit scenario, in which the supply chain entities make decision simultaneously. The model also shows that the product substitution gives a positive effect to the profit and become a competitive advantage of the offline channel.

1. Introduction

The speedy information technology development has significantly improved supply chain performance in every industry. The use of internet brings many opportunities to the companies [1]. The development enables a new supply chain structure which integrates conventional selling at retail store (offline) with internet based direct selling (online) called dual channel supply chain (DCSC) [2]. Simple DCSC structure usually consists of manufacturer, retail store, and online channel. The structure can be directly managed by manufacturer. Offline consumers demands are fulfilled by inventory in retailer, while online consumers demands are fulfilled by inventory in manufacturer [3].

In DCSC structure, online and offline channels can compete with each other to attract consumers, mostly by pricing strategy. In this structure, where the supply chain consists of several entities, there is a power hierarchy that influences the decision making process [4]. In DCSC structure which consists of manufacturer and retailers, the manufacturer, which has role as the leader, usually has a higher bargaining power that decides the pricing first. The retailers have the role as followers that decide the pricing based on manufacturer price [1]. This decision making scenario is known as Stackelberg Leadership. In this scenario, each player works hard to maximize their own profit without considering the whole profit of supply chain [5]. On the other side, there is a scenario where players make decision at the same time (simultaneous) [4] that we refer as the Total Profit scenario. In Total Profit scenario, players cooperate with each other to maximize profit of the whole supply chain [5].

Besides price, inventory also has a very important role, especially in fulfilling consumer demand to reach the required service level. The high level of inventory causes high saving cost and company's capital is stuck in inventory form, while if the number of inventory is too low, consumer demand cannot be fulfilled and it causes loss of sakes [6]. Therefore, the decision about the level of inventory becomes the main problem in minimizing inventory cost while it still maintains good service level [7].

In previous research [8], EQQ model with linear demand function is used to maximize profit from DCSC structure. It also introduced product substitution in offline channel. Product substitution plays important role especially when consumer think that main product price in both channels is too expensive and/or when there is empty inventory for the main product in both channels. The model used Stackelberg Leadership scenario. Coordination by sharing information within entities to maximize total profit of the whole supply chain, as in the Total Profit scenario explained earlier, is not considered yet. Hence this study developed pricing and inventory policy in DCSC structure by considering product substitution, and applying the two scenarios, Total Profit and Stackelberg Leadership.

2. Model

The development of the model is based on [8], which is a profit model of DCSC structure, considering product substitution, using Stackelberg Leadership scenario.

2.1. Conceptual model

The model considers a dual channel supply chain structure consisting of a central warehouse, one offline channel and one online channel (see figure 1). We assume that the product substitution only available in offline channel. Central warehouse gets its product supply from two manufacturers, Manufacturer A for the main product and Manufacturer B for the substitute product. The warehouse then distributes the product to the offline and online channel. There will be either one of the two possible scenarios which may occur. The first one is Stackelberg Leadership scenario, where central warehouse as a leader will make the prices decision first and announce the prices to offline and online channel, followed by online and offline channel sequentially. The second is Total Profit scenario, where all entities in the structure will make joint prices decision in order to maximize total profit of the structure.

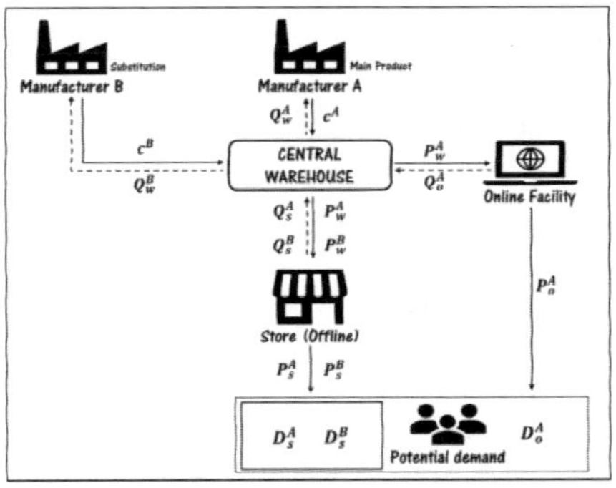

Figure 1. Dual channel supply chain structure with product substitution.

2.2. Notation

Parameters and variables used in this model are denoted by these following notations.

i : index for product type, A for main product and B for substitute product;

j : index for the structure's entity, w for central warehouse, s for offline channel and o for online channel;

β_s : price sensitivity in offline channel (unit/Rp);

γ_{so} : cross-price sensitivity of offline channel to substitute online channel (unit/Rp);

P_j^i : price of product i for entity j (Rp) (decision variable);

Q_{ij}^* : optimal order quantity for product i in entity j (unit) (decision variable);

T_{ij}^* : reorder point for product i in entity j (days) (decision variable);

ρ : proportion of customers who prefer online than offline channel relative to the total demand;

σ : proportion of customers who prefer substitute product over main product in offline channel;

β_o : price sensitivity in online channel (unit/Rp);

γ_{os} : cross-price sensitivity of online channel to substitute offline channel (unit/Rp);

γ_{BA} : cross-price sensitivity of product B to substitute product A in offline channel (unit/Rp);

D_{max} : maximum demand when the price is equal to zero (unit/period);

D_j^i : demand of product i for entity j (unit/period);

R_j : revenue of entity j;

c_i : cost of product i from manufacturer (Rp/unit);

π_j : profit of entity j;

h_j : holding cost for entity j (Rp/unit/period);

s_j : ordering cost for entity j (Rp/order);

O_{bj} : total purchasing cost for entity j (Rp);

O_{sj} : total ordering cost for entity j (Rp);

O_{hj} : total holding cost for entity j (Rp);

O_{Tj} : total inventory cost for entity j (Rp);

2.3. Model development

Referring to [7], linear demand functions of this model are as follow.

$$D_o^A = \rho D_{max} - \beta_o P_o^A + \gamma_{os} P_s^A \tag{1}$$

$$D_s^A = (1-\sigma)(1-\rho)D_{max} - \beta_s P_s^A + \gamma_{so} P_o^A \tag{2}$$

$$D_s^B = \sigma(1-\rho)D_{max} - \beta_s P_s^B + \gamma_{BA} P_s^A \tag{3}$$

2.3.1. Inventory model. We use a basic EOQ model as the inventory control method. All the assumptions in EOQ model is applied in this model. According to Bahagia [9], inventory cost in EOQ model consist of purchasing cost, ordering cost, and holding cost.

$$O_{Tw} = c_A(D_o^A + D_s^A) + (c_B \times D_s^B) + s_w\left(\frac{D_o^A + D_s^A}{Q_{Aw}^*} + \frac{D_s^B}{Q_{Bw}^*}\right) + h_w\left(\frac{Q_{Aw}^* + Q_{Bw}^*}{2}\right) \tag{4}$$

$$O_{Ts} = (P_w^A \times D_s^A) + (P_w^B \times D_s^B) + s_s\left(\frac{D_s^A}{Q_{As}^*} + \frac{D_s^B}{Q_{Bs}^*}\right) + h_s\left(\frac{Q_{As}^* + Q_{Bs}^*}{2}\right) \tag{5}$$

$$O_{To} = (P_w^A \times D_o^A) + s_o\frac{D_o^A}{Q_{Ao}^*} + h_o\frac{Q_{Ao}^*}{2} \tag{6}$$

Optimality of order quantity and reorder point are the decision variable of inventory policies in EOQ model. Formulation of optimal order quantity and reorder point for central warehouse can be expressed by the following equations.

$$Q_{Aw}^* = \left(\frac{2s_w(D_s^A + D_o^A)}{h_w}\right)^{-1} \tag{7}$$

$$Q_{Bw}^* = \left(\frac{2s_w D_s^B}{h_w}\right)^{-1} \tag{8}$$

$$T_{Aw}^* = \left(\frac{2s_w}{(D_s^A + D_o^A)h_w}\right)^{-1} \tag{9}$$

$$T_{Aw}^* = \left(\frac{2s_w}{(D_s^A + D_o^A)h_w}\right)^{-1} \tag{10}$$

Optimal order quantity and reorder point for offline channel can be expressed by the following equations.

$$Q_{As}^* = \left(\frac{2s_S D_S^A}{h_S}\right)^{-1} \quad (11)$$

$$Q_{Bs}^* = \left(\frac{2s_S D_S^B}{h_S}\right)^{-1} \quad (12)$$

$$T_{As}^* = \left(\frac{2s_S}{D_S^A \times h_S}\right)^{-1} \quad (13)$$

$$T_{As}^* = \left(\frac{2s_S}{D_S^B \times h_S}\right)^{-1} \quad (14)$$

Formulation of optimal order quantity and reorder point for online channel can be expressed by the following equations.

$$Q_{Ao}^* = \left(\frac{2s_O D_O^A}{h_O}\right)^{-1} \quad (15)$$

$$T_{Ao}^* = \left(\frac{2s_O}{D_O^A \times h_O}\right)^{-1} \quad (16)$$

2.3.2. Revenue model. Revenue of each entity can be obtained by multiplying demand of its product with the product's price and can be expressed by the following equations.

$$R_W = P_W^A (D_S^A + D_O^A) + (P_W^B \times D_S^B) \quad (17)$$

$$R_S = (P_S^A \times D_S^A) + (P_S^B \times D_S^B) \quad (18)$$

$$R_O = P_O^A \times D_O^A \quad (19)$$

2.3.3. Stackelberg leadership model. Central warehouse maximized its profit by the following equation.

$$\pi_W = (P_W^A - c_A)(D_S^A + D_O^A) + (P_W^B - c_B)D_S^B - s_W\left(\frac{D_O^A + D_S^A}{Q_{Aw}^*} + \frac{D_S^B}{Q_{Bw}^*}\right) - h_W\left(\frac{Q_{Aw}^* + Q_{Bw}^*}{2}\right) \quad (20)$$

Maximization of function (20) should consider these following constraints.

$$P_W^A \geq c_A \quad (21)$$

$$P_W^B \geq c_B \quad (22)$$

$$Q_{Aw}^*, Q_{Bw}^* \geq 0 \quad (23)$$

The profit of online channel can be expressed by the following equation.

$$\pi_S = (P_S^A - P_W^A)D_S^A + (P_S^B - P_W^B)D_S^B - s_S\left(\frac{D_S^A}{Q_{As}^*} + \frac{D_S^B}{Q_{Bs}^*}\right) - h_S\left(\frac{Q_{As}^* + Q_{Bs}^*}{2}\right) \quad (24)$$

The previous variables, P_W^A and P_W^B now become parameters in function (24), where their values were determined from equation (20). Function (24) should consider the following constraint.

$$Q_{Ao}^* \geq 0 \quad (25)$$

Using the prices at central warehouse (P_W^A and P_W^B), the price at online channel (P_O^A) and the demand at online channel (D_O^A), offline channel maximized its profit.

$$\pi_O = (P_O^A - P_W^A)D_O^A - s_O\frac{D_O^A}{Q_{Ao}^*} - h_O\frac{Q_{Ao}^*}{2} \quad (26)$$

2.3.4. Total profit model. In this scenario, all entities in the structure will make a joint decision to maximize the entire structure's profit. Total profit can be expressed by the following equation.

$$\pi_{total} = \{(P_S^A - c_A)D_S^A + (P_o^A - c_A)D_o^A + (P_S^B - c_B)D_S^B\} - \left\{s_W\left(\frac{D_o^A + D_S^A}{Q_{Aw}^*} + \frac{D_S^B}{Q_{Bw}^*}\right) + h_W\left(\frac{Q_{Aw}^* + Q_{Bw}^*}{2}\right) + s_S\left(\frac{D_S^A}{Q_{As}^*} + \frac{D_S^B}{Q_{Bs}^*}\right) + h_S\left(\frac{Q_{As}^* + Q_{Bs}^*}{2}\right) + s_o\frac{D_o^A}{Q_{Ao}^*} + h_o\frac{Q_{Ao}^*}{2}\right\} \quad (27)$$

3. Numerical example

Parameters in this model are adopted from [8]. There are some parameters values that were changed, price and cross-price sensitivity, because the data from previous research was not feasible in this model. Consider a DCSC structure with the following data: D_{max}=84 units/month, ρ=0.7, σ=0.5, β_o=0.1x10^{-3} unit/Rp, β_s=0.08x10^{-3} unit/Rp, γ_{os}=0.06x10^{-3} unit/Rp, γ_{so}=0.04x10^{-3} unit/Rp, γ_{BA}=0.05x10^{-3} unit/Rp, c_A=Rp320,000/unit, c_B=Rp295,000/unit, s_w=Rp95,000/order, h_w=Rp20,000/unit/month, s_s=Rp135,000/order, h_s=Rp22,000/unit/month, s_o=Rp110,000/order and h_o=Rp22,000/unit/month. Using LINGO 11, the optimal value for the objective function and each variable in both scenarios can be obtained. The result is presented in table 1.

Table 1. Numerical example results.

	Stackelberg Scenario					Total Profit Scenario				
	Central Warehouse		Offline Channel		Online Channel	Central Warehouse		Offline Channel		Online Channel
	A	B	A	B	A	A	B	A	B	A
P_j^i	502,143	383,839	552,143	452,589	789,286	338,732	297,976	477,144	393,215	664,286
Q_{ij}^*	13.78	8.15	0	7.01	11.42	14.46	6.89	3.50	7.83	14.49
T_{ij}^*	20.68	34.99	0	52.55	26.31	19.71	41.35	105.10	47	20.70
O_{Tj}	8,903,776		1,689,500		8,083,277	8,941,977		2,078,018		7,432,177
O_{Ttotal}			18.676.553					18.452.173		
π_j	3,825,956		120,857		2,177,437	0		365,196		6,517,823
π_{total}			6,124,250					6,883,019		

4. Analysis

It can be seen from table 1 that the total profit in Total Profit scenario is higher than in Stackelberg Leadership scenario. It is also found that the prices at each entity is lower in Total Profit scenario, which shows better competitiveness.

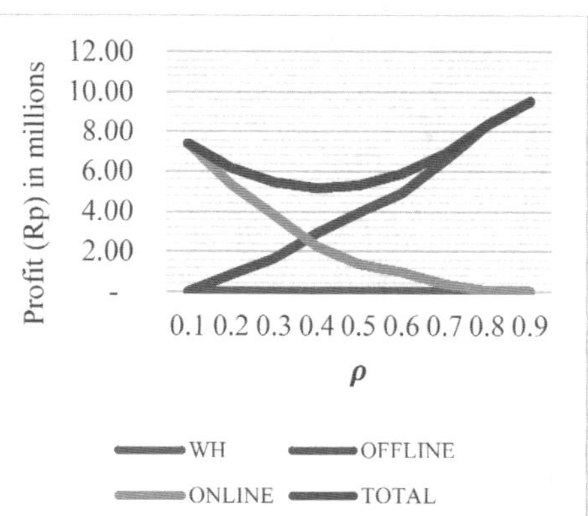

Figure 2. Impact of ρ to profit in Total Profit scenario.

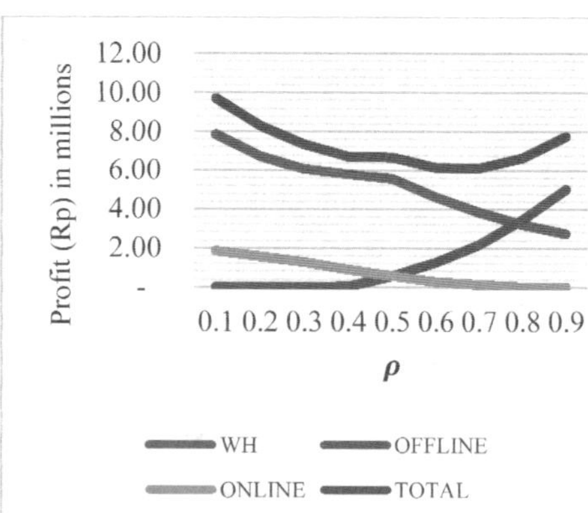

Figure 3. Impact of ρ to profit in Stackelberg Leadership scenario.

The consumer preference to online channel (ρ) influences the profits of each entity, and also the whole structure. The impact can be observed in figure 2 and figure 3. Parameter γ_{BA} measures cross-price sensitivity of product B to substitute product A in offline channel. In both scenarios, higher γ_{BA}

will give positive impact to the profit of offline channel. The impact can be observed in figure 4 dan figure 5.

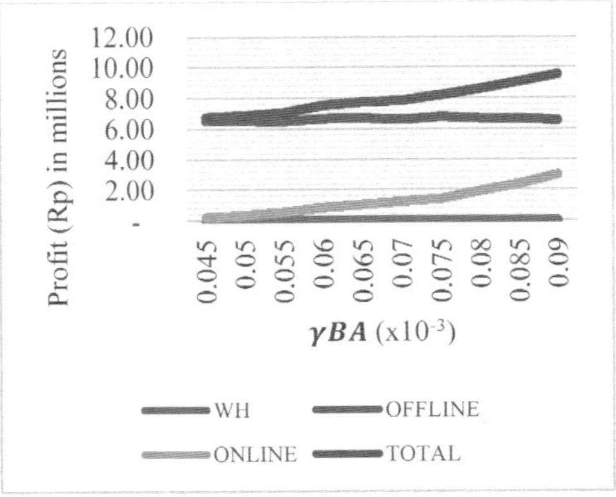

Figure 4. Impact of γ_{BA} to profit in Total Profit scenario.

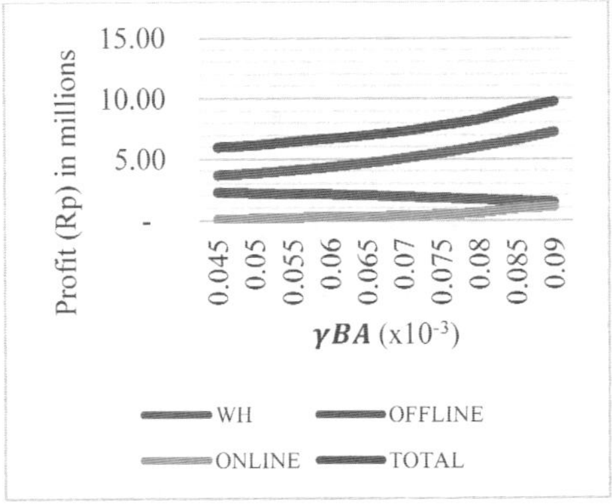

Figure 5. Impact of γ_{BA} to profit in Stackelberg Leadership scenario.

5. Conclusion and further research

This study has developed a model to accommodate price and inventory strategy in DCSC structure, implementing two scenarios namely Total Profit dan Stackelberg Leadership. The result shows that the total profit in Total Profit scenario is higher compared to Stackelberg Leadership scenario. There is a requirement to enable Total Profit scenario, which is coordination and information sharing, to get prices decision altogether. Hence, if all entities can collaborate and share information, Total Profit scenario is preferable. The study can be further developed by considering probabilistic demand and product lead time. There is also a possibility to develop application and the platform suitable, to enable the coordination. Application of the model to several different products is also interesting for further research.

References

[1] Yue X and Liu J 2006 Demand Forecasting Sharing in a Dual Channel Supply Chain *Eur. J. Oper. Res.* **174** 646.

[2] Widodo E, Takahashi K, Morikawa K, Pujawan N, and Santosa B 2011 Managing Sales Return in Dual Sales Channel: An Analysis of Its Product Substitution *Int. J. Ind. Syst. Eng.***9** 121.

[3] Chiang W K and Monahan G E 2005 Managing Inventories in a Two-Echelon Dual Channel Supply Chain *Eur. J. Oper. Res.***162** 325.

[4] Matsui K 2019 A Supply Chain Member Should Set Its Margin Later If Another Member's Cost is Highly Uncertain *Eur. J. Oper. Res.***275** 127.

[5] Modak N M and Kelle P 2019 Managing a Dual Channel Supply Chain Under Price and Delivery Time Dependent Stochastic Demand Eur. J. Oper. Res.**272** 147.

[6] Silitonga R Y H and Kawet M P 2017 Pengembangan Model untuk Aplikasi Pengendalian Persediaan Probabilistik Multi Item Single Supplier *Jurnal Telematika.* **12** 13.

[7] Limanjaya B and Silitonga R Y H 2018 Development of Multi Item Probabilistic Inventory Model by Considering Perishable and Purchase Bonus Factors *Jurnal Telematika* 13.

[8] Widodo E 2015 A Model Reflecting the Impact of Product Substitution in Dual Chanel Supply Chain Inventory Policy *Procedia Manuf.***4** 168.

[9] Bahagia S N 2006 *Sistem Inventori* (Bandung: Institut Teknologi Bandung)

IOP Conf. Series: Materials Science and Engineering **703** (2019) 012021 doi:10.1088/1757-899X/703/1/012021

Assessing materials from hoarded mobile phones: hidden e-waste subject for reverse logistics

R Siringo[1], H Herdiyansyah[1], R D Kusumastuti[2] and A E Lucianto[1]

[1]School of Environment, Universitas Indonesia
[2]Departemnet of Management, Faculty of Economics and Business, Universitas Indonesia

E-mail: herdis@ui.ac.id

Abstract. The advancement of communication technology is shortening the lifetime of mobile phones (MP), resulting in a phenomenal increase in mobile phones waste (MPW). However, instead of recycling their obsolete devices, customers shelved MPW for the personal data stored inside. This study aims to estimate secondary minerals contained in MPW stockpiled by Indonesians. "Consumption and used" method was employed to assess the quantity of MPW generated in this country based on the data of mobile subscribers. Considering that MP life span was around 3 years, and about 38% of obsolete MP were hoarded by their owners, this study revealed that about 339.69 million units of MPW were waiting to enter the reversed logistics. This volume of MPW may generate economic intrinsic at approximately USD 508 million from valuable metals such as 80,949 Kg of silver, 8,873 Kg of gold, and 2,547 Kg of palladium. It implied that hoarding MPW impedes the noble goal of recycling; that is to promote sustainable materials management (SMM) since mining precious metals from e-waste will reduce the demand of virgin minerals directly extracted from nature and reduce the environmental impact. This paper highlighted that it is imperative to develop a sustainable system to support the mineral supply chain.

1. Introduction

When mobile phones (MP) were firstly introduced in Indonesia market, not many residents could afford to buy this particular technology. However, the purchasing power in this country has improved along with its economic growth. As a result, the demand of these equipments escalates persistently. In facts, the penetration of medium brands with affordable price has made smartphones popularity increases among lower-medium income residents. They are attracted to the multiple functions embedded in these devices. Having diverse features, smartphones can assist people in many ways, for example, maintaining social relationship, reading news and doing many other activities.

The number of cellular customers in this country experienced enormous growth since it was the third biggest market of MP in Asia. In 2001, there were only 3.6 million subscribers then it multiplied almost a hundred times within 15 years (see figure 1); at a point that outnumbered its population. In 2015, this country was populated by 255 million inhabitants [1] but mobile cellular subscriptions reached more than 338 million [2]. This mobidensity (mobile connections per 100 citizens) of 132% was clear evidence that some residents possessed more than one active MP. It was supported by a survey conducted by Maheswari et al. [3] that there were 47.9% of Indonesians who had 2 or 3 active phones.

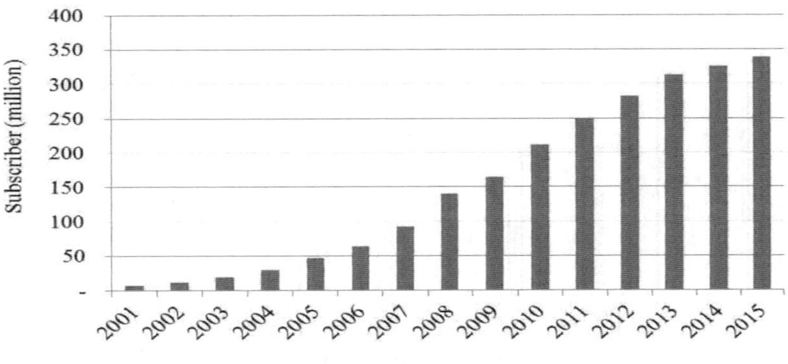

Figure 1. Mobile cellular subscriptions in Indonesia [2].

The increasing amount of MP ownership along with significant improvements in communication infrastructure explains how this country is in the period of a digital revolution [4]. It raises concerns because this proliferating number of MP demand has a positive correlation to the enormous number of MPW. The MPW quantity will continue to rise driven by technology enhancement that provokes people to replace their gadgets. Due to its short lifespan, literature shows that MPW dominated the number of e-waste in this country [5,6]. Recently, Santoso et al. [6] applied population balance method and estimated that in 2028, the amount of e-waste generated in Indonesia will be about 49.62 million units, including MPW of 40.78 million units or 82%. This number is alarming since the management of obsolete telecommunication devices in the country shows minimum progress. Besides, it is portraying that developing sustainable e-waste management is urgent.

Sustainable e-waste management needs a legal framework, collection mechanism, transportation scheme and processing infrastructure [7]. In this system, collection plays as a front gate, and it influences the whole process. The collection rate will be affected by consumer's willingness to recycle their old devices. However, for many reasons, users prefer to retain their end of used MP. Some people replace their old MP since they have been malfunctioned, and it is cheaper to buy a new one rather than bringing the broken handheld to service centres [8]. Others who are addicted to gadgets, change their MP merely because they are fascinated by the innovative features in the latest models [8, 9]. Moreover, people hesitate to recycle their MPW since their personal data are loaded inside. They are afraid of privacy disclosure [3]. As this type of e-waste is small enough, keeping them does not need much space in a room. Consequently, a large proportion of MPW are buried deep inside one's drawers; stockpiled by their owners. This has become a barrier to mineral resources sustainability as shown in figure 2. Indonesians who show this kind of behaviour are about 38% [3]. To compare, residents in India, China, and Spain who have a similar habit are 59%, 62% and 73% respectively [8, 9, 10].

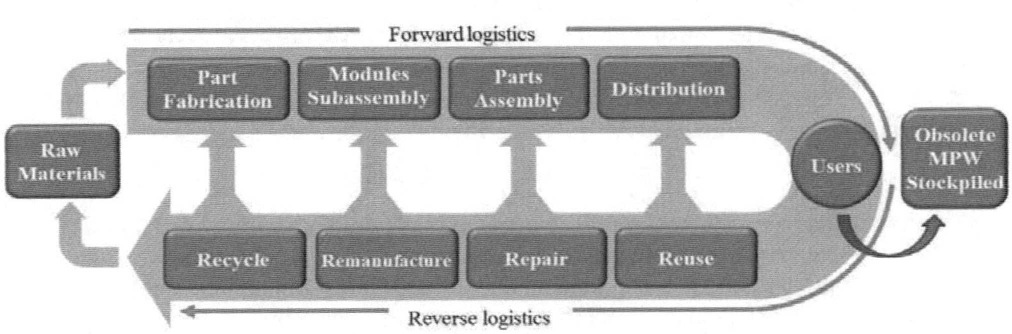

Figure 2. Minerals in stockpiled MPW outside the reverse logistics (modified from [11]).

Minerals recovery is imperative for MP since numerous metals including rare earth elements are embedded in them. Although in a tiny amount, they are indispensable. They make the device working. As MP production is increasing, so does demand for these minerals. However, natural reserves of some metals are reaching their scarcity level nowadays [12, 13]. It makes the supply vulnerable. Moreover, the market and regulation fail to support the mineral life cycle from MPW [14]. For this reason, promoting sustainable materials management through recycling is urgent because it can recover a lot of minerals. Effective recycling also supports reverse supply chain. It will lengthen the life cycle of critical metals. Secondary minerals extracted from e-waste can be used for manufacturing new products. The cost for processing metals from e-waste is also more effective than mining the virgin minerals [15]. Besides, having better life cycle can preserve virgin material extracted directly from nature and reduce destructive environmental impact [16]. A study by Szamalek and Galos [17] estimated that the quantity of MPW in the world hidden by their owners was around 14 billion units and half of them are smartphones. These hoarded telecommunication handhelds may generate precious minerals such as 3,892,000 Kg of silver, 378,000 Kg of gold and 140,000 Kg of palladium [17].

The amount of metals in MP is diverse for each model. In general, minerals in 4G models are different in comparison to their predecessor 2G models. It corresponds to their design and features. Material consumption in modern smartphones with their rich feature is dissimilar from older models that only have basic functions. The type and amount of materials covered in a touch screen smartphone is in contrast with MP with QWERTY keyboard. For example, the usage of indium (In) is related to smartphones with LCD display [18]. Although MP models are varied, it should be noticed that valuable metals can be found in every model; from the outdated to a modern one. These precious metals undoubtedly are driving factors for the high economic value of e-waste [16]. The average content of precious metals inside MP which are produced in three distinct generations; 2001, 2005 and 2010 are listed in table 1 [18, 19].

Table 1. Precious metals in mobile phones [18, 19].

	Mass (g)		
Metals	2001	2005	2010
Silver (Ag)	0.244	0.150	0.305
Gold (Au)	0.038	0.018	0.030
Palladium (Pd)	0.015	0.001	0.011

This study aims to assess the valuable metals that can be recovered from the end of life of MP, which passively lie in wait for entering reverse supply chain. It raises awareness among MP users that the recycling of materials embedded in MPW is related to preserving minerals in nature and extending the circular economy. It also portraits that stockpiling MPW will affect the supply chain and cost the environment dearly. Moreover, this research emphasized on how important for this country to have sustainable e-waste management with reliable collection systems. A trustworthy one so that it can motivate residents to recycle their MPW. The remainder of the paper structured as follows: the research method was presented in section 2, followed by result and discussion in Section 3 and conclusions in section 4.

2. Method

Two steps were employed to assess the valuable minerals from hoarded MPW. First, the number of MPW retired annually was calculated by adopting "consumption and used" method [20]. This method was also known as the approximation approach. It needed two input data; the number of active MP and their average lifetime. As the estimated proportion of stockpiled MPW, data presented by Maheswari et al. [3] in their research was utilized in this study. The calculation was conducted using the following equation:

$$Q_t = (N_t/\mathrm{l})\mathrm{H} \quad (1)$$

where Q_t (units/year) is the quantity of retired MPW in year t, N (units) is the number of active MP which is equivalent to mobile phone subscriptions in a particular year, l (years) is the average lifetime of MP, and H (%) is the proportion of MP stockpiled by customers. Maheswari et al. in their survey revealed that average useful lifespan of MP in this country is about 2.57 year [3]. By applying the Weibull distribution, Santoso et al. [6] found that the average lifetime of MP in this country is around 3.42 years. Based on their research, in this study, 3 years is assumed as the average replacement time of MP. For example, if one bought his phone in 2005, he would replace it in 2008. Besides, in accordance with the previous study [3], 38% is adopted as a proportion of hoarded MPW.

Finally, the secondary minerals contained in hoarded MPW were estimated based on the information presented in table 1. The quantity of MPW was grouped in three according to their generation. It was because every generation of MP had its own characteristics and may include a distinct amount of metals. The average mass of metals for MP produced around 2001 was a key to assess minerals for MPW generated from 2001 to 2007. Mostly they were phones with basic functions; making phone calls and sending text messages. As for metals covered in MP produced around 2005 was used for MPW generated from 2008 to 2012. These were for MP with several additional functions. For example, they had an embedded camera and speaker. Besides, they also had bigger capacity to save data. Finally, materials embedded in MP manufactured around 2010, were useful to estimate the valuable metals contained in MPW generated from 2013 to 2016. These were smartphones generation with touch screen models. Usually they had rich features and able to connect to the internet.

3. Result and Discussion

3.1. Retired mobile phones hoarded by owners

The annual quantity of MPW generated in Indonesia was assessed by applying equation 1. The results are presented in table 2. It shows that by the end of 2016, the total MPW generated in this country had already reached 893 million units. It included 339 million MPW that were passively waiting outside the reverse supply chain. This extensive amount was an accumulation effect of low recycling rate. Besides, the vacancy of appropriate e-waste treatment and the lack of environmental awareness also played significant parts [3].

Table 2. Estimation of retired and hoarded MPW.

Year	MPW retired	MPW hoarded
2001	2,173,649	825,987
2002	3,900,000	1,482,000
2003	6,165,084	2,342,732
2004	10,112,202	3,842,637
2005	15,636,657	5,941,930
2006	21,267,672	8,081,715
2007	31,128,960	11,829,005
2008	46,859,414	17,806,577
2009	54,558,987	20,732,415
2010	70,430,078	26,763,430
2011	83,268,540	31,642,045
2012	93,987,888	35,715,398
2013	104,408,971	39,675,409
2014	108,527,606	41,240,490

2015	112,982,780	42,933,456
2016	128,524,466	48,839,297
Total	893,932,956	339,694,523

As shown in table 2, MPW that was retired in 2001 but still in stored by their owners was about 825,987 units or 82.59 ton. It increased consistently every year. In 2016 the number reached 48,839,297 units or in terms of weight it was 4,883 ton. This number will grow persistently in the future as people will maintain their behaviour before trustworthy system to collect e-waste is available. The current finding had a small contrast with previous research performed by Santoso et al. [6] in which they estimated that the number of MPW in 2028 was about 40.78 million units or 4,078 ton. In total, the quantity of hidden MPW in the country was about 339 million units.

To answer the objective of this research, it was imperative to group the quantity of hoarded MPW in table 2 in three. The first one was for MPW generated from 2001 to 2007. Because in this period, the number of cellular subscribers was only about 93 million, this group had the lowest quantity; that was 34,346,005 units in total. The next group was for MPW generated from 2008 to 2012. The number of hoarded MPW from this generation was 132,659,865 units. The last group was MPW generated from 2013 to 2016 or smartphones models. With 172,688,653 units, the last group had the highest number compared to another two; almost half of the total quantity. Having megabytes capacity to save data, users found their smartphones bursting with personal data. It made them have a stronger emotional attachment. Consequently, they hesitated to recycle their old gadgets and preferred to retain them for a longer period instead. The result was in agreement with Szamalek and Galos [17], that half of the hoarded phones were smartphones.

3.2. Valuable minerals lie in wait for reverse supply chain

This section discussed the estimated amount of valuable materials contained in three distinct generations of hidden MPW and their economic value. To determine the amount of valuable metals contained in hoarded MPW, metals weight in each MP as listed in table 1 was multiplied by the MPW quantity in each group. The result is presented in table 3. It shows that in total, silver, gold, and palladium that can be recovered from hoarded MPW in this country were about 80,949 kg, 8,873 kg, and 2,547 kg respectively.

Table 3. Amount of materials in hoarded mobile phone.

	Mass (Kg)			
Metals	2001-2004	2005-2009	2010-2015	Total
Ag (Silver)	8,380.43	19,898.98	52,670.04	80,949.44
Au (Gold)	1,305.15	2,387.88	5,180.66	8,873.69
Pd (Palladium)	515.19	132.66	1,899.58	2,547.43

When this study was conducted, the average price index for Ag, Au, and Pd in the global market were (in USD/t oz) 14.50, 1,280 and 1,290 respectively [21]. The monetary values of precious metals in table 3 were determined based on those prices. The potential economic value for each identified metal was approximately USD 37.7 million for silver, 365.2 million for gold, and 105.6 million for palladium. Those three metals worth USD 508.5 million in total or in local currency they were worth IDR 7.3 trillion. However, this study was assessing the value of three precious minerals only. It should be noted that the number would be higher if all the types of minerals including the rare earths covered in MPW was also considered. On the other hand, the cost for reverse logistics, including processing secondary minerals, even though a lot more effective compared to traditional mining [15], will reduce the monetary benefit. This issue was not discussed in detail in this manuscript though.

4. Conclusion and Recommendation

This study assessed the quantity of precious minerals inside hoarded MPW and its economic value. Every MP contains various materials, which makes recycling imperative. Because sustainable mineral management has a positive correlation to sustainable e-waste management, consumer's awareness and behaviour play an essential role. Most people hesitate to bring their MPW to collection points; they hide these small devices in their home instead. The quantity of hoarded MPW in Indonesia was estimated at around 339.69 million units, and half of them were smartphones. Economic intrinsic in the form of valuable metals contained in this volume of MPW was approximately 80,949 Kg of silver, 8,873 Kg of gold and 2,547 Kg of palladium. One disheartening fact was that these metals worth of USD 508.5 million was waiting outside the reverse supply chain. Therefore, sustainable e-waste management need to be developed to save these mineral.

By reliable collection system, secondary minerals extracted from e-waste will find their way back in the manufacturing cycle. Some measures can be applied to collect MPW from the society. Government can provide a formal and permanent collection facility as part of sustainable e-waste management. It is useful for the country for long terms period. However, asking residents to donate their MPW in a campaign to solve a specific issue may show effective result. It helps the reverse supply chain. For example, when Australian zoos campaigned for gorilla conservation, they successfully collected 115,369 MPW from their visitors from 2009 to 2014 [22]. Recently, Japan's Olympic Committee is asking e-waste donation from society in order to create the entire medals for 2020 Tokyo Olympic. Those are some evidences that organized missions can influence people and raise their awareness.

Acknowledgments

This research is funded by Master Thesis Research or Penelitian Tesis Magister Universitas Indonesia-DRPM Ministry Research, Technology and Higher Education 2019, in the contract number NKB-1902/UN2.R3.1/HKP.05.00/2019.

References

[1] Statistics Indonesia 2015 Data of population
[2] International Telecommunication Union 2017 Data of mobile-cellular subscription
[3] Maheswari H, Yudoko G and Adhiutama A 2017 *IOP Conf. Ser.: Mater. Sci. Eng.* **277** 012062
[4] Das K, Gryseels M, Sudhir P and Tan K T 2016 *Unlocking Indonesia's Digital Opportunity* (McKinsey Indonesian Office)
[5] Andarani P and Goto N 2013 *J. Mater. Cycles Waste Manage.* **16** 306–20
[6] Santoso S, Zagloel Y M, Ardi R and Suzianti A 2019 *IOP Conf. Ser.: Earth Environ. Sci.* **219** 012006
[7] Honda S, Khetriwal D S and Kuehr R 2016 *Regional E-Waste Monitor: East and Southeast Asia* (Bonn: United Nations University, ViE-SCYCLE)
[8] Borthakur A and Govind M 2018 *J. Environ. Plann. Manage.* 1–24
[9] Tan Q, Duan H, Liu L, Yang J and Li J 2018 *J. Cleaner Prod.* **195** 1228–36
[10] Bovea M D, Ibáñez-forés V, Pérez-belis V and Juan P 2017 *Waste Manage.* 1-14
[11] Cerasis 2015 *Cerasis Whitepaper Reverse Logistics & Cerasis*
[12] Graedel T E, Harper E M, Nassar N T, Nuss P and Reck B K 2015 *Proc. Natl: Acad. Sci.U.S.A* **112** 4257–62
[13] World Economic Forum 2019 *A New Circular Vision for Electronics; Time for a Global Reboot* (Geneva; World Economic Forum)
[14] Evans C, *et al* 2014 *Case Study on Critical Metals in Mobile Phones* (Paris: OECD) 14
[15] X. Zeng, J. A. Mathews, and J. Li 2018 *Environ. Sci. Technol.* **52** 4835–41
[16] Vats M C and Singh S K 2015 *Waste Manage.* **45** 280–8
[17] Szamałek K and Galos K 2016 DOI.10.1515/gospo-2016-0037

[18] Buchert M, Manhart A, Bleher D and Pingel D 2012 *Recycling Critical Raw Materials from Waste Electronic Equipment* (Darmstadt: Oeko-Institut e.V.) 37-40
[19] Polák M and Drápalová L 2012 *Waste Manage.* **32** 1583–91
[20] UNEP 2007 *E-Waste Volume I - Inventory Assessment Manual* (Osaka: UNEP) 55-6
[21] Bloomberg 2019 Data of prices of precious and industrial metals
[22] Litchfield C A, Lowry R and Dorrian J 2018 *PLoS One* **13** 1–18

IOP Conf. Series: Materials Science and Engineering **703** (2019) 012022 doi:10.1088/1757-899X/703/1/012022

Optimisation of subtractive rapid prototyping process parameters using response surface methodology

T J Suteja[1] and M A Hadiyat[2]

[1]Manufacturing Engineering Department, Faculty of Engineering, University of Surabaya, Raya Kalirungkut, Surabaya, Indonesia

[2]Industrial Engineering Department, Faculty of Engineering, University of Surabaya, Raya Kalirungkut, Surabaya, Indonesia

E-mail: jayasuteja@staff.ubaya.ac.id

Abstract. Subtractive rapid prototyping machine is the most suitable tool to manufacture a polymer based prosthetic part because it is able to achieve a low surface roughness value for a complex and customised part. Many investigations have been conducted to explain the relation among the surface roughness value, the material rate removal, and the subtractive rapid prototyping process parameters. It is important to find the optimum process parameters in order to achieve the most efficient and productive process. However, none of the research found in the literature optimises the subtractive rapid prototyping process parameters in fabricating polycarbonate part. Therefore, this research aims to find the optimum process parameters to achieve the lowest arithmetic average of surface roughness value of the polycarbonate part in maximum material removal rate. In this research, the response surface methodology is implemented to optimised feed rate, step-over, and depth of cut of the subtractive rapid prototyping process. This research finds the feed rate, step-over, and depth of cut values that can be used to achieve the best result in manufacturing of polycarbonate material.

1. Introduction

Polycarbonate material has been applied for various prosthetic products due to the fact that it is strong, tough, and transparent thermoplastic material. A prosthetic product is mostly an intricate product and customized for each patient. As a result, the feasible process to fabricate the product is by using rapid prototyping. Rapid prototyping process can be performed by using subtractive or additive methods. Subtractive rapid prototyping is carried out by implementing high speed milling process to cut the raw material in order to produce the part. Meanwhile, additive rapid prototyping carried out by depositing materials layer by layer to build the shape of the part. For a certain type of prosthetic parts such as a socket of prosthetic leg that require a specific surface roughness and dimensional accuracy, the subtractive rapid prototyping is preferable to be implemented. The main purpose of subtractive rapid prototyping is to achieve the required minimum surface roughness and dimensional error in the maximum material rate removal. Therefore, the optimisation of subtractive rapid prototyping process parameters in fabricating polycarbonate materials is considered as a significant problem and needs to be tackled. This paper only describes the optimisation of subtractive rapid prototyping processes to achieve the maximum material removal rate and the minimum surface roughness. The novelty of this research lies in optimising the subtractive rapid prototyping process parameters to achieve the maximum material removal rate and half of the arithmetic average of surface roughness value resulted by using additive rapid prototyping process.

2. Literature review

Limited research found in the literature discussed rapid prototyping process of polycarbonate. The possibility to fabricate a thin fin of polycarbonate material by using subtractive rapid prototyping is investigated by Nieminem, I., et. Al [1]. However, they did not investigate the influence of the depth of cut and step-over on the surface roughness and the dimensional error of the polycarbonate material.

Other researchers conduct an investigation on the influence of high speed milling process parameters on metal materials. The research by Albertí, M., et. al., Vivancos, J., Urbanski, J.P., Oktem, H., et al., The, J.S., et. al., Ma, W., et.al., Zeroudi, N. and Fontaine, M., and Shimana, K., et. al. show that material removal rate, surface roughness, and dimension error of a material are affected by physical and mechanical characteristics of the material, depth of cut, step-over, feed rate, cutting speed, cut type, interpolation type, tool holder type, physical and mechanical characteristics of the tool, vibration, controller of the machine, and computer aided manufacturing software [2, 3, 4, 5, 6, 7, 8, 9].

Suteja, T.J. develops several models to show the influence of the depth of cut, feed rate, and step-over on the vertical and horizontal surface roughness, vertical length error, horizontal length error, and depth error of polycarbonate material in subtractive rapid prototyping [10, 11]. The models show that the vertical surface roughness is positively affected by the feed rate, the step-over, and the depth of cut. The parameter that has the most influence on the vertical surface roughness is the step-over. In addition, the depth of cut and the interaction between the step-over and the depth of cut has a positive influence on the horizontal surface roughness.

The expected surface roughness values to optimize the process parameters are determined based on literature review. As the additive rapid prototyping can also be used to fabricate a product made of polycarbonate, then a review of previous research is conducted to investigate the achieved surface roughness of part fabricated by using an additive rapid prototyping.

Ippolito, R., et. al. compared five methods of additive rapid prototyping [12]. According to Ippolito, R., et. al., the achieved arithmetic average of surface roughness of additive rapid prototyping process is varied in the range between 0.001 to 0.025 mm. In addition, the parallel arithmetic average of surface roughness is lower than the perpendicular feed rate direction.

Fox, J.C., et. al., conduct experiments performed on Laser Powder Bed Fusion type of additive rapid prototyping with the EOS M2701 system using the commercially available Stainless Steel powder [13]. According to their research, the value of arithmetic average of surface roughness increases as overhang angle decreases. The minimum and maximum surface roughness values achieved by using the additive rapid prototyping process are 0.015 to 0.045 mm respectively.

The purpose of this research is particularly for investigating the subtractive rapid prototyping of a polycarbonate prosthetic product. He, Y., et al. implemented Scanning Printing Polishing Casting to fabricate a prosthetic product [14]. Based on their research, the fabrication method can achieve a low arithmetic average of surface roughness (maximum 0.002 mm) and require low cost. However, the method requires more complex process and longer time.

Udroiu, R., et. al., investigate the surface quality for a polymer-based additive rapid prototyping [15]. The additive rapid prototyping machine used is EDEN 350 using PolyJet technology. Based on the result, the arithmetic average of surface roughness values for the PolyJet material jetting process with matte finish were in the range of 0.0005 to 0.015 mm. For the glossy finish, the arithmetic average of surface roughness values in the range of 0.0005 to 0.010 mm. Though PolyJet technology can achieve lower surface roughness, the cost of the process is still high.

The widely used additive rapid prototyping technology is Fused Deposition Modelling because the cost of the technology decreases into feasible price. The research by Kaji, F., et. al. proposes a model of surface roughness in Fused Deposition Modelling especially for the geometry of the cusp profile [16]. The result of the research shows that the layer thickness generally has a deep effect on the surface roughness. The minimum surface roughness value, which is achieved by using 0.125 mm layer thickness, is 0.015 mm.

Based on the literature review, the aim of this research is to optimize the process parameters in subtractive rapid prototyping to achieve the half of the arithmetic average of surface roughness values of the Fused Deposition Modelling, which is maximum of 0.0075 mm (7.5 μm), in the maximum material rate removal of polycarbonate material.

3. Methodology

Response Surface Methodology is implemented in this research to develop several mathematical models and validate the models. The shape and the dimensions of the polycarbonate material specimen fabricated by using the subtractive rapid prototyping process are shown in Fig. 1. Roland MDX 40 is used as the subtractive rapid prototyping machine. The machine is assisted by CAM Modela Player 4.0 software to generate the tool path from the STL format model. A carbide solid square end mill with 5 mm diameter is used as the cutting tool. In order to move the cutting tool, the software uses zigzag cut type.

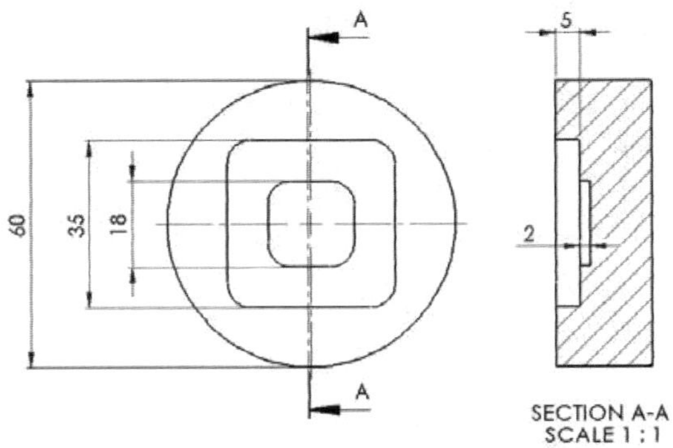

Figure 1. The specimen (in mm).

Different parameter values for roughing and finishing processes are determined based on the tools catalogue and interview with the expert. Table 1 shows the roughing parameter value of the subtractive rapid prototyping. Three levels of value for depth of cut, feed rate, and step-over are designed for finishing process. The value of each level for each parameter as shown in Table 2 is determined based on the machine specification, literature study, and the preliminary experiment. The subtractive rapid prototyping is performed under dry operating condition. The spindle for finishing process is 10000 rpm and the entry speed for finishing process is 4 mm/s. The polycarbonate material is assumed to be always homogeneous. Then, the cutting temperature is assumed to be always constant. Finally, the tool wear is assumed to occur after performing three roughing and finishing processes.

Table 1. Parameter value for roughing.

Feed Rate	: 12 mm/s
Entry Speed	: 4 mm/s
Spindle Speed	: 8500 rpm
Depth of Cut	: 0.37 mm
Step-over	: 1 mm

Table 2. Parameter value for finishing.

	Low	Middle	High
Depth of Cut [mm]	0.1	0.235	0.37
Feed Rate [mm/s]	12	14.5	17
Step-over [mm]	0.3	0.65	1

Material rate removal and arithmetic average of surface roughness are investigated in this research as the responses. To calculate the material rate removal, the time needed to fabricate the specimen is

IOP Conf. Series: Materials Science and Engineering **703** (2019) 012022 doi:10.1088/1757-899X/703/1/012022

measured by using a stopwatch. The arithmetic average of surface roughness is measured by using Mitutoyo SJ 210 with 0.01 µm of accuracy at Industrial Metrology Laboratory of University of Surabaya. Two direction of arithmetic average of surface roughness measurement is conducted, which are horizontal and vertical directions. After the measurement process, the measured data is analyzed by using MINITAB release 14 software.

4. Results and discussion

The design of the first order experiment are shown in Table 3. The first order experiment involves all two level factors using 2^3 factorial design with additional 5 center points. The result of the first order experiment is showed in Table 3. RaVer is vertical arithmetic average of surface roughness [µm], RaHor is horizontal arithmetic average of surface roughness [µm], and MRR is material rate removal [mm^3/s].

Table 3. First order experiment results.

Std Order	Run Order	Feed Rate [mm/s]	Step-over [mm]	Depth of Cut [mm]	RaHor [µm]	RaVer [µm]	MRR [mm^3/s]
11	1	14.50	0.65	0.235	3.13	7.25	1.221673
13	2	14.50	0.65	0.235	3.22	7.83	1.220947
2	3	17.00	0.30	0.100	2.53	4.46	0.266056
12	4	14.50	0.65	0.235	3.18	8.65	1.222763
7	5	12.00	1.00	0.370	4.88	12.39	2.607133
9	6	14.50	0.65	0.235	2.89	7.83	1.220637
3	7	12.00	1.00	0.100	3.64	12.01	0.704914
5	8	12.00	0.30	0.370	3.00	4.70	0.801036
10	9	14.50	0.65	0.235	3.55	8.33	1.217333
6	10	17.00	0.30	0.370	4.18	4.59	0.983487
1	11	12.00	0.30	0.100	2.81	3.60	0.215947
4	12	17.00	1.00	0.100	4.30	12.08	0.868179
8	13	17.00	1.00	0.370	4.49	12.01	3.213400

Based on the result of the validation step, the first order models are not adequate as a linear regression model. For that reason, the second order experiment for material removal rate, vertical arithmetic average of surface roughness, and horizontal arithmetic average of surface roughness must be conducted. The central composite design is used to determine the number of the second order experiment run. The design and result of the second order experiment is shown in Table 4.

By using the experiment result shown in Table 4, the prediction model of material removal rate, vertical arithmetic average of surface roughness, and horizontal arithmetic average of surface roughness are shown in Eq. 1, Eq. 2, and Eq. 3.

$$MRR = 0.462 - 0.0260 \text{ x } F - 0.901 \text{ x } S - 2.833 \text{ x } D - 0.00067 \text{ x } F^2 - 0.162 \text{ x } S^2 - 0.345 \text{ x } D^2 + 0.0761 \text{ x } F \text{ x } S + 0.2129 \text{ x } F \text{ x } D + 7.804 \text{ x } S \text{ x } D \quad (1)$$

$$Ra_{hor} = 2.61 + 0.007 \text{ x } F - 0.41 \text{ x } S + 1.31 \text{ x } D + 0.0038 \text{ x } F^2 + 1.450 \text{ x } S^2 + 12.46 \text{ x } D^2 - 0.067 \text{ x } F \text{ x } S - 0.485 \text{ x } F \text{ x } D + 3.73 \text{ x } S \text{ x } D \quad (2)$$

$$Ra_{ver} = -8.9 + 0.88 \text{ x } F + 17.65 \text{ x } S + 6.4 \text{ x } D - 0.0146 \text{ x } F^2 + 3.34 \text{ x } S^2 + 13.9 \text{ x } D^2 - 0.521 \text{ x } F \text{ x } S - 0.59 \text{ x } F \text{ x } D - 4.10 \text{ x } S \text{ x } D \quad (3)$$

where F is feed rate [mm/s], S is step-over [mm], and D is depth of cut [mm]. According to the result of the validation step for the second order experiment, the developed equations can be used as the best prediction model.

The aim of this research is to determine the feed rate, step over, and depth of cut in subtractive rapid prototyping to achieve the maximum material rate removal of polycarbonate material and the arithmetic average of surface roughness less than or equal to 7.5 μm. The multiple response optimizer based on desirability approach is used to maximize the material rate removal and minimize the horizontal and vertical arithmetic average of surface roughness [17]. The result shows that the optimum condition achieved when the feed rate, step-over, and depth of cut are set in 18.70 mm/s, 0.54 mm, and 0.46 mm respectively. The maximum material removal rate achieved by implementing these parameters is 2.3672 mm^3/s. The achieved horizontal and vertical arithmetic averages of surface roughness are 3.5842 μm and 7.4867 μm respectively.

Table 4. Second order experiment results.

Std Order	Run Order	Feed Rate [mm/s]	Step Over [mm]	Depth of Cut [mm]	RaHor [μm]	RaVer [μm]	MRR [mm^3/s]
18	1	14.5000	0.65000	0.235000	3.64	8.20	1.33510
19	2	14.5000	0.65000	0.235000	3.36	8.73	1.22121
10	3	18.7045	0.65000	0.235000	2.80	7.76	1.47727
15	4	14.5000	0.65000	0.235000	2.74	7.57	1.22204
20	5	14.5000	0.65000	0.235000	2.72	9.67	1.22152
13	6	14.5000	0.65000	0.007958	3.33	9.53	0.04138
14	7	14.5000	0.65000	0.462042	3.90	9.08	2.40249
5	8	12.0000	0.30000	0.370000	3.71	5.05	0.79896
11	9	14.5000	0.06137	0.235000	2.35	0.88	0.11581
8	10	17.0000	1.00000	0.370000	4.29	13.21	3.21431
6	11	17.0000	0.30000	0.370000	3.51	5.38	0.98372
2	12	17.0000	0.30000	0.100000	3.10	4.88	0.26597
9	13	10.2955	0.65000	0.235000	3.28	8.90	0.97840
7	14	12.0000	1.00000	0.370000	5.03	13.72	2.61059
12	15	14.5000	1.23863	0.235000	4.60	18.61	2.25131
4	16	17.0000	1.00000	0.100000	3.48	12.50	0.86892
3	17	12.0000	1.00000	0.100000	3,26	13,20	0.70532
16	18	14.5000	0.65000	0.235000	2,94	7,41	1.21149
1	19	12.0000	0.30000	0.100000	2,95	2,77	0.21595
17	20	14.5000	0.65000	0.235000	3,42	8,76	1.22183

5. Conclusions

The goal of this research is to optimize the subtractive rapid prototyping parameters for polycarbonate material by implementing the response surface methodology in order to achieve the maximum material rate removal of polycarbonate material and achieve the horizontal and vertical arithmetic average of surface roughness less than or equal to 7.5 μm at the same time. The optimized feed rate, step over, and depth of cut for the subtractive rapid prototyping are found to be 18.70 mm/s, 0.54 mm, and 0.46 mm respectively. By implementing these parameters, the achieved material removal rate is 2.3672 mm^3/s and the maximum achieved vertical arithmetic average of surface roughness is 7.4867 μm.

References

[1] Nieminen I, Paro J and Kauppinen V 1996 High-speed milling of advanced materials *J. Mater. Process. Technol.* **56(1-4)** 24-36

[2] Albertí M, Ciurana J and Rodriguez C 2007 Experimental analysis of dimensional error vs. cycle time in high-speed milling of aluminium alloy *Int. J. Mach. Tools Manuf.* **47(2)** 236-246

[3] Vivancos J, Luis C, Costa L and Ortız J 2004 Optimal machining parameters selection in high speed milling of hardened steels for injection moulds *J. Mater. Process. Technol.* **155** 1505-1512

[4] Urbanski J, Koshy P, Dewes R and Aspinwall D 2000 High speed machining of moulds and dies for net shape manufacture *Mater. Des.* **21(4)** 395-402

[5] Oktem H, Erzurumlu T and Kurtaran H 2005 Application of response surface methodology in the optimization of cutting conditions for surface roughness *J. Mater. Process. Technol.* **170(1-2)** 11-16

[6] The J, Candra S and Aquarista Y 2008 Optimasi proses pemesinan milling fitur pocket material baja karbon rendah menggunakan response surface methodology *Jurnal Ilmiah Teknik Mesin Universitas Kristen Petra* **10(1)** 1 -7

[7] Ma W, He G, Zhu L and Guo L 2016 Tool deflection error compensation in five-axis ball-end milling of sculptured surface *Int. J. Adv. Manuf. Technol.* **84(5-8)** 1421-1430

[8] Zeroudi N and Fontaine M 2015 Prediction of tool deflection and tool path compensation in ball-end milling *Journal of Intelligent Manufacturing* **26(3)** 425-445

[9] Shimana K, KondoE, Karashima H, Nakao M and Yamashita S 2016 An approach to real-time compensation of machining error using deflection of tool estimated from cutting forces in end-milling process *Journal of Advanced Mechanical Design, Systems, and Manufacturing* **10(2)** 1-10

[10] Suteja T 2016 The influence of depth of cut, feed rate and step-over on surface roughness of polycarbonate material in subtractive rapid prototyping *Lecture Notes in Electrical Engineering* **365** 409-414

[11] Suteja T 2017 The influence of depth of cut, feed rate, and step-over on dimensional accuracy in subtractive rapid prototyping of polycarbonate material *IOP Conf. Ser.: Mater. Sci. Eng.* **187** 012033.

[12] Ippolito R, Iuliano L and Gatto 1995 A benchmarking of rapid prototyping techniques in terms of dimensional accuracy and surface finish *CIRP Annals-Manufacturing Technology* **44(1)** 157-160

[13] Fox J, Moylan S and Lane B 2016 Effect of process parameters on the surface roughness of overhanging structures in laser powder bed fusion additive manufacturing *Procedia CIRP* **45** 131-134

[14] He Y, Xue G and Fu J 2014 Fabrication of low cost soft tissue prostheses with the desktop 3D printer *Scientific reports* **4** 6973

[15] Udroiu R, Braga I and Nedelcu A 2019 Evaluating the Quality Surface Performance of Additive Manufacturing Systems: Methodology and a Material Jetting Case Study *Materials* **12(6)** 995

[16] Kaji F and Barari A 2015 Evaluation of the surface roughness of additive manufacturing parts based on the modelling of cusp geometry *IFAC-PapersOnLine* **48(3)** 658-663

[17] Candioti L, De Zan M, Camara M and Goicoechea H 2014 Experimental design and multiple response optimization using the desirability function in analytical methods development *Talanta* **124** 123-138

IOP Conf. Series: Materials Science and Engineering **703** (2019) 012023 doi:10.1088/1757-899X/703/1/012023

Green dynamic capability for enhancing green innovations performance in a manufacturing company: a conceptual framework

R Amaranti[1,2], R Govindaraju[1], and D Irianto[1]

[1]Department of Industrial Engineering, Institut Teknologi Bandung, Jalan Ganesa 10, Bandung, 40132, Indonesia
[2]Department of Industrial Engineering, Universitas Islam Bandung, Jalan Tamansari 1, Bandung, 40116, Indonesia

E-mail: reniamaranti2709@yahoo.com

Abstract. Green innovation has become a necessity for companies in line with the demands of environmental requirements. Green innovation is the activity of developing a product or process that is carried out by considering environmental aspects (such as energy savings, pollution prevention, waste recycling, reduction of resource consumption) to reduce the impact of industrial activities on the environment. One of the factors associated with green innovation performance is the company's dynamic capabilities. This paper proposes a conceptual framework that illustrates the role of companies' dynamic green capabilities on the performance of green innovation in manufacturing companies in Indonesia.

1. Introduction

The industrialization has become a major driver of economic growth and increased the welfare of countries in the world. On the other hand, industrial activities also contribute to the reduction of natural resources, destruction of ecosystems, and the threat of climate change. Limited natural resources, global warming, waste management, and other environmental issues are challenges for industrial growth. In addition, international environmental regulations, various environmental protection conventions, and increasing environmental awareness of consumers have an impact on the pattern of business competition and how industrial companies conduct their business [1, 2]. This encourages industries throughout the world to increase attention and awareness of environmental issues. Therefore, the company's environmental management plays an important role in business today.

Environmental management and green practices are related to firm innovativeness [3]. Therefore, green innovation is one way to achieve sustainable development in the manufacturing industry that is in line with increasing environmental awareness among consumers. Green innovation is innovation that is carried out related to environmental aspects (energy-saving, pollution prevention, waste recycling, reduction of resource consumption) to reduce the impact of industrial activities on the environment. Green innovation is a topic that has been widely studied by researchers in recent years. If examined from various literature on innovation, research on green innovation can be broadly grouped into several groups. The first group is the literature that discusses green design, such as the influence of environmental problems in the R&D process [4]; the influence of market performance on green innovation and the development of new products [5]; influences from consumers and environmental regulations on green product innovation [6]. The second group consists of literature that discusses drivers and barriers to green innovation. This group includes those who discuss internal and external barriers [7] and challenges for green innovation [8]. Third, is a literature group that discusses the impact of green innovation on performance. Included in this group is the literature on how green innovation impacts the company's competitive advantage [1] and the influence of green innovation on company performance [9, 10, 11, 12]. The last group is the literature that discusses the performance of green innovation, which is related to dynamic capabilities [13]; absorptive capacity [14]; regulatory and customer pressure [15]; and evaluation of green innovations [16, 17].

Green innovation performance depends not only on existing capabilities but also on how to change these capabilities to build new capabilities. The dynamic capability perspective is one concept that provides a theoretical view to examine the practices of innovation management at the organizational level. Dynamic capabilities allow companies to combine and rearrange resources to respond to environmental challenges. The dynamic capability perspective also explains that dynamic capabilities help companies to sense and capture opportunities and to reconfigure important knowledge at various levels of the organization [18, 19]. In addition, a key factor for achieving sustainable innovation is knowledge [20]. Innovation is associated with high-level variation and exploration. This requires new knowledge that is different from existing knowledge and specific combinations of new knowledge [21]. When companies innovate, they not only process existing information but also create new information and knowledge to re-create a new corporate environment. The creation of company knowledge becomes important when companies are going to innovate [22]. Therefore, knowledge management becomes an important activity for companies especially to be able to innovate and adapt to environments changes. According to [23], knowledge management also has relevance to dynamic capabilities. This is supported by [24] and [25] which shows that the achievement of dynamic capabilities is influenced by knowledge-based resources. The availability of company knowledge is a prerequisite for the flow of new knowledge in the learning process to create dynamic capabilities [26].

Based on the literature, it can be seen that the dynamic capability and ability of the company to absorb knowledge are related to green innovation performance. However, none of the literature has comprehensively discussed how dynamic capabilities and knowledge play a role in green innovation performance. Therefore, this paper will explain a conceptual framework that illustrates how the relationship between a company's ability to acquire and create knowledge, green dynamic capabilities, and green innovation performance.

2. Literature review

2.1. Green Innovation

Green innovation is also known as eco-innovation or environmental innovation. Eco-innovation, in general, can be defined as innovation for reducing negative impacts on the environment. According to [27], eco-innovation activities can be analyzed based on three dimensions, namely 1) targets, are areas of focus on eco-innovation: products, processes, marketing methods, organizations, and institutions; 2) mechanism, is the way in which changes are made in the target: modification, redesign, alternatives, and creations); and 3) impacts (the impact of eco-innovation on the environment). Green innovation is generally defined as activities related to the development of new products and processes that can

IOP Conf. Series: Materials Science and Engineering **703** (2019) 012023 doi:10.1088/1757-899X/703/1/012023

reduce negative impacts on the environment [1]. Green innovation is categorized into technology, management functions, product design, and production processes. Green innovations carried out during product life cycle assessments include the process of modifying existing product designs to produce environmentally friendly products. Whereas green innovation in the process is done by modifying production methods so as to reduce consumption of natural resources, water, and energy, and reduce waste and pollution.

2.2. Green dynamic capability

Dynamic capabilities are needed by companies to adapt to changes in the environment. Dynamic capability is different from organizational capabilities. Dynamic capabilities allow companies to innovate beyond their current routines, while organizational capabilities provide the means to configure organizational resources for company routines [18]. This capability consists of high-level organizational practices that transform organizational and member expertise into products and services. Dynamic capability is the ability of companies to reconfigure their competencies both internal and external competencies to deal with environmental changes [19]. In this definition, organizational competence shows managerial and organizational processes or "current patterns of practice and learning". In line with [18], [21] then redefining dynamic capabilities as a corporate process that uses resources to match and even make market changes. This definition describes dynamic capabilities as processes within organizations to integrate and reconfigure resources. Dynamic capabilities are widely regarded as combining processes that enable organizations to maintain superior performance over time [28].

The term green dynamic capability used in this study is adapted from [29]. In this case, green dynamic capabilities are defined as the ability of companies to use existing resources and knowledge to renew and create green organizational capabilities. With this capability, the company can change its products and processes to meet the requirements of environmental conditions and adapt to dynamic environmental changes. Dynamic capability consists of sensing capability, seizing capability, and reconfiguring capability.

- Sensing capability is the company's ability to see, interpret, and pursue opportunities related to "green" aspects [13, 19, 28, 30, 31]. Sensing capability describes the intensity of company activities for scanning, search, exploration, and observing green best practices in the same industry.
- Seizing capability is the company's ability to respond to "green" opportunities through new products, processes, or services [19, 28].
- Reconfiguring capability is the ability of companies to carry out sustainable alignment, reorganize, and protect assets (tangible and intangible) to adapt to environmental changes [19, 28].

2.3. Absorptive Capacity

Based on [32], absorptive capacity is defined as the ability of a company that is indicated by a series of routines and processes in the company to acquire, assimilate, transform, and exploit knowledge. Absorptive capacity consists of potential absorptive capacity and realized absorptive capacity [32, 33].

- Potential absorptive capacity is the company's ability to acquire and assimilate knowledge from the external environment. Knowledge acquisition describes the company's ability to carry out activities to identify and obtain external knowledge that is considered important for the company. Assimilation is the company's ability to analyze, process, interpret, and understand the knowledge that has been acquired.
- Realized absorptive capacity describes the ability of companies to utilize external knowledge that has been absorbed. This capability illustrates how companies combine existing knowledge with

newly acquired and assimilated knowledge, and how companies develop and improve their competencies by incorporating modified knowledge.

3. Conceptual framework

The conceptual framework that will be discussed in this paper refers to [13], [14], [34], and [35]. Green innovation is used to improve environmental management performance in terms of meeting environmental protection requirements [1]. Green innovation performance is a company's performance derived from innovation in processes and products that pay attention to environmental aspects (energy savings, pollution prevention, waste recycling, green product design, or environmental management) so as to reduce negative impacts on the environment [1,13,14,35]. Green innovation performance consists of green product innovation performance and green process innovation performance [1, 13, 17,35].

Based on the perspective of dynamic capabilities, companies need dynamic capabilities to adapt to change. The company's dynamic capability is something that can transform resources into performance [24, 36]. Green innovation performance does not only depend on existing capabilities but also the ability to change these capabilities so that the company can build new capabilities. Dynamic capability helps companies to perceive and capture opportunities and to reconfigure important knowledge at various levels of the organization [18, 19]. Dynamic capability is an important mechanism for changing the ability to manage knowledge into performance at the company level. Dynamic capability is the company's ability to create and utilize resources in the company to achieve sustainable competitive advantage [31]. In this case, the company combines individual skills and knowledge, physical and technical resources, structure and culture to stimulate the dynamism of knowledge within the company.

Knowledge is a source of dynamic capability through various activities in the company such as technical training provided to partners, informal information exchange, identification of market information through direct communication between individuals, and systematic collection of information from central companies. The ability to recognize opportunities depends on individual abilities and existing knowledge. Various forms of knowledge, especially when transmitted through social interaction, can act as a source of dynamic capabilities. However, the relationship between organizational knowledge and dynamic capabilities is not always positive and is very dependent on the broader organizational context [23]. The ability to absorb knowledge is also needed for knowledge creation in companies which also influences the company's dynamic capabilities [34]. Referring to [32], absorptive capacity is divided into two namely potential absorptive capacity and realized absorptive capacity. Potential absorptive capacity is the ability of companies to assess and obtain knowledge originating from outside the company (acquisition and assimilation of knowledge), while the realized absorptive capacity is the ability of companies to utilize the knowledge that has been absorbed (transformation and exploitation of knowledge). Potential absorptive capacity and realized absorptive capacity both affect the company's dynamic capabilities. According to [14], the ability to absorb knowledge has a positive effect on the performance of innovation performance both green product innovation and green process innovation.

Based on the literature review described in the previous section, a conceptual framework is developed that illustrates how manufacturing companies can improve their green innovation performance through the company's green dynamic capability. The conceptual framework is illustrated in Figure 1.

4. Methodology

The conceptual framework described in this paper is part of the ongoing research on green innovation performance. The conceptual framework was built based on the literature review and preliminary observations in several manufacturing companies. The next step is to develop the conceptual framework into a conceptual model by identifying research variables and relationships between research variables. Furthermore, the conceptual model will be empirically tested. Data collection will be carried out by distributing questionnaires to manufacturing companies in Indonesia that have implemented environmental management (such as companies that have obtained ISO 14000 certificates, have followed the PROPER standard assessment). Empirical data testing will be carried out using PLS-SEM. The analysis and validation phase of the research model will be carried out with multi-case studies to obtain a complete and in-depth analysis of how green innovation can be achieved through a green dynamic capability.

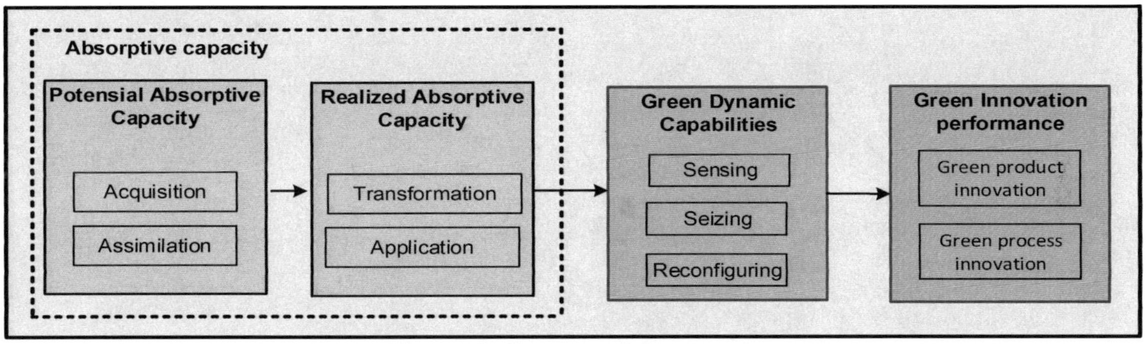

Figure 1. Conceptual framework

5. Conclusion and further study

Based on the conceptual framework, there are three important aspects that are considered related to achieving green innovation performance, namely the ability to absorb knowledge, the company's dynamic capabilities, and green innovation. This conceptual framework will be developed into a conceptual model by determining research variables and relationships between variables. Data collection with surveys will be conducted to test the model empirically. For this purpose, the development of measurement instruments, data collection and analysis will be carried out. Measuring instruments will be designed by adopting and modifying measurement items from the relevant literature that already exists.

References

[1] Chen Y, Lai S, and Wen C 2006 The influence of green innovation performance on corporate advantage in Taiwan *J. Bus. Ethics.* **67** 331–339
[2] Chen Y and Chen Y 2008 The driver of green innovation and green image–green core competence *J. Bus. Ethics.* **81** 531–543.
[3] Aragón-Correa J A 1998 Strategic proactivity and firm approach to the natural environment *Acad. Manag. J.* **41**(5) 556–567
[4] Foster C and Green K 2000 Greening the innovation process *Bus. Strateg. Environ.* **9** 287–303
[5] Pujari D 2006 *Technovation* **26** 76–85
[6] Yalabik B and Fairchild R J 2011 *Intern. J. Prod. Econ.* **131**(2) 519–527
[7] Abdullah M, Zailani S, Iranmanesh M and Jayaraman K 2015 Barriers to green innovation initiatives among manufacturers: The Malaysian case *Rev. Manag. Sci.*
[8] Hall J and Vredenburg H 2003 The challenges of innovating for sustainable *MIT Sloan Manag Rev.* **45**(1) 61–68

[9] Sezen B and Çankaya S Y 2013 Effects of green manufacturing and eco-innovation on sustainability performance *Procedia-Soc. Behav. Sci.* **99** 154–163.

[10] Ar IM 2012 The impact of green product innovation on firm performance and competitive capability: The moderating role of managerial environmental concern *Procedia - Soc. Behav. Sci.* **62** 854–864.

[11] Fernando Y and Xin W 2017 The impact of eco-innovation drivers on environmental performance: Empirical results from the green technology sector in Malaysia *Sustain. Prod. Consum.* **12** 27–43

[12] Xie X, Huo J, and Zou H 2019 *J. Bus. Res.* 1–10

[13] Albort-Morant G, Leal-Millán A and Cepeda-Carrión G 2016 The antecedents of green innovation performance: A model of learning and capabilities *J. Bus. Res.* **69** 4912–4917

[14] Albort-Morant G, Henseler J, Cepeda-Carrión G, and Leal-Rodríguez A L 2018 Potential and realized absorptive capacity as complementary drivers of green product and process innovation performance *Sustainability* **10**(381) 1–20

[15] Huang X X, Hu Z P, Liu C S, Yu D J, and Yu L F 2016 The relationships between regulatory and customer pressure, green organizational responses, and green innovation performance *J. Clean. Prod.* **112** 3423–3433

[16] Wang W, Yu B, Yan X, Yao X, and Liu Y 2017 Estimation of innovation's green performance : A range-adjusted measured approach to assess the unified efficiency of China's manufacturing industry *J. Clean. Prod.* **149** 919–924

[17] Tseng M, Wang R, Chiu A S F, Geng Y, and Hsu Y 2013 Improving performance of green innovation practices under uncertainty *J. Clean. Prod.* **40** 71–82

[18] Teece D J, Pisano G, and Shuen A 1997 Dynamic capabilities and strategic management *Strateg. Manag. J.* **18**(7) 509–533

[19] Teece D J 2007 Explicating dynamic capabilities: The nature and microfoundations of (sustainable) enterprise performance *Strateg. Manag. J.* **28**(13) 1319–1350

[20] Liao S, Fei W, and Chen C 2007 Knowledge sharing, absorptive capacity, and innovation capability: An empirical study of Taiwan's knowledge-intensive industries *J. Inf. Sci.* **33**(3) 340–359

[21] Eisenhardt K M and Martin J A 2000 Dynamic capabilities: What are they? *Strateg. Manag. J.* **21** 1105–1121

[22] Nonaka I, Umemoto K and Senoo D 1996 *Technol. Soc.* **18**(2) 203–218

[23] Prieto I M and Easterby-Smith M 2006 *Eur. J. Inf. Syst.* **15** 500–510

[24] Wang C L and Ahmed P K 2007 Dynamic capabilities: A review and research agenda *Int. J. Manag. Rev.* **9**(1) 31–51

[25] Zollo M and Winter S G 2013 Deliberate learning and the evolution of dynamic capabilities *Organ. Sci.* **13**(3) 339–351

[26] Ambrosini V and Bowman C 2009 What are dynamic capabilities and are they a useful construct in strategic management? *Int. J. Manag. Rev.* **11**(1) 29–49

[27] OECD 2009 *Eco-innovation in industry: Enabling green growth* OECD Publishing 1-279

[28] Wilden R, Gudergan S P, and Nielsen B B 2013 Dynamic capabilities and performance: Strategy, structure, and environment *Long Range Plann.* **46** 72–96

[29] Lin Y H and Chen Y 2016 *Qual Quant.*1–23

[30] Pavlou P A and El-Sawy O A 2011 Understanding the elusive black box of dynamic capabilities *Decis. Sci. J.* **42**(1) 239–273

[31] Tseng S M and Lee P S 2014 *J. Enterp. Inf. Manag.* **27**(2) 158–179

[32] Zahra S A and George G 2002 Absorptive capacity: A review, reconceptualization, and extension *Acad. Manag. J.* **27**(2) 185–203

[33] Kang M and Lee M 2016 Absorptive capacity, knowledge sharing, and innovative behavior of R&D employees *Technol. Anal. Strateg. Manag.* 1-14

[34] Pai F Y and Chang H F 2013 The effects of knowledge sharing and absorption on

organizational innovation performance–A dynamic capabilities perspective *Interdiscip. J. Information, Knowledge, Manag.* **8** 83–97
[35] Huang J W and Li Y H 2017 Green innovation and performance: The view of organizational capability and social reciprocity *J. Bus. Ethics.* **145**(2) 309–324

Combined structural equation modelling – artificial neural networks model for predicting customer loyalty

M A Hadiyat

Department of Industrial Engineering, University of Surabaya, Raya Kalirungkut, Surabaya, 60293, Indonesia

E-mail: arbi@staff.ubaya.ac.id

Abstract. Customer loyalty becomes considerations by service providers to maintain for reducing the churn rate. Many studies propose factors that are significantly influencing customer loyalty, and apply them for predicting it. Based on mathematical models, loyalty prediction methods are developed, and it involves new approaches including machine learning. This research aim is predicting customer loyalty using the combination of structural equation model (SEM) and artificial neural networks (ANN). The methodology starts by applying SEM for selecting statistically significant factors affect the loyalty. The linear SEM model ensures this relationship by fulfilling statistical hypothesis and fulfilled assumptions. Once selected factors are found, they are treated as inputs for ANN modelling. ANN is selected because of its ability in nonlinear modelling to enhance its prediction. ANN then learns the relationship between those inputs and the loyalty in real time as any additional observation recorded in. Based on trained ANN, prediction of customer loyalty based on input factors could be done. A case study was conducted at a Hotel by asking 130 customers. SEM inputs includes tangibles, facility, and staff attitudes, while loyalty scores become output. Combination of SEM-ANN has successfully predicted the customer loyalty and brought up chances for improvement strategies.

1. Introduction

Customer loyalty becomes one of important targets in marketing strategies. Many studies discuss about how to improve customer loyalty by practicing customer relationship management. The main reason why this loyalty should be noticed by product or service provider is about its effect on frequent customer purchase exclusively [1], and of course it leads to domination of the market. Once the providers identify factors influencing customer loyalty, then customer service and relationship can be designed [2]. Some quantitative methods have been developed in measuring customer loyalty (see [3]), and some of them focus on predicting customer loyalty. Service or product provider uses this prediction to gain information about what customer relationship strategies should be deployed based on significant influencing factors (see [4]). Moreover, the prediction model can retrieve customer perception data taken from periodic survey to improve individual loyalty prediction, as done by [5]. Some researchers apply mathematical model to predict the loyalty, such as logistic regression [6], discriminant analysis [7], and artificial neural networks [8]. Some of them even combine the prediction model with data mining techniques (see [5]).

However, applying only mathematical model to determine factors influencing the customer loyalty has limitation in updating the fitted model with periodically survey data. Thus, the researcher needs to re-fit the model with new data and re-analyse it based on statistically significant factors. On the other hand, implementing only artificial neural networks (ANN) model for predicting customer loyalty leads to ineffectiveness, where this model involves all the factors without considering their statistical significance. The absence of these such statistical hypothesis also leads the ANN to black-box modelling which is hard to be interpreted by researchers.

The objective of this research is to propose alternative framework in combining the mathematical structural equation model (SEM) for determining factors influencing the customer loyalty with prediction techniques of it based on artificial neural networks (ANN). This framework applies iterative ANN fitting based on updated customer survey data to improve the prediction accuracy. Along with the amount of data collected by survey periodically, this iterative procedure forms simple data mining technique in loyalty prediction.

2. Literature review

2.1. Customer loyalty

In [9], it is mentioned that customer loyalty is a customer emotion that comes to the product or service provider, in spite of the presence of another competitor, more financially lucrative offers on the market. Providers should create strategies to improve this loyalty by determining specific factor, action, and activities to increase the indicator values of loyalty. Predicting customer loyalty could be useful for designing customer relationship management and improving customer perception about benefits they received from service or product providers [10].

2.2. Structural equation modelling (SEM)

Since firstly proposed by Sewall Wright at 1921, structural equation modeling (SEM) has become important tools for understanding relationship between latent and indicator variables in terms of reflective and formative causality.[11]. An example of simple SEM is shown in Figure 1.

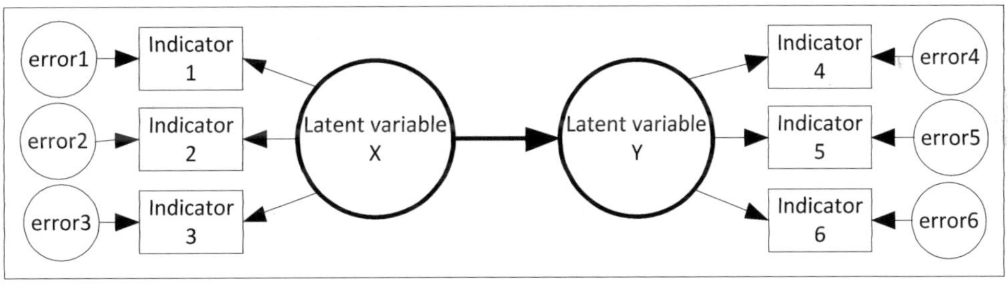

Figure 1. Simple SEM model

In this research, SEM model takes part in determining significant factors or variables that affect the customer loyalty. Indicator variables are measured by valid and reliable questionnaire taken from survey activity. Theses significant factor are then treated as input in ANN model for loyalty prediction.

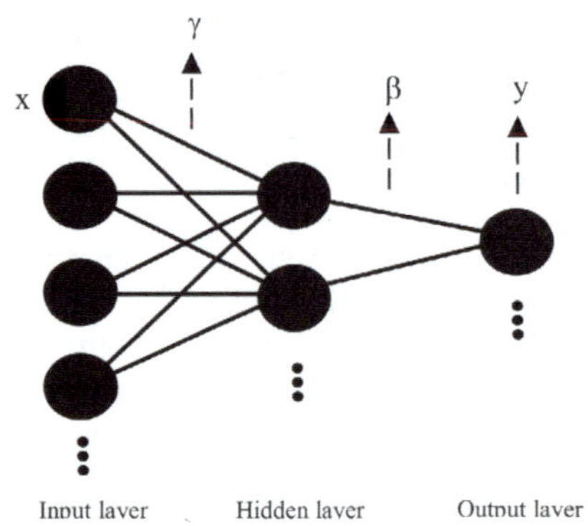

Figure 2. Single hidden layer ANN model

2.3. Artificial neural networks model (ANN)

Many studies apply ANN for several purposes, and some of them use it as part of data mining system (see [12]). It is a complex mathematical model that has flexibility for modelling nonlinear causality among variables, and some literatures categorize it as an artificial intelligence model [13]. However, ANN didn't provide any statistical hypothesis testing for their input variables, and modelling process lead to black-box analysis [13]. General ANN architecture with single hidden layer that is used in this research shown in Figure 2. Each node in a layer is connected (as weights) with another node at the closest layer. A general mathematical model of ANN is shown in equation (1)

$$\boldsymbol{y} = \sum \boldsymbol{\beta}_j \boldsymbol{f}\left(\sum \boldsymbol{\gamma}_{ij} \boldsymbol{x}_i\right) + \boldsymbol{\varepsilon} \tag{1}$$

Where i represents the number of inputs, j expresses the number of hidden nodes, and function f represents pre-determined activation function which commonly uses logistic function. The ANN learning or training process estimates the weights γ_{ij} and β_j based on observed input (x) and output y (target) variables. Once the weights are estimated, then this model uses new observed input data to predict its output. Some papers have successfully combined the SEM and ANN to predict such targeted variable based on questionnaire responses (see [14] and [15])

3. Proposed framework

The proposed customer loyalty prediction framework consists of four steps including iterative ANN weight updating process. Charted steps are shown in Figure 3.

- First step: research on customer loyalty should be conducted based on ordinary survey using pre-determined variables potentially influence the loyalty. Next, all observed survey data leads the fitted SEM model to find significant X variables that affect customer loyalty (Y variable).
- Second step: all significant X variables obtained from SEM are then treated as input for ANN model. ANN learning process also involves Y variable as response or targeted output. The ANN produces weights which is used to predict of customer loyalty.
- Third step: as new customer comes and fills the questionnaire, the provider could predict the loyalty based on his/her answer using ANN prediction. Every new recorded data from new customer gives information for the ANN to update its weights and improve prediction accuracy
- Fourth step: ANN would update the weights by using re-learning process as new customer data recorded. Once the weights updated, ANN uses these new weights to predict loyalty of another new customer, and then again update the weights iteratively.

The framework involves iteratively updating ANN weights, the more data from new customer will then result to more accurate loyalty prediction. These steps form simple data mining process for customer loyalty prediction based on combination between SEM and ANN model.

4. Implementation result

4.1. SEM modelling

The framework in figure 2 takes case study on a digital printing service provider. Twenty variables are shown in Table 1, measured in Likert scale questionnaire filled by 100 customers that have been using this service. All predictor variables are involved in SEM model to find the significance ones. The SEM path model is shown in figure 4. Based on common analysis in SEM on some statistical goodness of fit test including Akaike's Information Criterion (see [16]), significance predictors are then chosen. ANN modelling needs these predictors as input to predict customer loyalty.

4.2. ANN modelling

As significance predictor variables was determined by SEM modelling, ANN started the learning process by first created its architecture (see figure 5). As shown in Table 1, there are 8 predictor which should be involved in ANN, including two dimensions. This information is needed by ANN to set the input layer with 8 nodes, and to set the single hidden layer with two nodes. The learning process was then started with back-propagation algorithm [13], and the weights result is shown in Table 3. The Akaike's Information Criterion (AIC) was also used to compare model performance between ANN and SEM.

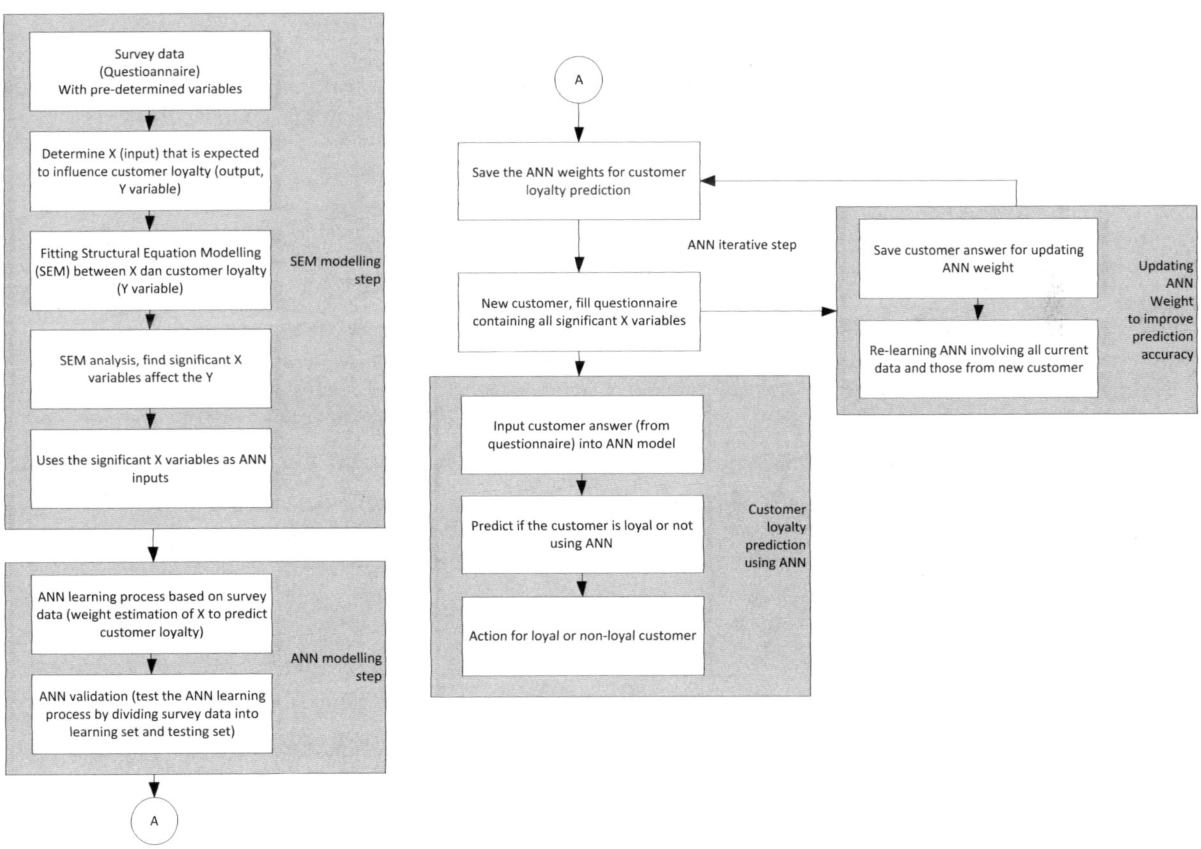

Figure 3. Proposed SEM-ANN framework

IOP Conf. Series: Materials Science and Engineering **703** (2019) 012024 doi:10.1088/1757-899X/703/1/012024

Table 1. Predictor and response variable in SEM

ID	Variable	Code	Predictor Significance	Variable dimension	Dimension Significance	Goodness of fit
1	Various service	M1	No			
2	Latest technology equipment	M2	Yes			
3	Price according to quality	M3	Yes			
4	Competitive price	M4	No			
5	Friendly staffs	M5	No			
6	Experienced staffs	M6	Yes			
7	Responsive staffs	M7	No			
8	Effective promotion	M8	No	Marketing mix dimension (predictors)	Significant	
9	Easy access information	M9	No			Akaike's Information Criterion (AIC) for SEM model -19.378
10	Modern interior design	M10	Yes			
11	Clean and comfort environment	M11	No			
12	Strategic location	M12	No			
13	Easy to find location	M13	No			
14	Simple order procedure	M14	No			
15	Monitored order progress	M15	Yes			
16	Satisfaction for service provided	S1	Yes			
17	Fulfilling customer expectation	S2	Yes			
18	Customer comfort and trust	S3	Yes	Satisfaction dimension (predictors)	Significant	
19	Empathy and care	S4	No			
20	Excellence service	S5	No			
21	Response (target) variable	T1		Customer loyalty		

4.3. Customer loyalty prediction

New three customer with new answer then asked with only statistically significant variables, because SEM model has selected them for ANN inputs, and remove insignificant ones. Answers shown in Table 2. These new data then becoms input in learned ANN to predict the loyalty (see Table 3).

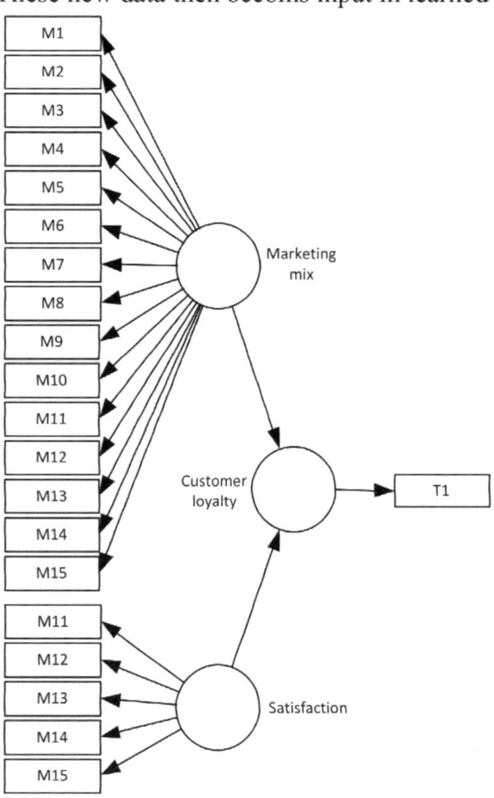

Figure 4. SEM path model

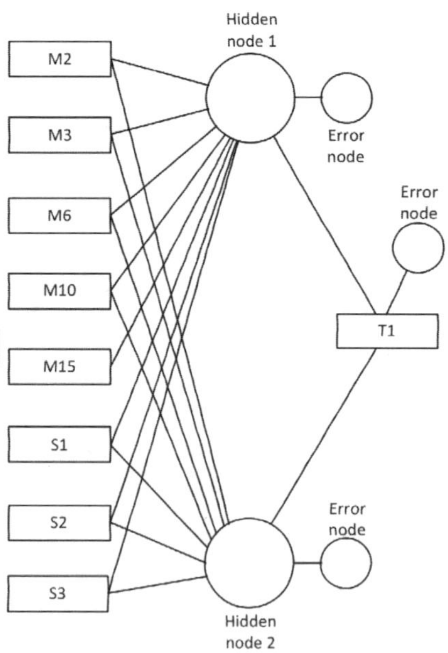

Figure 5. ANN path model

Table 2. New customer loyalty prediction by ANN *(8 input nodes, 2 hidden nodes)*

New Customer 1 questions	Answer (Likert scale)	New Customer 2 questions	Answer (Likert scale)	New Customer 3 questions	Answer (Likert scale)
M2	3	M2	3	M2	5
M3	3	M3	3	M3	4
M6	4	M6	4	M6	4
M10	4	M10	4	M10	4
M15	5	M15	4	M15	5
S1	4	S1	5	S1	5
S2	4	S2	5	S2	5
S3	4	S3	5	S3	5
Predicted loyalty	3.08 (loyal)	Predicted loyalty	3.83 (loyal)	Predicted loyalty	4.42 (loyal)

Table 3. Updated ANN weights

ANN path	Previous weights	Weights after updating	ANN path	Previous weight	Weights after updating	Goodness of fit
M2 → hidden unit 1	1.419	0.045	M2 → hidden unit 2	0.834	0.745	AIC for ANN model -48.9398
M3 → hidden unit 1	0.267	0.551	M3 → hidden unit 2	-1.258	0.465	
M6 → hidden unit 1	2.217	-0.23	M6 → hidden unit 2	-0.338	-0.33	
M10 → hidden unit 1	1.409	0.223	M10 → hidden unit 2	-0.571	0.674	
M15 → hidden unit 1	3.153	0.211	M15 → hidden unit 2	2.079	0.326	
S1 → hidden unit 1	1.843	-0.341	S1 → hidden unit 2	0.325	0.152	
S2 → hidden unit 1	0.799	-0.315	S2 → hidden unit 2	-0.028	0.571	
S3 → hidden unit 1	1.228	-0.027	S3 → hidden unit 2	-1.863	0.298	
Hidden unit 1 → T1	0.787	0.218	Hidden unit 2 → T1	-0.472	0.912	

4.4. Updating ANN weights

These data from new customer then become new raw input for ANN in updating the weights. Using similar learning process, ANN then updates the weights and produces new ones, as shown in Table 3. Every new raw data involved in ANN would then update the weights. This iterative procedure is repeated as new customer continuously comes to the provider, and ANN prediction becomes better with more data inputted in. The AIC criterion also shows better (smaller) result than SEM model.

5. Concluding remark

The framework proposed in Figure 3 has been successfully implemented, including iterative procedures to improve prediction accuracy. With additional programming techniques, this framework becomes a simple data mining system to predict customer loyalty, as there are steps for saving customer answer and updating the ANN weights continuously and automatically.

References

[1] Hayes B 2008 *Measuring Customer Satisfaction and Loyalty3rd Ed* (Milwaukee: Quality Press, American Society for Quality) 79

[2] Hassan R S, Nawaz A, Lashari M N, and Zafar F 2015 Effect of customer relationship management on customer satisfaction *Procedia Economics and Finance* **23** p 563

[3] Szwarc P 2005 *Researching Customer Satisfaction and Loyalty* (London: Kogan Page) p 55

[4] Chou P F, Lu C S, and Chang Y H 2014 Effects of service quality and customer satisfaction on customer loyalty in high-speed rail services in Taiwan *Transpormetrica A: Transport Science* **10** p 917

[5] Lu G, Lu P, and Qi X 2011 Customer loyalty prediction and implementation using data mining *Proc. International Conference on Computer Science and Service System*(Nanjing: IEEE) p 3120

[6] Tamaddoni A, Stakhovych S, and Ewing M 2016 Comparing Churn Prediction Techniques and Assessing Their Performance: A Contingent Perspective *J. Service Research* **19** p 123
[7] Khodabandehlou S and Rahman M Z 2017 Comparison of supervised machine learning techniques for customer churn prediction based on analysis of customer behavior *J. Systems and Information Technology* **19** p 65
[8] Tiwari A, Hadden J, and Turner C 2010 A New Neural Network Based Customer Profiling Methodology for Churn Prediction *Lecture Notes in Computer Science Volume 6019* (Berlin: Springer)
[9] Kiseleva E M, Nekrasova M L, Mayorova M A, Rudenko M N, and Kankhva V S 2016 The theory and practice of customer loyalty management and customer focus in the enterprise activity *Int. Review of Management and Marketing* **6** p 95
[10] Figueroa C J 2010 Predicting customer loyalty labels in a large retail database *Data Mining, Annals of Information SystemVol 8* (Boston: Springer)
[11] Pearl J 2012 The causal foundations of structural equation modelling *Handbook of Structural Equation Modelling* (New York: Guilford Press) p 68
[12] Hadiyat M A and Prilianti K R 2012 Comparing statistical feature and artificial neural networks for control chart pattern recognition: acase study *Proc 3rdInt Conference on Technology and Operation Management* (Bandung: SBM ITB) p 83
[13] Bishop C 1999 *Neural Networks for Pattern Recognition* (Oxford: Clarendon Press) p 116
[14] Sharma S K, Gaur A, Saddikuti V, and Rastogi A 2017 Structural equation model (SEM)-neural network (NN) model for predicting quality determinants of e-learning management systems *Behavior & Information Technology* **36** p 1053
[15] Ahani A, Rahim N Z A, and Nilashi M 2017 Forecasting social CRM adoption in SMEs: a combined SEM-neural network method *Comput. Hum. Behav.***75** p 560
[16] Suhartono and Hadiyat M A 2003 Building neural network model for time series as statistical modeling *Proc. The SEAMS-GMU Int. Conf. of Mathematics and its Applications* (Yogyakarta: UGM Press)

IOP Conf. Series: Materials Science and Engineering **703** (2019) 012025 doi:10.1088/1757-899X/703/1/012025

The use of consumer behavior to identify the flow mapping of waste cooking oil: A finding from Semarang, Indonesia

S Hartini, D P Sari and A A Utami

Department of Industrial Engineering, Diponegoro University, Semarang 50275, Indonesia

E-mail: annisautami.amalia@gmail.com

Abstract. Management of waste cooking oil is expected to receive more concern nowadays. As a daily consumed good, cooking oil generates a huge amount of waste. Besides, the potency of WCO to cause environmental damages such as water pollution is also inevitable. Adopting the concept of a circular economy, WCO is actually convertible into other value-added products such as soap or fuel. However, in practice, WCO is at scattered points. This paper identified the flow mapping of WCO in Semarang Indonesia as a baby step before stakeholders arrange the further move. Based on the survey that is conducted to 347 households and 146 culinary enterprises, it is known that 90% of households and 67,6% of culinary enterprises disposed the WCO to drains, land, or trash. There is also a small number of participants who give or sell the WCO to 3rd parties such as garbage collectors, oil collectors, household assistants, relatives, and others. WCO-recycling is only carried out by 0,9% of household and none of culinary enterprises.

1. Introduction

In the last two decades, industries began to put more concern in initiating the practice of sustainable supply chain management (SSCM) [1], [2]. The international trend now is challenging industries to focus on SSCM in achieving competitive advantage [3] by applying this concept in all product life cycle including the end-of-life (EOL) [4]. One of the products with poor EOL management is cooking oil. The issue of inappropriate management of waste cooking oil (WCO) is currently attracting public concern and serious attention in recent years [5]. This is because the production of WCO is always there, noticing that cooking oil is a daily consumed good. In China, which is the largest producer of WCO, fast food restaurants in large cities can produce 15 liters of WCO every day [6]. In Mediterranean countries, one person produces 3-5 kg of WCO every year [7].

Besides the huge amount, another factor that makes WCO kept being discussed is the serious impact caused by the inappropriate management of WCO. High-temperature frying can momentarily eliminate the edible and nutritional value [8], so that re-consuming cooking oil will only endanger food safety and human health [9]. Improper handling of WCO can cause environmental pollution especially for water and soil [10]. On a larger scale, the mistreatment of WCO can also damage the aquatic community as the oil layer in water will cover the surface and inhibit oxygen to diffuse[11].

Seeing the adverse effects on humans and the environment resulted from the inappropriate handling of WCO, there should be procedures for proper management of cooking oil waste. WCO can actually be collected and then used as a biofuel production unit, soap, detergent, paint or lubricant [12]. Not to mention, collecting and recycling this waste is a contribution to solve three problems at once, which are reducing waste, reducing dependence on fossil fuel energy, and reducing pollutant emissions [13]. This concept adopts the concept of the circular economy (CE). The CE concept is when material flow continues to circulate and does not enter the biosphere, except biological nutrition [14]. Today, industries are beginning to see this concept as a mechanism to increase competitive advantage [15]. One of the three CE principles is to keep the product or material used.

Before the concept of CE can be implemented, it is necessary to ensure that the WCO can be collected first. Because in practice WCO is at scattered points such as from households, restaurants, and hotels, so a collecting chain is needed at the very first place [16]. Thus, this paper will contribute in identifying the flow of waste cooking oil as a system portrayal to be considered in formulating the further move.

2. Theoretical Background

2.1. Waste Cooking Oil

WCO is a waste produced from food frying activities using vegetable cooking oil [17]. Some examples of plants that can produce vegetable cooking oil are palm oil, corn, sunflower, olives, soybeans, and peanuts [18]. During frying activity, vegetable oil will be heated at 160-200 ° C for a relatively long time. The condition when frying is done will certainly result in changes both physically and chemically in the cooking oil.

Discharging WCO to the soil or Final Disposal Site is not the right form of handling. Such treatment will only cause water and soil pollution, disrupt aquatic ecosystems, and endanger human health [18]. One liter of WCO discharged into the water stream can contaminate 1000 water tanks containing 500 liters. When it reaches a water source, WCO will form a layer on the surface of the water which inhibits oxygen exchange and leads to changes in the ecosystem. Not only that, the drain pipe will also be blocked by cooking oil waste that accumulated so that it will disrupt the flow of the wastewater and cause unpleasant odors or various diseases [19].

WCO is potential to be converted into soap, as a raw material for oil or oleochemical processing industries [20], as energy sources such as biodiesel, bio methanol, hydrogen, H2 / CO, and low molecular weight hydrocarbons [21]. Seeing the business opportunity behind this waste stream, some companies have even invested in collecting WCO to be further processed [13].

2.2. System Description

Based on the Report on the Distribution of Indonesian Edible Oil Commodities published by the Central Bureau of Statistics in 2016, it is known that the circulation of cooking oil in the province of Central Java is the largest in households. Because 89.55% of cooking oil from distributors will continue to retailers, and 69.37% are sold to households [22].

Unlike the distribution of cooking oil, the flow of WCO's EOL in Central Java is no recapped yet. This is because no stakeholder who is responsible for the WCO collection. Unlike the domestic waste that has a National Waste Management Information System managed by the Ministry of Environment and Forestry of the Republic of Indonesia, the data on waste cooking oil produced has also not been recorded. Organizations related to waste cooking oil such as the Association of WCO's Collector for

Indonesian Renewable Energy (APJETI) also do not provide much information. Even the Semarang branch of APJETI can no longer be contacted.

3. Research Design

The objective of this research is to perform a flow mapping of WCO in Semarang, Indonesia. A survey was conducted in Semarang, Indonesia. It is the capital city of Central Java Province. This city currently has not implemented WCO collection yet and is considered as one of the cities that is believed to have a rapid development associated with the food business. The questionnaire that was used in this survey has two sections consisting of the respondent's personal information in the first section and the respondent's behavior in managing the WCO in the second section. Respondents of the survey are household and culinary enterprise's owners. The questionnaire is designed differently for household and culinary enterprises. For the first part of the household, the information conveyed is the respondent's name, gender, age, educational background, marital status, income (in IDR) per month, and the location they live. Meanwhile, for the culinary enterprise, the information includes the kind of food they sell, monthly income, and the location of the food trucks. In the second section for both culinary and household. The participants were asked about the intensity of reusing the oil, the treatment for the WCO, the amount of WCO generated, and their understanding of the harm. To generate the mapping, the percentage of each answer is first counted.

4. Findings and Discussion

The survey was performed to 347 households and 164 culinary enterprises in Semarang. 47% of household reuses the same cooking oil twice, 23.3% once, 17.9% do not reuse, 10.4% 3 times and 1,4% more than 3 times. This information showed in Table 1. In the after-use phase, 97,7% of household produces WCO, while the other 2,3% does not produce WCO. Of the 97,7% of the waste produced, the most common behavior is to dispose the waste cooking oil with the details 66% to the drains, 8,6% to the land, and 15,8% to the trash. If it is proportionated to the amount of WCO generated in Semarang, there are 252.271 liters of WCO ended up in water. This is because not everyone understands the impact caused by WCO's mistreatment. Only 54.9% of the participants understand the dangers of disposing WCO to landfills. In addition, many households who dispose the WCO to drains claim that it is the easiest and most practical way because they only need to pour the leftover cooking oil directly from the frying pan into the sink when doing the dishes. The little amount of leftover oil makes them think that the impact will not be significant. In fact, if that little amount of WCO is collected for the whole month, one household produces 0,973 liters of WCO.

Another behavior is to give the WCO to other parties. Other parties that receive the WCO are varied, such as garbage collectors, household assistants, culinary enterprises, friends or family, Waste Bank (Bank Sampah), and Woman Association (PKK). Besides being given, some, such as culinary enterprises, sell the WCO to other parties. The cooking oil that was sold was only used for one-time cooking. Recycling the WCO was only carried out by 0,9% of the participant. Some of these wastes are used as fuel to turn on kerosene lights, make chicken feed, or become fence lubricants. This is described in Figure 1.

26,6% of culinary enterprises claimed that they never reuse the cooking oil. 24,4% reuses 3 times, 22,6% 2 times, 18,3% more than 3 times, and 8,5% 1 time. In contrast to the domestic sector, in the culinary sector, more participants do not produce any leftover oil after-use, which is 13.4%. This is described in Figure 2.

After-use behavior of cooking oil between culinary enterprises and households is more likely the same. The most common behavior is to dispose of cooking oil waste. 31,7% dispose it into the drains, 11,6% throw it to the land, and 24,4% dispose it in the trash. There is no culinary enterprise who can afford to recycle the WCO. Aside from being disposed, 9,8% sell it to other culinary enterprises, oil sellers or collectors, military academy (KODAM), a traditional market, friends, family, and other 9.1% gave to oil collectors, garbage collectors, friends, family or using it again for cooking at home. The data used in the mapping is entirely sourced from the results of the questionnaire, so that it is prone to being subjective. For example, in the amount of cooking oil waste produced, respondents' answers often still use estimation such as referring to the size of mineral water bottles without any field observation used to find out the amount of cooking oil waste produced.

Table 1. The Result of The Survey

Household			Culinary Enterprise		
Intensity of cooking oil re-usage			Intensity of cooking oil re-usage		
Never reuse	62	17.87%	Never reuse	43	26.22%
1 time reuse	81	23.34%	1 time reuse	40	24.39%
2 times reuse	163	46.97%	2 times reuse	37	22.56%
3 times reuse	36	10.37%	3 times reuse	14	8.54%
>3 times reuse	5	1.44%	>3 times reuse	30	18.29%
Behavior of WCO management			Behavior of WCO management		
Dispose to drains	229	65.99%	Dispose to drains	52	31.71%
Dispose to trash	55	15.85%	Dispose to trash	40	24.39%
Dispose to land	30	8.65%	Dispose to land	19	11.59%
Recycle	3	0.86%	Give to other parties	14	8.54%
Give to other parties	21	6.05%	Sell to other parties	16	9.76%
Sell to other parties	1	0.29%	Do not produce leftover	23	14.02%
Do not produce leftover	8	2.31%			

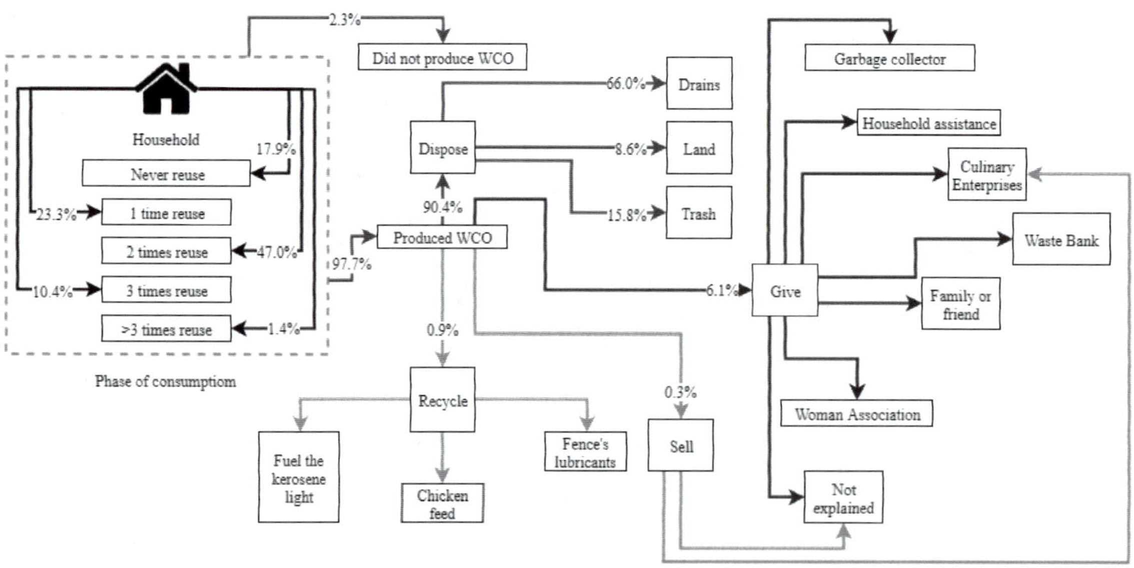

Figure 1. The WCO Flow Mapping of Households

Besides, specifically in culinary enterprises, there appears to be a difference between the information from the respondents and the results of the observations. In some cases when the frying pan can be seen, for example when conducting an interview with a Sempolan seller, the oil used for frying looks dark brownish, but when filling out the questionnaire, business owners said that they only re-use the cooking oil once. There are two things that are thought to be the cause of this. First, because the respondent misinterprets the question, or because the respondent feels that this should not be known to the public so the answers given do not really reflect the actual conditions.

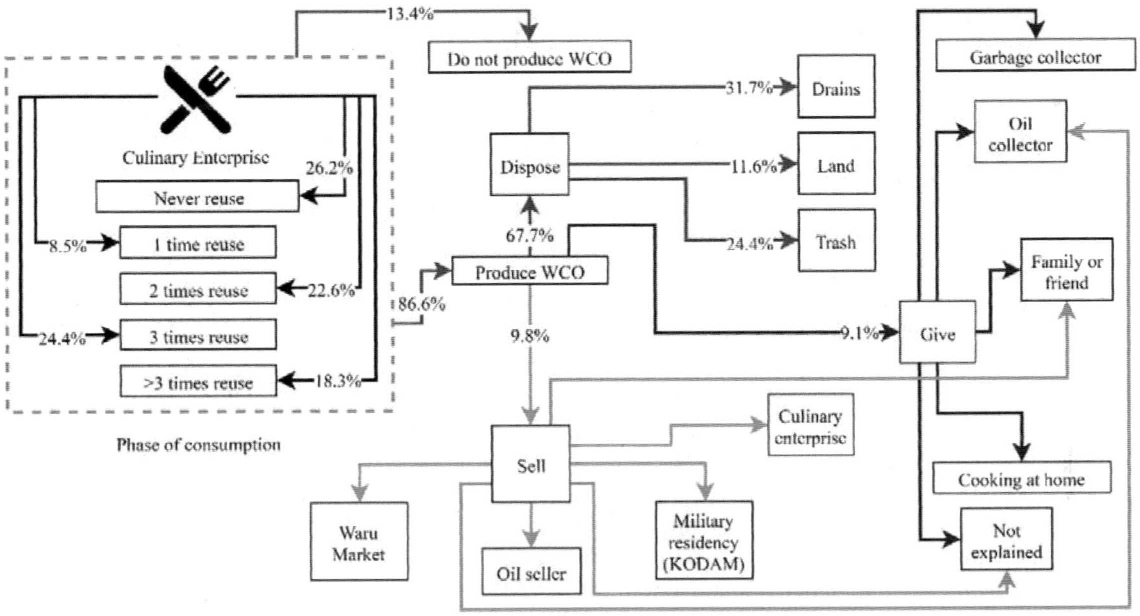

Figure 2. The WCO Flow Mapping of Culinary Enterprises

5. Conclusions, Limitations, and Future Research Directions

The objective of this study was to identify the flow mapping of WCO in Semarang. Based on a practical perspective, this study could give such insight to the stakeholders, for example, local government of Semarang in formulating the further treatment of WCO since proper handling of WCO is considered as essential [10], [11]. Based on the theoretical point of view, this study could give such a contribution toward the related literature of waste flow mapping especially for waste cooking oil in Indonesia.

The result of this research stated that most of the households or culinary enterprises have not managed the WCO well. For households, 66% discharges WCO to the drains, 8.6% to the land, and 15.8% to the trash. This number is as much as 252.271 litres of WCO that ends up in water source. Meanwhile from culinary enterprises, 31.7% discharges WCO to the drains, 11.6% to the land, and 24.4% to the trash.

The limitations of this study are twofold. First is related to the accuracy of the culinary enterprises' answer as there is no access to observe the cooking and waste managing activity. For the upcoming research, a field observation is expected so it can confirm the answer from the respondent. The second limitation is related to the area (location) of this research, i.e., the study was conducted in Semarang, Indonesia. As suggested for future research is a research with a larger sample size so that more detailed analysis would be obtained. In addition, a more nationally representative sample, not only geographically but also demographically, could provide higher generalizability. Therefore, to generalize, the upcoming research should broaden and expand the research area.

References

[1] L. Preuss 2011 *Supply Chain Manag. An Int. J.* **14** 213–23.
[2] T. Abdallah, A. Diabat, and J. Rigter 2013 *Int. J. Prod. Econ.* **146** 465–77.
[3] L. Shen, L. Olfat, K. Govindan, R. Khodaverdi, and A. Diabat 2013 *Resources, Conserv. Recycl.* **74** 170–9.
[4] J. Ma and G. E. O. Kremer 2014 *J. Clean. Prod.* **108**. 289-300.
[5] R. Yang, W. Tang, R. Dai, and J. Zhang 2018 *J. Clean. Prod.* **201** 61–77.
[6] N. H. Abdullah, S. H. Hasan, and N. R. M. Yusoff 2013 *Int. J. Mater. Sci. Eng.* **1** 94–99.
[7] A. Mannu, G. Vlahopoulou, V. Sireus, G. L. Petretto, G. Mulas, and S. Garroni 2018 *Nat. Prod. Commun.* **13** 1934578X1801300523.
[8] S. Nanda, R. Rana, H. N. Hunter, Z. Fang, A. K. Dalai, and J. A. Kozinski 2018 *Chem. Eng. Sci.* **195** 935-45.
[9] T. Liu *et al.* 2018 *J. Clean. Prod.* **204** 636–42.
[10] K. Hanisah, S. Kumar, and A. Tajul 2013 *Heal. Environ. J.* **4** 76–81.
[11] L. C. Lange and A. F. M. Ferreira 2017 *Waste Manag.* **61** 269–75.
[12] A. Talebian-Kiakalaieh, N. A. S. Amin, and H. Mazaheri 2013 *Appl. Energy* **104** 683–710.
[13] T. R. P. Ramos, M. I. Gomes, and A. P. Barbosa-Póvoa 2013 *Waste Manag.* **33** 1691–1703.
[14] Ellen MacArthur Foundation, "Ellen MacArthur Foundation," 2012.
[15] Ellen MacArthur Foundation, "What is the circular economy 100?," 2013. [Online]. Available: https://www.ellenmacarthurfoundation.org/ce100. [Accessed: 24-Nov-2018].
[16] Y. Jiang and Y. Zhang 2016 *Transp. Res. Procedia* **12** 938–49.
[17] L. Iglesias, A. Laca, M. Herrero, and M. Díaz 2012 *J. Clean. Prod.* **37** 162–71.
[18] M. Carlini, S. Castellucci, and S. Cocchi 2014 *Energy Procedia* **45** 198–206.
[19] C. A. Guerrero F., A. Guerrero-Romero, and F. E. Sierra 2011, "Biodiesel Production from Waste Cooking Oil," *Biodiesel - Feed. Process. Technol.*
[20] C. Sheinbaum, M. V. Balam, G. Robles, S. Lelo De Larrea, and R. Mendoza 2015 *Waste Manag. Res.* **33** 730–39.
[21] D. C. Panadare and V. K. Rathod 2015 *Iran. J. Chem. Eng.* **12** 55–76.
[22] Badan Pusat Statistik, "Distribusi Perdagangan Komoditas Minyak Goreng Indonesia," 2016.

Perceived kansei and performance-based usability impact on satisfaction for web-based applications

M Hartono

Department of Industrial Engineering, University of Surabaya, Raya Kalirungkut, Surabaya, 60293, Indonesia

E-mail: markus@staff.ubaya.ac.id

Abstract. Kansei Engineering has been applied intensively for product design and development considering the deepest layer of customer needs, known as Kansei, which is related to emotional need or affect. This method has been extended into service design and innovation. It has been shown that Kansei's impact on customer loyalty is equally important as that of cognition or performance. A recent research in investigating the proportional effect of Kansei and performance on usability measurement for e-services has been conducted. It shows that both Kansei and performance is found to be equally significant as well. Then, a question is raised, whether they are antecedent of customer satisfaction. This study is proposed to investigate both perceived Kansei-based and performance-based usability measures on satisfaction. Both theoretical and practical implication is highly expected due to the growth of more human-computer interaction (HCI) features and services nowadays. A proposed applicative framework followed by an empirical study on how the perceived Kansei and performance-based usability impact on satisfaction for web-based applications is discussed.

1. Introduction

Concerning to web service and web-based applications, usability analysis focuses on structure and behaviour rather than technical aspects. A web-based application is a program which can be accessed through a network connection using HTTP and runs inside a certain web browser. The usability of web-based application is highly expected. Both the impression and performance of the web-based application are significantly expected by the user. A recent study shows that both cognitive and affective process is associated with usability [1]. Cognitive deals with human capabilities, limitations, and tendencies; whereas, affective is known as Kansei (in Japanese) concerns on human feelings and impression. Affect complements the whole cognitive evaluation, according toa study by Hartono and Tan [2]. This study highlights the importance of emotional satisfaction in services, and how the service provider should take the impact of emotional needs on particular service attribute performance for the improvement prioritization.

In the area of Human-Computer Interaction (HCI), both cognitive and affective aspects for usability have been implemented into electronic products, medical equipment, mobile phones and e-commerce sites [1,3,4]. In the case of HCI-based websites, the cognitive aspect includes ease of use, content, identification, delay, and trust. More specifically, in usability study for e-commerce, its cognitive process covers navigation, interaction, layout, merchandise and information quality [5]. A recent study on the role of affect and cognition on customer satisfaction and loyalty shows that effect has the proportional weight to cognition which leads to customer satisfaction and loyalty [6]. However, effect has received little intention for the usability evaluation [1]. Hence, a study on how to investigate the impact of effect- and cognition-based usability on satisfaction is highly prospective.

There is still no comprehensive conclusion about the relationship between effect and usability and how its impact on user satisfaction. Some studies on the affect-based usability have been conducted [7]. However, some studies could not show that relationship [1]. Therefore, a study on the combination of cognitive and affective processes is highly expected to deliver more comprehensive information for user satisfaction.

A case study on a web-based application is selected, or it is deemed related to the website. The development of usable websites is important for e-business and e-commerce success today as online customers and users will feel, search, and experience products or services mainly through websites. A full usable website has been deemed to be a media for delivering positive attitude toward online shops, boosting engagement and revisit rates and also stimulating online purchase [8]. It is also stated that *difficult-to-understand* website content, inconsistent format, disorientation, poor interaction, and reliability and ill-defined help functions are regarded as common problems for commercial websites [9].

Given this above background, this study has two objectives. First, it is to build an applicative research framework of perceived Kansei and performance-based usability impact on satisfaction for web-based applications. Second, it is to conduct a simple empirical test to investigate the effect of web-based application usability constructs on satisfaction.

This paper is systematically structured as follows. After the introduction section, it is followed by a brief Literature Review. Afterwards, it continues to the Research Methodology and Framework, Result and Discussion, and is wrapped up with Conclusion and Further Research Recommendation.

2. Brief Literature Review

2.1 Usability for website and web-based application

The mutual interaction between human and computer is highly needed in designing and implementing a variety of computing systems. According to ISO 9241 [10], usability is defined as the extent to which a product can be utilized by particular users to obtain effectiveness, efficiency, and satisfaction. Related to website or web-based application, it refers to the extent to which web can be used by specific users in achieving goals in terms of efficiency, effectiveness and satisfaction in a specified context of website or web-based application use.

According to Lee & Kozar [9], effectiveness represents the accuracy where the online users can achieve their goals (e.g., purchase, information gathering); efficiency is related to quick time and less cognitive effort to achieve goals (e.g., website with simple and consistent design features to make the pages easy to read and learn); and satisfaction is defined as the comfort and acceptability of a website to its users (e.g., real-time chat menu with customer support personnel).

2.2 Kansei, Performance and Satisfaction in web-based application usability

In website design and application, affect refers to impression or delight (known as Kansei in Japanese) whereas performance deals with cognition, according to the previous study on emotional needs and Kansei Engineering [1]. Delight, firmness, and convenience are found to be determinants of customer satisfaction and loyalty toward the site [11]. Delight explains the effect, whereas, firmness and convenience are dominantly describing the performance or cognitive processes. Another study proposes a set of usability dimensions such as download delay, navigability, site content, interactivity and responsiveness which bring impact on the customer satisfaction, return possibility and frequency [12].

About various constructs used in various usability studies, Prastawa et al [1] proposes two main factors (i.e., cognitive and affective processes) which bring significant impact on usability. Even though this approach has simplified the understanding of usability performance drivers, an extended approach which is to investigate the proportional impact of effect and cognition on user satisfaction in a case of usable web-based application or website is prospective.

3. Research Framework and Hypothesis

Referring to what has been discussed in Section 1 and 2 above, this study starts with the perceived performance-based and Kansei-based usability measurement, and then followed by the overall satisfaction measurement, as shown in figure 1. The measure variables were adopted from Prastawa et al. [1]. The details of them are provided in the Appendix. According to Helander and Khalid [13], perceived performance-based is related to cognition, which is something more on analytical and rational thinking, whereas Kansei-based is related to emotions and attitudes. Variables of both constructs explain how usability can be derived from both rationale and emotional aspects. For instance,"readability" explains the performance-based usability, it may relate to sentences and steps to operate the website that are easy to understand. For the Kansei-based usability, for instance, "appeal" is proposed. Appeal is related to the appearance of the web is interesting and eye-catching.

Apart from performance-based and affective-based usability constructs, the third construct to consider is overall satisfaction, which is only measured by 1 variable called as "overall satisfaction". The relationship among the three constructs is shown in figure 1.

There are two hypotheses defined, i.e., H_1: there is a significant impact of perceived performance-based usability on overall satisfaction, and H_2: there is a significant impact of perceived Kansei-based usability on overall satisfaction.

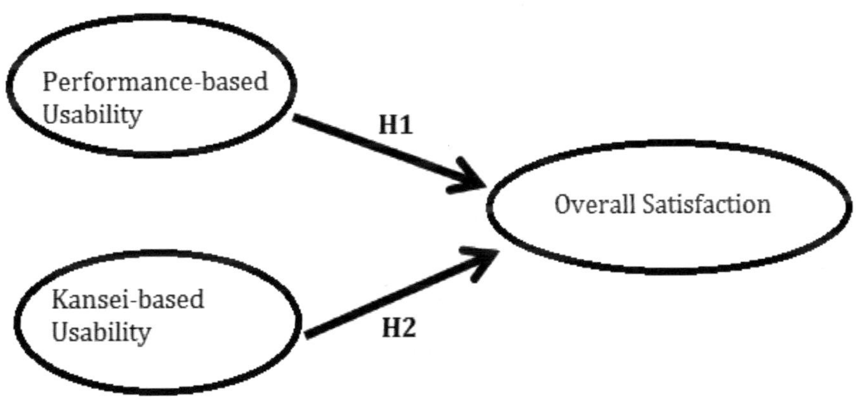

Figure1. Conceptual framework of Performance- and Kansei-based usability for web-based application.

4. Result and Discussion

A case study has been conducted in one of the famous web-based application operated in Indonesia, namely Tokopedia. The targeted subjects were those who have experienced the Tokopedia services in the last six months (October 2018 – March 2019) through the purposive sampling method. It is a non-probability sampling method when the researchers believe that they can achieve a representative sample through a sound judgment which may save time, effort and money. There were 35 respondents involved during the survey using the questionnaire (20 males and 15 females) with the age range from 20 to 40 years. One copy of the questionnaire was given to represent one group of participants (e.g., either one family or one single respondent). The questionnaire consists of three different parts, i.e., Performance-based Usability, Affective-based Usability, and Overall Satisfaction. It took about 10 minutes to complete a questionnaire.

4.1 Validity and reliability test for constructs

The analysis is performed using SmartPLS 3 [14]. It is Partial Least Squares Structural Equation Modeling (PLS-SEM) which focuses on the analysis of variance with condition of small sample size and little availability of tested theory. Confirmatory factor analysis (CFA) has been carried out for testing how well the measured variables represent the number of constructs. In other words, CFA is used to confirm or reject the proposed measurement theory. Model was deemed fit, as it satisfied the following criteria: root mean square residual (< 0.1), d_ULS and d_G (< the upper bound of the confidence interval), and normal fit index (> 0.7 and it is closer to 1). Reliability and validity tests (i.e., convergent validity, discriminant validity and reliability) have been conducted and are showed in table 1and figure 2. In terms of convergent validity, all constructs were deemed acceptable since: (i) all item factor loadings in all constructs exceed 0.6, (ii) composite reliability in all constructs exceeds 0.7, and (iii) AVE in all constructs exceeds 0.5.

Table 1. Measurement of constructs.

Construct	AVE*	Composite Reliability	Cronbach's Alpha
Performance-based Usability	0.571	0.922	0.905
Kansei-based Usability	0.580	0.917	0.896
Overall Satisfaction	1	1	1

*average variance extracted

Referring to discriminant validity, these three constructs above were considered valid as the square root of AVE for each construct is greater than its corresponding inter-construct correlation coefficient [15]. Lastly, in terms of reliability, all three constructs show Cronbach's alpha values which exceed 0.6, thus, they were deemed acceptable.

4.2 Path coefficients and hypothesis testing

It was found that Kansei-based (known as affect or emotion) and performance-based (referred to cognition or rationale) usability account for 47.7% and 50.5% of their impact on overall satisfaction, respectively**,** as shown in table 2.

Table 2. Parameter significance values.

Construct	Original Mean	Sample Mean	Standard Deviation	T Statistics	p values
Performance-based Usability	0.505	0.505	0.091	5.538	<0.001
Kansei-based Usability	0.477	0.479	0.089	5.379	<0.001

The first hypothesis (H_1) that perceived performance-based usability is positively related to overall satisfaction is supported by the results (T Statistics = 5.538, p-value < 0.001). The second hypothesis (H_2) is also supported. Kansei-based usability is significantly positively correlated with overall satisfaction (T Statistics = 5.379, p-value < 0.001). This study shows that both Kansei and performance-based usability have a significant role and impact on overall satisfaction. They cannot be separated; they are a set of human-based experience evaluation in a service encounter. The high value of R square of 0.918 explains both Kansei and performance are regarded as the significant predictors of overall satisfaction. Their impact on customer overall satisfaction is relatively equally balanced.

More specifically, the critical usability variables for both constructs (performance-based and Kansei-based) were defined according to high factor loadings (i.e., factor loadings ≥ 0.8). The most critical performance-based usability measures were interactivity (P3), logical navigation (P8) and easy back option (P9), whereas Kansei-based usability measures were salience (A4), appeal (A5) and reliability (A7).

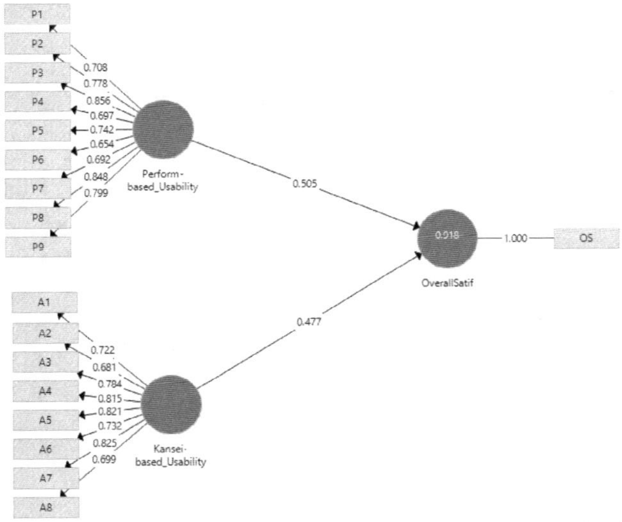

Figure2. Structural model and path coefficients for web-based application case study.

5. Conclusion and Further Research

Kansei(which is related to affect or emotion) was found to be equally important as performance (known as cognition) in usability measurement which brought significant impact on overall satisfaction. This study complements the previous study on usability in web-based application by taking into account both performance and Kansei impact on overall satisfaction. This study is limited due to relatively small sample size and a single service setting. Future study should consider that concern.

6. Appendix

Performance-based and Kansei-based usability variables for web-based application, adopted from [1]

Performance-based usability variables	Kansei-based usability variables
P1. Error prevention a [usability – multiple similar tasks to be done easily]	A1. Colour [usability – conceptual image of the app]
P2. Error prevention b [usability – warning message to prevent possible errors]	A2. Brightness [usability – conceptual image of the app]
P3. Interactivity, Feedback, and Help [usability – multiple opportunities for interaction and communication]	A3. Harmoniousness [usability – compatibility and alignment of the app]
P4. Learnability [usability – appropriate information related to attributes sold]	A4. Salience [usability – convenient use of the app]
P5. Readability [usability – sentences that are easy to understand]	A5. Appeal [usability – appearance of the app is interesting]
P6. Memorability [usability – steps that are easy to remember]	A6. Comfort [usability – convenient use of the app]
P7. Easy Navigation [usability – guide to easy to operate]	A7. Reliability [usability – reliable and trusted app]
P8. Logical Navigation [usability – guide to logical to run]	A8. Attractiveness [usability – attractive and pleasing app]
P9. Easy Back Option [usability – return to the previous page]	

References

[1] Prastawa H, Ciptomulyono U, Singgih M L and Hartono M 2019 The effect of cognitive and affective aspects on usability *Theor. Issues in Ergon. Sci.* TIES **20** p 507-531

[2] Hartono M and Tan K C 2011 How Kano Model Contributes to Kansei Engineering in Services *Ergon.* **54** p 987-1004

[3] Han S H, Yun M H, Kwahk J and Hong S W 2001 Usability of Consumer Electronic Products *Int. J. of Ind. Ergon.* **28** p 143–151

[4] Seva R R, Gosiaco K G T, Crea M, Santos E D and Pangilinan D M L 2011 Product Design Enhancement Using Apparent Usability and Affective Quality *Appl. Ergon.* **42** p 511–517

[5] Li G, Ji S, Li C and Feng J 2011 Efficient Fuzzy Full-Text Type-Ahead Search *The VLDB J.* **20** p 617 – 640

[6] Hartono M and Raharjo H 2015 Exploring the mediating role of affective and cognitive satisfaction on the effect of service quality on loyalty *Total Qual. Manage. & Bus.Excellence* **26** p971-985

[7] Lavie T and Tractinsky N 2004 Assessing Dimensions of Perceived Visual Aesthetics of Web Sites *Int. J. of Hum.-Comput. Stud.* **60** p 269–298

[8] Nielsen J 1993 *Usability Eng.* (London: Academic Press)

[9] Lee Y and Kozar K 2012 Understanding of Website Usability: Specifying and Measuring Constructs and Their Relationships *Decis. Support Syst.* **52** p 450–463

[10] ISO 9241-11 2018 *Ergon. of Hum.-Syst. Interact. – Part 11: Usability: Definition & Concept* (Geneve: International Organization for Standardization)

[11] Kim H K., Hana S H, Park J, Park J 2016 Identifying Affect Elements Based on a Conceptual Model of Affect: A Case Study on a Smartphone *Int. J. of Indust. Ergon.* **53** p 193–204

[12] Palmer J W 2002 Web Site Usability, Design, and Performance Metrics *Inf.Syst. Res.* **13** p 151–167

[13] Helander M G and Khalid H M*Handbook of Hum. Factors & Ergon*, 3rdedSalvendy G (New York: Wiley Interscience) p 543 – 572

[14] Ringle C M, Wende S and Becker J M 2015 *SmartPLS 3*(Boenningstedt: SmartPLS GmbH, http://www.smartpls.com)

[15] Gefen D and Straub D 2005 A practical guide to factorial validity using PLS-graph: tutorial and annotated example *Commun. of the Assoc. for Inf. Syst.* **16** p 91–109

IOP Conf. Series: Materials Science and Engineering **703** (2019) 012027 doi:10.1088/1757-899X/703/1/012027

Measurement of student satisfaction and loyalty using service quality model for higher education (HedQual) at industrial engineering department University of Pelita Harapan

N Hartono, Laurence and B F Tjahjadhi

Department of Industrial Engineering, University of Pelita Harapan, Tangerang, 15811, Indonesia

E-mail: natalia.hartono@uph.edu

Abstract. The Department of Industrial Engineering (IE) at Pelita Harapan University (UPH) strives to continually improve the quality of its services. Previous studies were conducted in 2010 and 2015 using service quality model from Parasuraman. IE UPH wants to increase their student satisfaction; therefore, IE UPH needs to measure it with a customized model because of the distinctive nature of the service. This study develops the ServQual Model for IE UPH, which is named HedQual. The HedQual is predicted to have a positive influence on the students' satisfaction and loyalty. The most important construct of service quality for IE UPH Students was also measured using consensus. Data collection was done using a questionnaire and distributed to all IE UPH students. The model was tested using PLS-SEM (Partial Least Square – Structural Equation Modeling). The final model shows that HedQual consists of Academic and Non-academic. HedQual, Reputation and Learning Outcome have a significant and positive effect on Student Satisfaction. Student Satisfaction has a significant and positive effect on Student Loyalty. The most important service quality for IE student from the consensus calculation is the attitudes and sincere interests of lecturers in helping the students.

1. Introduction

University of Pelita Harapan (UPH) is a private university in Indonesia that adheres to Christianity principles. By providing good and complete facilities, UPH tries to provide the best education for the students. Industrial Engineering (IE) is one of the study programs offered by UPH Karawaci, Tangerang which always want to improve its service quality.

IE UPH conducted research to assessits service quality (ServQual) in 2010 and 2015 to improve the service quality provided for the students. The research in 2010 used general ServQual model that consists of 5 dimensions as follows: Tangible, Reliability, Responsiveness, Assurance, and Empathy. The results shows that student expectations were higher than perceived levels of service quality provided by IE UPH [1]. Those studies only measure the gap or the difference between expectations and perceptions. Another research combined ServQual and Kano models into Quality Function Deployment (QFD) to improve the service quality of IE UPH [2]. Both of these studies used general ServQual model. Studies to improve the service quality should be tailor made because of the uniqueness of service, which differs between providers.

Research by Hartono (2008) concerns with the development of a service quality measurement model (ServQual) developed specifically for high school called HisQual with 7 dimensions, namely Tangibility, Reliability, Responsiveness, Assurance, Empathy, Learning Outcomes, and Personality Development [3]. Hartono et al. in 2017 also conducted research on important factors that influence students in choosing Industrial Engineering and the continuation of this research is needed [4]. Therefore, this study continues the research of Hartono et al. This study aims to create a ServQual model for IE UPH called HedQual, to test the predictive relationship between constructs in the HedQual, Student Satisfaction and Loyalty and to find out the most important part of the service for IE students based on the model.

2. Methodology

Service Quality was first introduced by Parasuraman in 1988 and has gained popularity until now for its generic ability to assess the quality of service. In this paper, ServQual that was introduced by Parasuraman is called general ServQual model that uses 5 dimensions (Tangibility, Reliability, Responsiveness, Assurance, and Empathy) because of researcher around the world use this ServQual model to assess a different kind of service. However, the service itself has specific characteristics and it is not always appropriate to use general ServQual model. Several researchers have modified the general ServQual model to assess service in education. Table 1 shows the position of this research compared with other ServQual model in education with the addition of two well-known ServQual models from Parasuraman and Grönross [3,4,5,6,7,8,9,10,11,12,13,14].

Table 1. Research position.

Researcher and year	Instrument	Research area	Data source	Note
Parasuraman et al. (1988)	ServQual with 5 dimension: Reliability, Empathy, Tangibles, Assurance, Responsiveness	General Service		Published
Owlia dan Aspinwall (1998)	4 dimension: Academic Resources, Competence, Attitude, Content	Higher Education (Engineering) in England	College student	Published
Grönroos (2000)	7 dimension: Professionalism and skills, attitude and behavior, accessibility and flexibility, reliability and trustworthiness, service recovery, serviscape, reputation and credibility	General		Published
Munaf F (2002)	Integration Servqual (general model) with QFD	Higher Education (Engineering) in Indonesia	College student	Unpublished
Abdullah (2005)	Development of ServQual Model for Higher Education	Higher Education in Malaysia	College student	Published
Hartono J (2005)	Model ServQual Parasuraman et al. (general model)	Highschool in Bandung, Indonesia	Highschool student	Unpublished
Mahapatra and Khan (2007)	5 dimension: Learning Outcomes, Responsiveness, Physical Facilities, Personality Development, Academics	Higher Education (Engineering) in India	Student, Alumni, Parent, Employer	Published
Hartono N (2008)	Integration ServQual, Kano and QFD (development of ServQual Model)	Highschool in Bandung, Indonesia	Student, Parent, Teacher	Unpublished
Annamdevula & Bellamkonda (2012)	Development of ServQual Model, HiEdQual	Higher Education in India	College student	Published
Gupta & Kaushik (2018)	ServQual Dimension Comparison in Higher Education	Higher Education		Published
Hartono N and Laurence (2017)	ServQual (development ServQual model)	Higher Education (Engineering) in Indonesia	College student	Unpublished
Verma & Prasad (2017)	Development of SerQual Model (MeQual)	Higher Education (Management)	College student	Published
Mattah, Kwarteng, and Mensah (2018)	Development of SerQual Model	Higher Education in Ghana	College student	Published
This research	Development of ServQual Model for Industrial Engineering University of Pelita Harapan, called HEdQual	Higher Education (Engineering) in Indonesia	College student	

The steps in developing HedQual compare service quality dimension in education as presented in table 1. The items in the questionnaire were developed from those previous studies and added with several items from interview to IE UPH students. The final item was 63 with HedQual 55 item, Student Satisfaction 4 item and Student Loyalty 4 item. HedQual results in 5 dimension: Reputation, Tangibility, Academics, Non-academic Aspect, Learning Outcomes. The respondents are from the population of IE UPH active students, in which from 130 students, there were only 57 usable filled questionnaires.

The validity test using the Pearson correlation shows that all items are valid because all Pearson Correlation values are > r table value (0.2609). R table value is obtained from the degree of freedom (df) which is 55 and the level of significance 0.05 is 0.2609. The reliability test result which is 0.976 shows good reliability. The data shows that the population is 70.18% male and 29.82% is female. The class of 2015 is the biggest percentage of response, shown by 38.6% followed by class of 2018, class of 2016, class of 2017 and 2014 which are 26.3%, 17.54%, 14%, and 3.5% respectively.

The initial conceptual HedQual model consists of Reputation (R), Tangibility (T), Academics (A), Non-academic (N) and Learning Outcome (L) which form a second order construct for HedQual. HedQual is an independent variable, while Student Satisfaction (SS) and Student Loyalty (SL) are dependent variables. The initial model is shown in figure 1. In this initial model, there are 8 hypotheses to be tested as follows:

H1: Academics is a dimension (second-order construct) of HedQual
H2: Learning Outcome is a dimension (second-order construct) of HedQual
H3: Non-academic is a dimension (second-order construct) of HedQual
H4: Reputation is a dimension (second-order construct) of HedQual
H5: Tangibility is a dimension (second-order construct) of HedQual
H6: HedQual has a significant and positive influence on Student Loyalty
H7: HedQual has a significant and positive influence on Student Satisfaction
H8: Student Satisfaction has a significant and positive influence on Student Loyalty

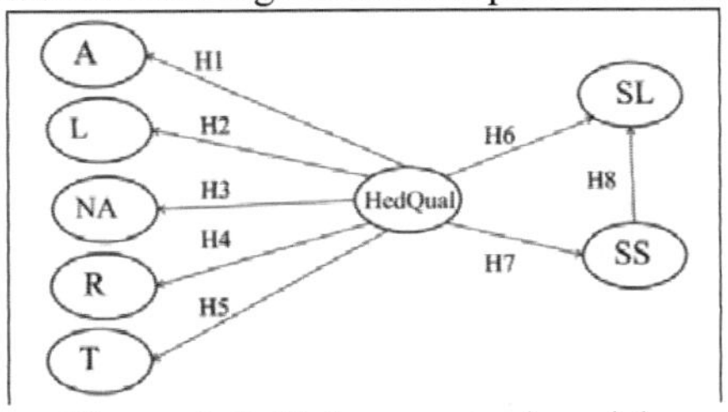

Figure 1. Initial conceptual model.

The PLS-SEM has been a subject of debates whether to use or not to use PLS [15]. The PLS path modeling is a full-fledged SEM method that can handle both factor models and composite models for constructing measurement, estimating recursive and non-recursive structural models, and conducting tests of model fit [16]. Sarstedt Hair points out that researchers have to clarify the difference between a conceptual measurement approach and statistical estimation perspective to choose specific SEM method and its appropriateness [15]. The PLS has advantages for exploratory research and important statistical tool for management, organizational research and other social sciences disciplines [17]. This research uses the PLS-SEM because developing a model using exploratory research that fits with IE UPH and not for generalization of the model.

The steps in PLS-SEM include importing data in the form of CSV file and constructing Path diagram. The model then evaluates after determining the number of bootstraps samples, in this research using 1000 as the number of bootstraps sample. Evaluation of the initial model consists of 2 stages: evaluation of measurement model (outer model) and structural model (inner model).

3. Results

The results of output outer loading show that several indicators have a value below 0.7 in the factor loading and have to be eliminated from the model. Results of PLS shows the indicators that must be eliminated are A1, A2, A3, A4, A5, A6, A7, A8, N1, N2, N4, and N13. All indicators in the construct of Reputation (code R), Tangibility (code T), and Learning Outcome (code L) are also eliminated.

After re-testing, it can be seen that all construct indicators have a value of loading factor > 0.7. The AVE value of all constructs is > 0.5, thus meet the convergent validity requirements. Based on the values of Cronbach's Alpha and Composite Reliability, all constructs have value > 0.7 that shows good reliability. Result of construct validity and reliability depicted in Table 2 shows that the evaluation of the outer model is good. Evaluation of the inner model (structural model) is from the R-square value and significance test. Table 2 also shows the R-square for the initial model which shows that the prediction of the structural model is moderate to high. The significance test with significance level 0.05 shows that construct indicator with t-statistics > 1.96 model is valid. The hypothesestest results are in table 3.

Table 2. Construct validity, reliability and R square of initial model after deletion.

Variable	Average Variance Extracted (AVE)	Cronbach's Alpha	Composite Reliability	R Square Adjusted
Academics	0.662	0.927	0.940	0.873
HedQual	0.657	0.971	0.973	
Non-academic	0.751	0.966	0.971	0.949
Student Loyalty	0.817	0.925	0.947	0.617
Student Satisfaction	0.765	0.896	0.928	0.619

Table 3. Hypothesis test result.

Hypothesis	t-statistics	Note	Hypothesis	t-statistics	Note
H1	47.827	Proven	H5	-	Not proven
H2	-	Not proven	H6	0.275	Not proven
H3	131.005	Proven	H7	14.919	Proven
H4	-	Not proven	H8	7.134	Proven

4. Discussion

Based on the results of the initial model evaluation, the model needs to be modified. The modification is based on other research and exclude the variables that are not proven in the hypothesis test. The initial model is modified by pulling out Reputation and Learning Outcome from HedQual. The Reputation construct is connected with Student Satisfaction and Student Loyalty, while the Learning Outcome is connected with Student Satisfaction. Research by Eryilmaz (2016) said that students' intentions for collaboration in the future are influenced by the reputation and image of universities [18]. This finding is supported by research conducted by Barusman (2014) which states that perceptions of reputation and image of higher education have a positive and significant influence on student satisfaction and loyalty [19].

Tangibility construct is retained as a second-order HedQual construct to see whether this construct affects HedQual after the Reputation and Learning Outcome are eliminated from HedQual. In addition, the relationship between the construct HedQual and Student Loyalty is removed because the initial model hypothesis test shows that the relationship between the two constructs is not proven. These findings lead to the formation of the following hypothesis:

H1: Academics is a dimension (second-order construct) of HedQual
H2: Non-academic is a dimension (second-order construct) of HedQual
H3: Tangibility is a dimension (second-order construct) of HedQual
H4: HedQual
H5: Learning Outcome has a significant and positive influence on Student Satisfaction
H6: Reputation has a significant and positive influence on Student Loyalty
H7: Reputation has a significant and positive influence on Student Satisfaction
H8: Student Satisfaction has a significant and positive influence on Student Loyalty

Figure 2 shows the modification model. The steps of PLS-SEM is the same as the previous steps.

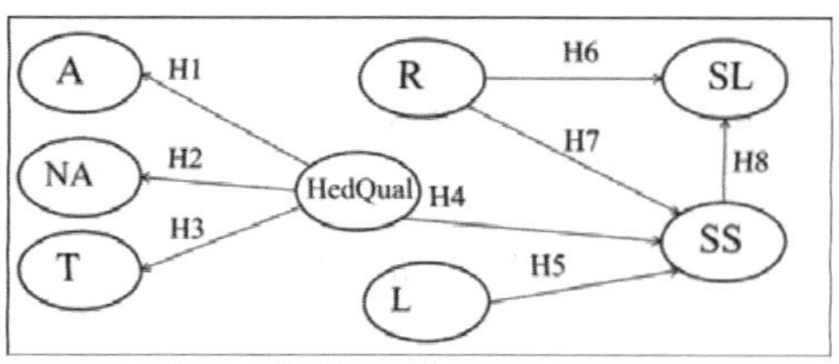

Figure2. Modification model.

The final modification model shows that all indicators of constructs have a loading factor > 0.7 and AVE > 0.5 that shows that the convergent validity requirement is met. Based on the values of Cronbach's Alpha and Composite Reliability, all constructs have value >0.8 that shows good reliability. The outer model evaluation is in good category that is shown from composite reliability and Cronbach's Alpha > 0.8 is depicted in table 4. table 4 also shows the R-square and the hypotheses test results are in table 5.

Table 4. Construct validity, reliability, and R-square of final modification model.

Variable	Average Variance Extracted (AVE)	Cronbach's Alpha	Composite Reliability	R Square Adjusted
Academics	0.662	0.927	0.940	0.873
HedQual	0.657	0.971	0.973	
Learning Outcome	0.644	0.815	0.878	
Non-academic	0.751	0.966	0.971	0.949
Reputation	0.666	0.875	0.909	
Student Loyalty	0.817	0.925	0.947	0.630
Student Satisfaction	0.764	0.896	0.928	0.795

Table 5. Hypothesis test result.

Hypothesis	t-statistics	Note	Hypothesis	t-statistics	Note
H1	51,881	Proven	H5	4,575	Proven
H2	134,539	Proven	H6	1,384	Not proven
H3	-	Not proven	H7	2,146	Proven
H4	3,706	Proven	H8	8,160	Proven

Based on the results of the R-Square, it can be seen that the constructs of Academics, Non-academic, and Student Satisfaction have values that fall into the strong category in predicting, while Student Loyalty which is included in the moderate category in predicting. This shows that most variations in the construct of Student Loyalty are explained by 37% factors not included in the model. The next stage is the significance test using the significance level of 5% using the bootstrapping procedure where t-value > 1.96 states that the hypothesis is proven.

The modification model results outperform the initial model. The interesting part is that tangibility is deleted both in initial and modification model because all indicators of tangibility have a loading factor < 0.7. Thus, it can be said that IE UPH students do not consider the Tangibility construct such as the appearance of buildings, facilities, and the university environment as a factor that influences the quality of services. In addition, the overall value of R-Square in the modification model tends to be the same or higher than the initial model. Thus, the evaluation results in the modification model show that the model predicts constructs better than the initial model. The final model shows that HedQual which consists of Academic and Non-academic, HedQual together with Reputation and Learning Outcome has a significant and positive effect on Student Satisfaction, and Student Satisfaction has a significant and positive effect on Student Loyalty.

The consensus is carried out to know what the most important factor in service quality is for IE UPH students. The consensus is a general agreement from the respondents and calculated using the equation from Tastle and Wierman [20]. The consensus ranges between 0-1 where the number closest to one shows the highest level of agreement of the respondents. The calculation results for the item based on the Level of Importance from the questionnaire are listed in Table 6 for the top five highest scores.

5. Conclusion

IE UPH service quality is valued by its students from Academic and Non-academic qualities. HedQual, Reputation and Learning Outcome affect positively to Student Satisfaction and thus also affect positively to Student Loyalty. The consensus shows that IE UPH students consider lecturers as the most important part in service quality, shown by the positions of A15,A10, A9,A16,A14 in the first, second, fifth, sixth and ninth in level of importance. The most important service quality for IE student from the consensus calculation is the attitudes and sincere interests of lecturers in helping the students. The reputation is also considered important, as shown by R2, R1 and R4 at the fourth, seventh and eight positions in the level of importance. Lecturers and Reputation should become major concerns to improve for IE Department to give the best service quality for their students.

Table 6. Consensus-based on level of importance.

No	Code	Questions	Consensus
1	A15	When students have a problem, the lecturer shows genuine interest in helping to solve it	0.7650
2	A10	The lecturer showed a positive attitude towards students	0.7648
3	L3	Students get the skills to use software (statistics & techniques)	0.7582
4	R2	IE UPH has a high reputation	0.7530
5	A9	The lecturer handles students in a caring and polite manner	0.7497
6	A16	Availability of lecturer for guidance and advice	0.7485
7	R1	IE UPH has a professional image	0.7473
8	R4	IE UPH has good accreditation	0.7467
9	A14	Lecturers understand the academic needs of students	0.7459
10	N12	Students was treat equally and respectfully by staff	0.7431

Acknowledgment
The writers express their gratitude to the University of Pelita Harapan that fully supports the research and permission to publish the result of this research.

References

[1] Sjukur F U 2010 *Satisfaction Analysis of Industrial Engineering Students to UPH Industrial Engineering Department Tangerang* (Tangerang: University of Pelita Harapan)
[2] Tampi C A, Silalahi R V, and Panjaitan A M 2015 *Integration of SERVQUAL and KANO Model Into Quality Function Deployment In An Attempt to Improve Service Quality at Industrial Engineering UPH* (Tangerang: University of Pelita Harapan)
[3] Hartono N 2008 *Development of Servqual Model for High School and its Application with Integration of Servqual, Kano Model, and QFD, Case Study in "Z" High School Bandung* (Bandung: Institute Technology of Bandung)
[4] Hartono N and Laurence 2017 *Factors That Influence Students in Choosing Industrial Engineering Major at University of Pelita Harapan* (Tangerang: University of Pelita Harapan)
[5] Parasuraman A, Leonard B, and Zeithaml V A 1988 SERVQUAL: A Multiple-Item Scale for Measuring Consumer Perceptions of Service Quality *J. of Retailing* **64** p 12–40
[6] Grönross C 2000 *Service Management and Marketing: A Customer Relationship Management Approach* (England: John Wiley and Sons)
[7] Owlia M S and Aspinwall E M 1998 A Framework for Measuring Quality in Engineering Education *Total Quality Management* **9** p 501–518
[8] Abdullah F 2005 HEdPERF versus SERVPERF: The Quest for Ideal Measuring Instrument of Service Quality in Higher Education Sector *Quality Assurance in Education* **13** p 305–328
[9] Hartono J 2005 *Measurement and Analysis Consumer Satisfaction Level toward education services, case study at BPK Penabur 1 High School Bandung* (Bandung: Maranatha Christian University)
[10] Mahapatra S S and Khan M S 2007 A Neural Network Approach for Assessing Quality in Technical Education: An Empirical Study *Int. J. of Productivity and Quality Management* **2** p 287–306
[11] Annamdevula S and Bellamkonda R S 2012 Development of HiEdQUAL for Measuring Service Quality in Indian Higher Education Sector *Int. J. of Innovation, Management and Technology* **3** p 412-416
[12] Gupta P and Kaushik N 2018 Dimensions of Service Quality in Higher Education - Critical Review (Students' Perspective) *Int. J. of Educational Management* **32** p 580-605

[13] Verma S and Prasad R K 2017 The MEQUAL Scale: Measure of Service Quality in Management Education *Int. J. of Comparative Education and Development* **19** p 193–206

[14] Mattah P A D, Kwarteng A J, and Mensah J 2018 Indicators of Service Quality and Satisfaction among Graduating Students of a Higher Education Institution (HEI) in Ghana *Higher Education Evaluation and Development* **12** p 36–52

[15] Sarstedt M, Hair J F, Ringle C M, Thiele K O, and Gudergan S P 2016 Estimation Issues with PLS and CBSEM: Where the bias lies! *J. of Business Research* **69** p 3998-4010

[16] Henseler J, Hubona G, and Ray P A 2015 Using PLS path modelling in new technology research: updated guidelines *Industrial Management & Data Systems* **116** p 2-20

[17] Henseler J, Dijkstra T K, Sarstedt M, Ringle C M, Diamantopoulos A, Straub D W, Ketchen Jr D J, Hair J F, Hult G T M, and Calantone R J 2014 Common Beliefs and Reality About PLS: Comments on Ronkko and Evermann (2013) *Organizational Research Methods* **17** p 182-209

[18] Eryilmaz M 2016 *Impact of Higher Education Institution's Reputation and Image on Students' Intention of Prospective Collaborations*. Proc. of ISERD Int. Conf. (Tokyo) p 8-12

[19] Barusman A R P 2014 *Student Satisfaction as a Mediating Variable between Reputation, Image and Student Loyalty* Proc. ITMAR-2014 p 1414–1436

[20] Tastle W J and Wierman M J 2007 Consensus and dissention: A measure of ordinal dispertion. *International Journal of Approximate Reasoning* **45** p 531-545

IOP Conf. Series: Materials Science and Engineering **703** (2019) 012028 doi:10.1088/1757-899X/703/1/012028

Expertise-based decision makers' importance weights for solving group decision making problems under fuzzy preference relations

E Herowati

Department of Industrial Engineering, University of Surabaya, Raya Kalirungkut, Surabaya, 60293, Indonesia

E-mail : evy@staff.ubaya.ac.id

Abstract. The quality of a decision is influenced by the level of expertise of the Decision Makers (DMs). In Group Decision Making, alternatives' scores are obtained by integrating the DMs opinions and the importance weights of the DMs greatly affect the resulted value. Expertise level is defined as the ability to differentiate consistently and expressed as the CWS-Index, a ratio between the Discrimination and Inconsistency. The DMs give their evaluations in pairwise comparison of Fuzzy Preference Relations (FPR) and the additivity property of FPR generates the estimators needed to get the CWS-Indexes and the expertise-based ranking of DMs. The weights of the DMs are obtained by using Induced Ordered Weighted Averaging (IOWA) operator and Basic Unit Monotonic Increasing functions and the resulted weights are used to evaluate the available alternatives to get the best one based on Fuzzy Majority and IOWA operators. This paper proposes an expertise-based weight allocation method for DMs and a numerical example is discussed to illustrate this expertise-based model to get the best alternative and it is concluded that the higher the DMs' expertise level, the higher his/her weight would be, and these weights affect the alternatives' score and the rank of the alternatives.

1. Introduction

Experts have the ability to think differently and this ability underlies the high quality of their decision making [1]. In Group Decision Making (GDM), a group of Decision Makers (DMs) with different expertise, interest and background have to evaluate several alternatives to be chosen in a selection process. The value of each alternative is obtained from group assessment that synthesizes all the DMs' individual values by aggregating their values mathematically [2]. Since the DMs have different expertise, knowledge and background, their individual values should be aggregated by using different importance weights. Assigning inappropriate importance weights of DMs can lead to incorrect results in obtaining the final selection [3], therefore studies related to GDM and DMs' Importance weight are increasingly developing [4] .

Studies related to the DMs' importance weight have been carried out previously. In the beginning, the DMs' importance weight were determined subjectively based on people's views on the DMs, for example an influential person determining the DMS weight directly [5] and a group of assessors who

IOP Conf. Series: Materials Science and Engineering **703** (2019) 012028 doi:10.1088/1757-899X/703/1/012028

were considered experts assessing the DMs to obtain their importance weights [6]. The subjectivity and difficulty of the assessors to evaluate DMs correctly has the potential to lead to biased decisions.

The DMs' importance weight can be determined objectively by observing the results of their assessment on a set of alternatives. The DMs' weight can be determined based on group consensus [7], in the sense that DM whose opinion increases consensus is given a larger weight. In addition, DM who assigned larger weight is the DM with a smaller deviation or smaller distance [8] from group composite value. The consistency of DM evaluations towards alternatives was also used to determine the weight of the related DM in several researches [9]. The DMS' weights that are obtained objectively are considered better than the DMs' weights that are determined subjectively, but there are still some decision biases that need to be considered, for example opinion that are approved by most people are not necessarily the right opinion and all of the above studies have not considered the expertise level of the DMs where the more expert a DM is, the better the quality of the decision will be.

There are several researches related to expertise level, for example expertise-based DM ranking [10-12] and expertise-based DMs' importance weights [13]. In these studies, DMs assessed the alternatives in the form of Fuzzy Preference Relations (FPR) and the DMs' weights obtained are objective and based on their expertise, but these weights should be revised since those researches have not considered reciprocal relations of FPR since an expert can certainly compare the value of two alternatives smartly. If he judges that the first alternative is better than the second alternative by some degree, then the reciprocal relation ensures that there is no need for him to re-evaluate by comparing the second alternative to the first one. This study proposes the revised version of expertise-based DMs' importance weights considering reciprocal relations and illustrated by examples used in prior expertise-based researches.

2. Literature reviews

2.1. Expertise level of the decision makers

An expert in a specific field is someone who has a strong background in this field. Shanteau et al [14] defines expertise as the ability to differentiate consistently, expressed as CWS-Index as shown in (1).

$$\text{CWS-Index} = \frac{\text{Discrimination}}{\text{Inconsistensy}} = \frac{\text{variance of different alternatives' values}}{\text{variance of the same alternative's values}} = \frac{\dfrac{\sum_{j=1}^{n} r(M_{j} - GM)^2}{n-1}}{\dfrac{\sum_{j=1}^{n}\sum_{i=1}^{r} (M_{ij} - M_{j})^2}{n(r-1)}} \qquad (1)$$

Where:

r	:	The number of replications
$M_{.}$	:	The average of individual valuesfor case-j
GM	:	The grandmean of all individual values
n	:	The number of different cases
$M_{..}$	:	The individual value for replication-i case-j

The evaluated DMs must repeat their assessment to obtain a measure of inconsistency.

2.2. The FPR and additive consistency property

Chiclana et al [15] proposed GDM model where a group of m experts $E = \{e_1, e_2, ...e_m\}$, $m \geq 2$ evaluate a set of n alternatives $X = \{x_1, x_2, ..x_n\}$, $n \geq 2$in FPR $P \subset XxX$ having a membership function $\mu_p : XxX \rightarrow [0,1]$ and represented by means of the n x n matrix $P = (p_{ij})$. $p_{ij}$is the preference degree of $x_i$over x_j. $p_{ij} > ½$ means $x_i$is preferred to x_j, $p_{ij} = ½$ means indifference between $x_i$and x_j.

The FPR P has a reciprocal relation as shown in (2) and the matrix P as shown in (3) and Additive Consistency (AC) property of FPR among three alternatives x_i, x_j, $x_k (i,j,k = 1,2,\dots,n)$ yields 3 formulas (εp_{ik}^{jg}, g=1,2,3) to estimate p_{ik} as in (4), (5) and (6).

$$p_{ij} + p_{ji} = 1 \quad (2)$$

$$P = \begin{bmatrix} 0.5 & p_{12} & p_{13} & p_{14} \\ 1-p_{12} & 0.5 & p_{23} & p_{24} \\ 1-p_{13} & 1-p_{23} & 0.5 & p_{34} \\ 1-p_{14} & 1-p_{24} & 1-p_{34} & 0.5 \end{bmatrix} \quad (3)$$

$$\varepsilon p_{ik}^{j1} = p_{ij} + p_{jk} - \tfrac{1}{2},\ j \neq i,k \quad (4)$$

$$\varepsilon p_{ik}^{j2} = p_{jk} - p_{ji} + \tfrac{1}{2},\ j \neq i,k \quad (5)$$

$$\varepsilon p_{ik}^{j3} = p_{ij} - p_{kj} + \tfrac{1}{2},\ j \neq i,k \quad (6)$$

For every element of the matrix FPR p_{ij}, there are 3x(n-2) replications (3 formulas and $j \neq i,k$) and has the range of [-0.5, 1.5]. If the reciprocal relation is considered, the 3 formulas are just one formula and the number of replications is reduced to (n-2). To keep the range of estimated value of p_{ik} between [0,1], all of the element p_{ik} should be transformed using $f(x) = \frac{x+a}{1+2a}$ and the range changes from $[-a,\ 1+a]$, $a > 0$ to $[0,1]$ [15].

2.3. *The induced OWA operator*

Yager [16] proposed an n-dimensional OWA Operator $F: I^n \rightarrow I$ where $F_w(a_1,a_2,\dots,a_n) = \sum_{j=1}^{n} w_j b_j$, b_j is the jth largest element of input arguments $(a_1,a_2,\dots,a_n)$, w_j is the ordered weight, and $w_j \geq 0$, $\sum_{j=1}^{n} w_j = 1$. The Induced OWA Operator is an OWA Operator using the order inducing variable as the ordering of the input arguments. An n dimensional Induced OWA Operator $F: I^n \rightarrow I$ where $F_w(\langle u_1,a_1\rangle,\langle u_2,a_2\rangle,\dots,\langle u_n,a_n\rangle) = \sum_{j=1}^{n} w_j b_j$, u_i is the order inducing variable, a_i is the input argument, w_j is the order weights and $w_j \geq 0$, $\sum_{j=1}^{n} w_j = 1$, b_j is the input argument with the j-th largest the order inducing variable value. The OWA weights can be obtained by using Basic Unit Monotonic Function (BUM) $Q:[0,1] \rightarrow [0,1]$, $Q(0)=0; Q(1)=1; Q(x) \geq Q(y)\ if\ x \geq y$ as follows [16]:

$$w_j = Q(R_j) - Q(R_{j-1}) \quad (7)$$

where $R_j > R_{j-1}$, $w_j \geq 0$, $\sum_{j=1}^{n} w_j = 1$. The accumulative value of the DMs' weights is associated with the BUM as in Figure1, Hence the DMs weights will never be negative ($w_j \geq 0$) with the total weights of one.

2.4. *Quantifier guided dominance degree*

Quantifier Guided Dominance Degree (QGDD) quantify the dominance of alternative-x_i over all of the others in a fuzzy majority as in (8).

$$QGDD_i = F_q(p_{ij}^{Group}, j = 1,2,\dots n, j \neq i) \quad (8)$$

The ranking of alternatives is determined based on ranking of their QGDD.

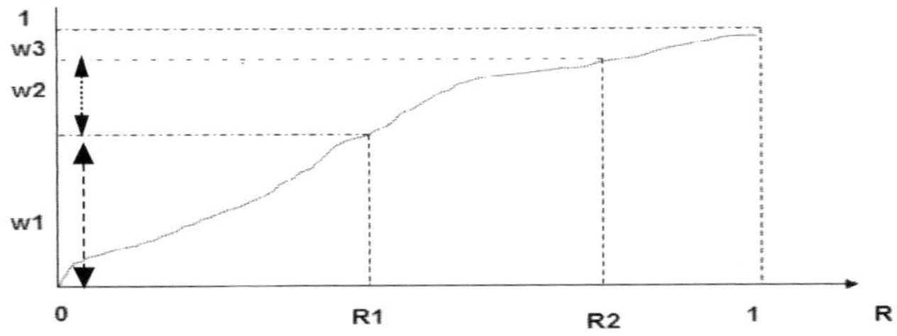

Figure 1. The OWA Weights from BUM

3. The proposed method

The steps used to select alternatives proposed in this study are as follows:

1) Combine Shanteau's expertise level and the AC property of FPR. The DMs' assessment must be given in FPR and we do not need repetition to measure consistency in expertise, and we get the CWS-Index for FPR in equation (9) and yield the CWS-Indexes and DMs' rank [12].

$$\text{CWS-Index} = \frac{\text{Variance of different 'pairwise comparisons between two alternatives'}}{\text{Variance of the same 'pairwise comparisons between two alternative's'}} \quad (9)$$

2) Combine the ordered CWS-Indexes and BUM, the accumulated logarithmic ordered CWS-Indexes are used as the horizontal axis of the BUM. Herowati et al [13] used BUM $Q(R)=R^n$, n ≤ 1 and this research revises the BUM to be a straight line, Q(R)=R to ensure that the DMs with the same expertise level will assigned the same weight and there are only (n-2) replications in this study instead of 3(n-2) in [13] since the previous study did not consider the reciprocal relation as in equation (2).
3) Aggregate the individual FPR of all DMs by using the weights generated from the previous step to form an FPR group and this value is used in the selection process using the concepts of QGDD.

4. Illustrative example

A numerical example is provided to illustrate the proposed method. Suppose 5 DMs $E=\{e_1,e_2,e_3,e_4,e_5\}$ assess a set of four alternatives $X=\{x_1,x_2,x_3,x_4\}$in form of Matrix FPR P as follows:

$$P_1=\begin{bmatrix}0.50&0.40&0.40&0.40\\0.60&0.50&0.40&0.70\\0.60&0.60&0.50&0.70\\0.60&0.30&0.30&0.50\end{bmatrix} P_2=\begin{bmatrix}0.50&0.30&0.40&0.45\\0.70&0.50&0.65&0.60\\0.60&0.35&0.50&0.45\\0.55&0.40&0.55&0.50\end{bmatrix} P_3=\begin{bmatrix}0.50&0.90&0.60&0.70\\0.10&0.50&0.30&0.40\\0.40&0.70&0.50&0.60\\0.30&0.60&0.40&0.50\end{bmatrix} P_4=\begin{bmatrix}0.50&0.40&0.80&0.70\\0.60&0.50&0.60&0.35\\0.20&0.40&0.50&0.10\\0.30&0.65&0.90&0.50\end{bmatrix} P_5=\begin{bmatrix}0.50&0.70&0.80&0.60\\0.30&0.50&0.65&0.55\\0.20&0.35&0.50&0.40\\0.40&0.45&0.60&0.50\end{bmatrix}$$

1) Estimate each element Matrix P_4 by using one of the formulas 1, 2 or 3 to generate 2 estimated value as shown in column 3 and 4 of Table 1. For each element there are 3 values (r=3), i.e. the original value and 2 estimated values. Transform the values outside the range of [0,1] by using transformation function $f(x)=\frac{x+0.1}{1+2x0.1}$ as shown in column 5,6,7 of Table 1.

Discrimination= $\frac{\sum_{j=1}^{n} r(M_j-GM)^2}{n-1}=\frac{1.01042}{(12-1)}=0.09186$;Inconsistency = $\frac{\sum_{j=1}^{n}\sum_{i=1}^{r}(M_{ij}-M_j)^2}{n(r-1)}=\frac{0.93750}{12x(3-1)}=0.03906$

CWS-Index for DM – 4 = $\frac{0.09186}{0.03906}$= 2.3518

CWS-Indexes for DM-1, DM-2, DM-3, DM-4 and DM-5 subsequently are 3.8395, 22.039, 52.1267, 2.3518 and 18.757 and the DMs rank is DM-3, DM-2, DM-5, DM-1 and DM-4.

2) The rank of the DMs obtained in step 1) and the CWS-Indexes are ordered as in Table 2, then the ordered CWS-Indexes were transformed by using the logarithmic function, accumulated and normalized. The normalized (Accumulated(Log(CWS-Index))) is the horizontal axis of the straight line BUM and we get the DMs' importance weights as in the last row of Table 2.
3) Aggregate the individual FPR of all DMs by using the importance weights to form a FPR group and calculate the QGDD by using fuzzy majority function 'Most' which has the weights to compare with the other three alternatives (1/15, 10/15, 4/15).

$$P_{Grup} = \begin{bmatrix} 0{,}500 & 0{,}609 & 0{,}589 & 0.579 \\ 0{,}391 & 0{,}500 & 0{,}505 & 0{,}517 \\ 0{,}411 & 0{,}495 & 0{,}500 & 0{,}490 \\ 0{,}421 & 0{,}483 & 0{,}510 & 0{,}500 \end{bmatrix}$$

p_{12}calculation to illustrate the aggregation process from individual FPRs to group FPR:

$$p_{12} = 0{,}105x0{,}4 + 0{,}2539x0{,}3 + 0{,}3247x0{,}9 + 0{,}0702x0{,}6 + 0{,}2407x0{,}7 = 0{,}609$$

The first row of the FPR group is the group value of the first alternative, the second row are the value for the second alternative. The QGDD for all the alternatives are calculated by using fuzzy majority 'Most' with the weight (1/15, 10/15, 4/15) and the element matrix values have to be ordered before the calculation. For example QGDD for alternative-2 are as follows:

$$QGDD_2 = F_q(p_{2j}^{Grup}, j = 1,2,\ldots,n, j \neq 2) = \tfrac{1}{15}x0{,}517 + \tfrac{10}{15}x0{,}505 + \tfrac{4}{15}x0{,}391 = 0{,}4755$$

Table 1. Example of CWS-Index calculation for DM-4

Before Transformation				After Transformation					
Element Matrix	Original Value	Estimated values		Original Value	Estimated values		M_j	$r(M_j - GM)^2$	$\sum^{r}$ [illegible]
p_{12}	0.40	0.70	0.85	0.4167	0.6667	0.7917	0.6250	0.0469	0.0729
p_{13}	0.80	0.50	**1.10**	0.7500	0.5000	1.0000	0.7500	0.1875	0.1250
p_{14}	0.70	0.25	0.40	0.6667	0.2917	0.4167	0.4583	0.0052	0.0729
p_{21}	0.60	0.15	0.30	0.5833	0.2083	0.3333	0.3750	0.0469	0.0729
p_{23}	0.60	0.75	0.90	0.5833	0.7083	0.8333	0.7083	0.1302	0.0313
p_{24}	0.35	0.80	0.20	0.3750	0.7500	0.2500	0.4583	0.0052	0.1354
p_{31}	0.20	**- 0.10**	0.50	0.2500	0	0.5000	0.2500	0.1875	0.1250
p_{32}	0.40	0.10	0.25	0.4167	0.2917	0.1667	0.2917	0.1302	0.0313
p_{34}	0.10	0. 25	0.40	0.1667	0.2917	0.4167	0.2917	0.1302	0.0313
p_{41}	0.30	0.60	0.75	0.3333	0.5833	0.7083	0.5417	0.0052	0.0729
p_{42}	0.65	0.80	0.20	0.6250	0.7500	9,2500	0.5417	0.0052	0.1354
p_{43}	0.90	0.75	0.60	0.8333	0.7083	0.5833	0.7083	0.1302	0.0313
				TOTAL				1.0104	0.9375

Table 2. The DMs' importance weights calculation

	DM-3	DM-2	DM-5	DM-1	DM-4
CWS- Index	52,1267	22,0390	18,7570	3,8395	2,3518
Log(CWS-Index)	1,7171	1,3432	1,2732	0,5843	0,3714
Accumulated(Log(CWS-Index))	1,7171	3,0603	4,3334	4,9177	5,2891
R=Normalized(Accumulated(Log(CWS-Index)))	0,3246	0,5786	0,8193	0,9298	1,0000
$Q(R) = R^{\alpha} = R$	0,3246	0,5786	0,8193	0,9298	1,0000
DMs' Importance Weights	0,3246	0,2539	0,2407	0,1105	0,0702

Table 3. Comparison of the results of the selection process

	Expertise-based DMs' weights (0,11, 0,25, 0,33, 0,07, 0,24)		DMs' Average Weight (1/5 , 1/5, 1/5, 1/5 , 1/5, 1/5)		Inverse Expertise-based DMs' weights (0,25, 0,11, 0,07, 0,33, 0,24)	
	QGDD	Rank	QGDD	Rank	QGDD	Rank
Alternative -1	0,588	1	0,564	1	0,556	1
Alternative -2	0,475	2	0,504	2	0,516	2
Alternative -3	0,469	3	0,439	4	0,392	4
Alternative -4	0,468	4	0,471	3	0,475	3

Table 3 shown that expertise-based DMs' importance weights yields alternatives ranking 1, 2, 3, 4. If the aggregation process uses DMS' average weights or for all the DMs, then the alternatives ranking are 1, 2, 4 and 3. The use of average as DMs' importance weights does not have any effect if we want to choose one best alternative. But if we want to eliminate one unwanted alternative, the results will be different from results based on expertise. The third case, inverse expertise-based uses of DMs' importance weights in reverse manner where the more expert a DM is, the smaller the weight and yield the same result with the second case. In this illustrative example, the expertise of the DMs are assumed to be unknown and not determined by the people's subjective view of these DMs. Expertise from DMS are determined based on their assessment of alternatives by using the Shanteau's concept, experts are those who can distinguish alternatives consistently and expressed by the CWS Indexes and the integration of DMs opinion using expertise-based weightsyield different rank of alternatives and it is expected to produce a better decision.

5. Conclusion

This paper revises expertise-based DMs' importance weights by considering reciprocal relations since it is more realistic when assessed DMs are carrying out their assessment task.The application of expertise-based DMs weights in this paper shows that DMs weights affects the alternatives' score and the alternatives' rank and it is expected that expertise-based DMs' weights leads to a better decisions.

References

[1] Malhotra V, Lee M D and Khurana A 2007 Domain experts influence decision quality: Towards a robust method for their identification *J. Pet. Sci. Eng.* **57** 181–94.

[2] Cooke RM 1991 *Experts in uncertainty opinion and subjective probability in science* (New York/Oxford: Oxford University Press)

[3] Emrah K and Kabak O 2019 Deriving decision makers weights in group decision making: An overview of objective methods *Inf. Fusion* **49** 146–60.

[4] Kabak O and Ervural B 2017 Multiple attribute group decision making: a generic conceptual framework and a classification scheme *Knowledge-Based Syst.* **123** 13–30.

[5] Herrera F, Herrera-Viedma E and Verdegay J L 1998 Choice processes for non-homogeneous group decision making in linguistic setting *Fuzzy Sets Syst.***94** 287–308.

[6] Ramanathan R and Ganesh L S 1994 Group preference aggregation methods employed in AHP: An evaluation and an intrinsic process for deriving members' weightages *Eur. J. Oper. Res.***79** 249–65.

[7] Kacprzyk J, Fedrizzi M, and Nurmi H 1992 Group decision making and consensus under fuzzy preferences and fuzzy majority *Fuzzy Sets Syst.***49** 21–31.

[8] Chuan Y 2017 Entropy-based weights on decision makers in group decision making setting with hybrid preference representations *Appl. Soft Comput.* **60** 737–49.

[9] Alonso S, Cabrerizo F, Chiclana F, Herrera F, and Herrera-Viedma E 2009 Group decision making with incomplete fuzzy linguistic preference relations *Int. J. Intell. Syst.* **24** 201–22.

[10] Herowati E, Ciptomulyono U, Parung J, and Suparno 2013 Competent-based experts ranking at fuzzy preference relations on alternatives. *Proc. Int. Conf. on Industrial Engineering and Service Science* (Surabaya:Sepuluh Nopember Institute of Technology, Indonesia)

[11] Herowati E, Ciptomulyono U, Parung J, and Suparno 2014 Expertise-based experts ranking at multiplicative preference relations on alternatives. *Proc. Int. Conf.Asia Pacific Industrial Engineering & Management Society* (Jeju Island: South Korea)

[12] Herowati E, Ciptomulyono U, Parung J, and Suparno 2017 Expertise-based ranking of experts: An assessment level approach *Fuzzy Sets Syst.* **315** 44–56.

[13] Herowati E, Ciptomulyono U, Parung J, and Suparno 2014 Expertise-based experts importance weights in adverse judgment *ARPN J. Eng. Appl. Sci.* **9** 1428-35.

[14] Shanteau J, Weiss D J, Thomas R P, and Pounds J C 2002 Performance-based assesmentof expertise: How to decide if someone is an expert ? *Eur. J. Oper. Res.***136** 253–63.

[15] Chiclana F, Herrera F, and Herrera-Viedma E 1998 Integrating three representation models in fuzzy multipurpose decision making based on fuzzy preference relations *Fuzzy Sets Syst.* **97** 33–48.

[16] Yager R R 1988 On ordered weighted averaging aggregation operators in multi-criteria decision making *IEEE Trans. Syst. Man Cybern. Part B Cybern.* **18** 183–90.

IOP Conf. Series: Materials Science and Engineering **703** (2019) 012029 doi:10.1088/1757-899X/703/1/012029

Organic-inorganic nanocomposite membranes for molecular separation and bioapplications

J Hou[1,2], P D Sutrisna[3], L Li[4] and V Chen[1]

[1]School of Chemical Engineering, University of Queensland
[2]Department of Materials Science and Metallurgy, University of Cambridge
[3]School of Chemical Engineering, University of Surabaya, Indonesia
[4]School of Chemical Engineering, University of New South Wales, Australia

E-mail: Jingwei.hou@uq.edu.au

Abstract. Novel porous materials like metal organic frameworks have shown good promise in catalysis, separation, sensing and adsorption. Compared with the conventional polymeric membranes, metal organic framework membranes usually exhibit higher separation efficiency. However, there are still many challenges which need to be addressed, like the interfacial compatibility within the mixed matrix membrane, defective engineering and stability of the framework stability. We will discuss our strategies to solve these problems towards better membranes. On top of the conventional separation membranes, metal organic framework materials can be applied to construct the artificial cellular membranes for bioentities and even living cells. The porous crystal provides exoskeleton to protect the soft biomolecules and their combination can enable a series bioapplications like catalysis, drug delivery and drug release.

1. Introduction

Metal organic framework materials have been recognized as one of the most promising candidates for separation and energy applications. It has a large surface area, controllable porous structure and chemical properties, and surface functionalization capability. Currently, the total structure of metal organic framework materials has reached over 60,000 of different types. However, its application, especially as a separation device, is still limited due to its difficulty in assembly into practical devices without introducing extra defects, which is extremely important for the membrane separation process as the presence of any defects can impose significant negative effect on the separation efficiency. The selectivity can be compromised by the microdefects on the membrane. In this contribution, we would like to introduce some of our recent contribution in this line of research, by regulating the interfacial compatibility between metal organic frameworks and polymeric, continuous supports, to fabricate better membrane devices toward their potential application in water treatment, gas separation, and energy-related applications [1, 2].

2. Compositing zeolite imidazolate framework-8 with polymeric supports

Zeolite imidazolate framework is usually constructed by the Zn, Co metal ions, interconnected by imidazolate ligands. The framework structure is very similar to that of the inorganic silica-based porous zeolite material. The pore size of the zeolite imidazolate framework can be regulated at the size range of gas molecules, e.g. CO_2 and N_2, thus it has been regarded as a promising candidate for molecular separation for gaseous mixtures. One of the early attempts from us, is to blend one type of zeolite imidazolate framework (ZIF-8) with the polymeric support (Figure 1) [3, 4], and the resultant membranes were tested for CO_2 and N_2 separation. Though the loss of selectivity was observed after the incorporation of ZIF-8 (i.e. the membrane has the lower capability to separate CO_2 from N_2), the membrane exhibited a significantly improved permanence (gas flow rate). In addition, one of the common problems for the polymeric membrane is its instability under the industrial-related operational conditions, like high temperature and elevated pressure. The inherent flexible, porous polymeric chain can experience compaction and rigidification, leading to the loss of separation efficiency. What we discovered through our research is that due to the satisfactory interfacial compatibility between ZIF-8 nanocrystals and their surrounding polymeric matrix, the formed hydrogen and pi-pi interaction can help stabilize the polymeric membrane structure, even at challenging operational conditions.

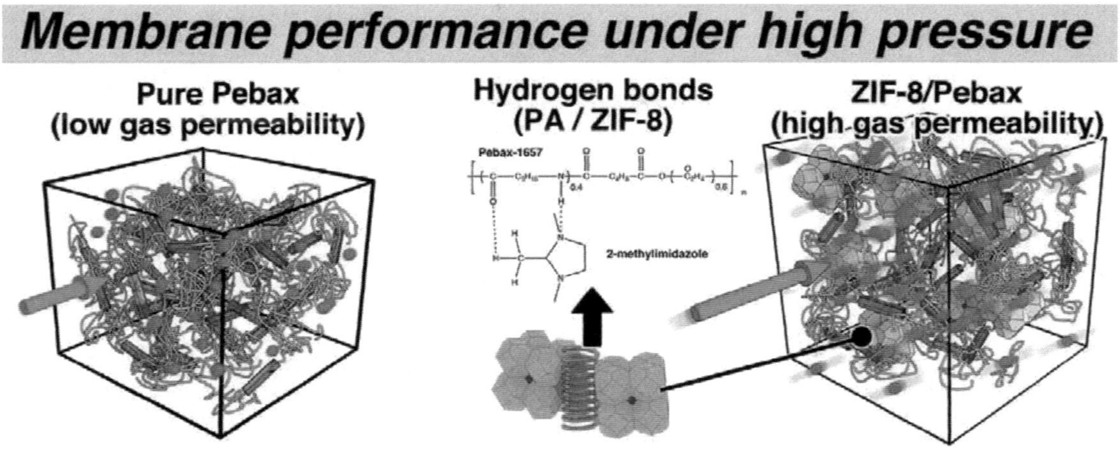

Figure 1. Schematic diagram of the ZIF-8/polymer mixed matrix membrane, and the mechanism of their improved operational stability under an elevated pressure condition.[3]

Since the presence of high porosity of the ZIF-8 nano-crystals, it is ideal to assembly the nano-crystal to form a coherent, continuous film to realize efficient separation. However, the as-synthesized ZIF-8 materials are usually in the form of discrete powder form, and it is difficult to assemble it into practical devices: the process ability of the nano-crystal is low, as it does not dissolve in solvents without compromising its porous structure. Recent advance in metal organic framework hot pressing can be inapplicable for membrane fabrication as well due to the difficulty to form a thin layer for separation. Ideally, a coherent film can be formed via *in-situ* heterogeneous crystallization and crystal growth of ZIF-8 of top of polymeric support. However, due to the limited compatibility between the crystal phase and polymeric phase, the heterogeneous nucleation efficiency can be limited, restricting the formation of a high-quality film. We proposed a concept to solve this problem, by introducing an interfacial mediate layer, composed of TiO_2 nanoparticles, to promote the growth and formation of ZIF-8 on a polymeric support. The polymeric porous support was firstly coated with a layer of nano-sized TiO_2 nanoparticles, followed by surface functionalization of amine groups to promote the heterogeneous nucleation of ZIF-8. The fabricated composite membrane has a thickness of 400 nm for the ZIF-8 layer, with its good alignment of the crystal structure with the pristine ZIF-8 powder (as suggested by the powder XRD results). Then the coherent, macroscope defect-free ZIF-8 layer showed satisfactory molecular sieving behaviour, exhibiting highly efficient H_2 and CO_2 separation efficiency. The practical membrane performance has good agreement with the theoretical estimations of the pure ZIF-8 layer, suggesting the separation is achieved via the ZIF-8 layer [5].

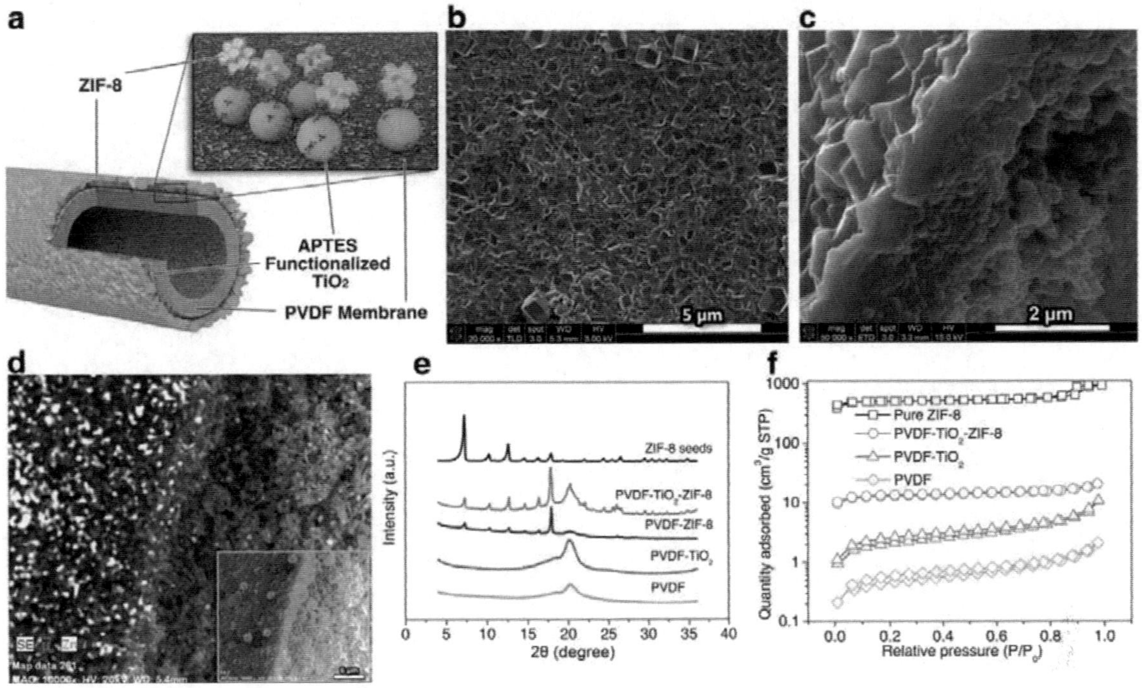

Figure 2. Formation of the coherent ZIF-8 coating layer on polymeric membrane surface. (a) Structure of the composite membrane, composed of polymeric PVDF support, surface coated TiO_2 layer and ZIF-8 top coating layer. (b-c) Surface and cross-sectional SEM image for the ZIF-8 membranes. (d) Elemental mapping results of the composite membranes, showing the presence of Zn elements on the ZIF-8 layer. (e-f) Crystal structure analysis and gas adsorption profiles of the membranes subject to different treatment [5].

3. Interfacial engineering to improve the membrane performance

As stated above, the presence of micro defects for membrane can lead to the loss of separation efficiency. However, the defect engineering for metal organic frameworks has been attracting increasing attention as it can induce extra functions for the materials. For example, the loss of metal ion or organic ligands can generate under-coordinated metal clusters, which further promote the catalytic efficiency and adsorptive capacity. In order to incorporate the defective yet functional metal organic framework materials into the membrane, we specifically engineered a type of topologically engineered metal organic framework material, with the presence of surface mesoporous structures, on top of the inherent microporous framework (Figure 3). The resultant materials exhibited a significantly improved water uptake from brine, showing good potential as pervaporation desalination membranes. After subsequent blending of the engineered crystal into substrates, we managed to fabricate a series of composite membranes, showing an improved brine desalination performance: compared with benchmark counterparts, the composite pervaporation membrane exhibits a nearly 100 % improvement in terms of the water productivity, and it maintained nearly 100 % rejection of the salts. This membrane technology has been considered one of the most promising candidates for challenging brine (e.g. 300 g/L NaCl solution) treatment (Figure 4).

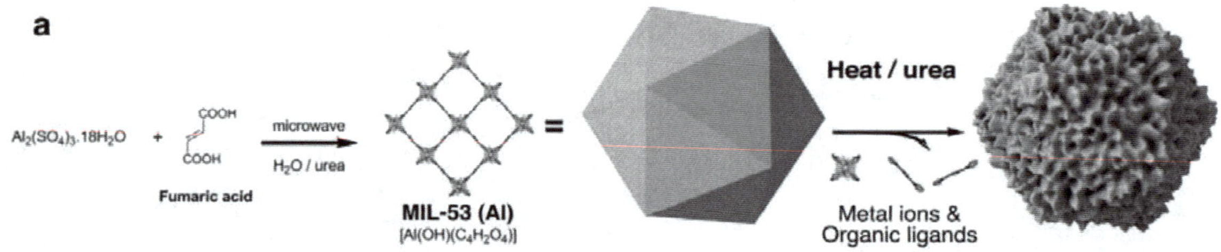

Figure 3. The fabrication of surface engineered metal organic framework materials with mesoporous structure [6].

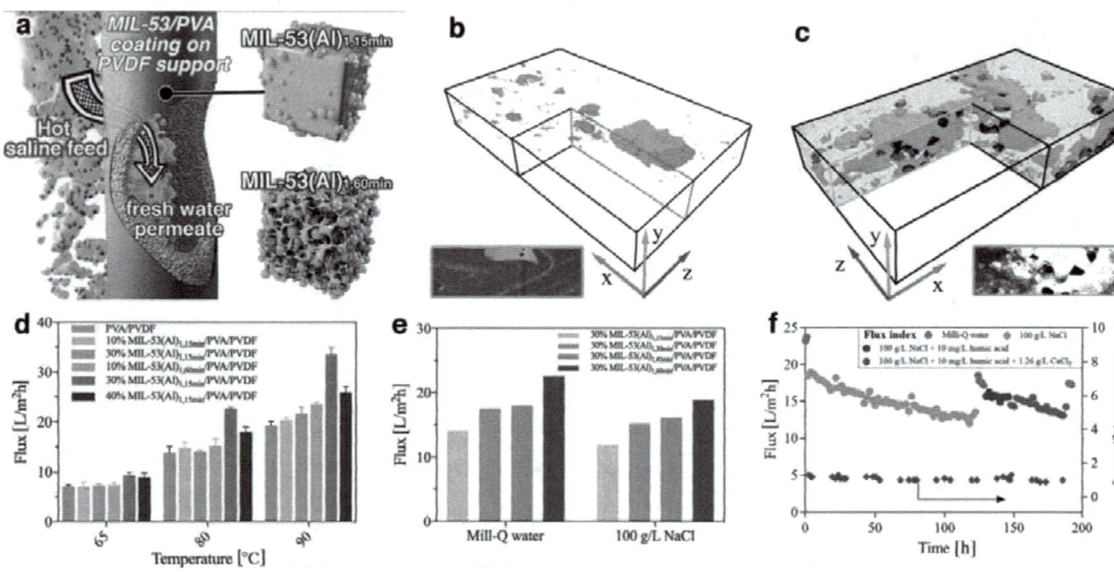

Figure 4. Nanocomposite membrane with surface engineered defective crystals for pervaporation desalination. (a) Schematic diagram of the composite membrane containing defective crystals. (b-c) FIB-SEM image of the composite membrane containing original and defective crystal materials. (d) Water productivity for the pervaporation membrane flux. (e) Membrane operational flux with milli-Q water and salty water. (f) Membrane long term operational flux with challenging brine as feed [6].

4. Bioapplications

Recently, we have demonstrated that the MOF materials can composite with biomolecules, providing an extra protective layer for bioentities against challenging environments. However, their practical application is still hampered by the limited chemical stability of MOF, due to the relatively weak coordinative bond between the metal ions and organic ligands. We have developed a bonding hybrid approach, to incorporate a carboxylate group into the pristine MOF structure, and the composite materials exhibit a significant improved stability against various challenging environments.

5. Conclusion

The development of material science provides numerous opportunities enabling the development of nanocomposite materials to realize advanced functions. This contribution summarizes the recent progress in this line of research in our group, through which we hope to shed light on the future development of a metal organic framework based nanocomposite membranes, and the development of the crystal materials at the same time.

References

[1] Ameloot R, Vermoortele F, Vanhove W, Roeffaers M B J, Sels B F and De Vos D E 2011 Interfacial synthesis of hollow metal–organic framework capsules demonstrating selective permeability *Nat. Chem.* **3** 382–7

[2] Banerjee R, Phan A, Wang B, Knobler C, Furukawa H, O'Keeffe M and Yaghi M 2008 High-Throughput Synthesis of Zeolitic Imidazolate Frameworks and Application to CO_2 Capture. *Sci.* **319** 939–43

[3] Sutrisna P D, Hou J, Li H, Zhang Y and Chen V 2017 Improved operational stability of Pebax-based gas separation membranes with ZIF-8: A comparative study of flat sheet and composite hollow fibre membranes *J. Membr. Sci.* **524** 266–79

[4] Sutrisna P D, Hou J, Zulkifli M Y, Li H, Zhang Y, Liang W, D'Alessandro D M and Chen V 2018 Surface functionalized UiO-66/ Pebax-based ultrathin composite hollow fiber gas separation membranes *J. Mater. Chem.* A **6** 918–31

[5] Hou J, Sutrisna P D, Zhang Y and Chen V 2016 *Formation of Ultrathin, Continuous Metal–Organic Framework Membranes on Flexible Polymer Substrates* Angew. Chem. Int. Ed. **55** 3947–51

[6] Liang W, Li L, Hou J, Shepherd N D, Bennett T D, D'Alessandro D M and Chen V 2018 Linking defects, hierarchical porosity generation and desalination performance in metal–organic frameworks *Chem. Sci.* **9** 3508–16

[7] Gao S, Hou J, Deng Z, Wang T, Beyer S, Buzanich A G, Richardson J J, Rawal A, Seidel R, Zulkifli M Y, Li W, Dennett T D, Cheetham A K, Liang K and Chen V 2019 improving the Acidic Stability of Zeolitic Imidazolate Frameworks by Biofunctional Molecules *Chem* **5** 1597-1608

Tensile Properties of Kenaf Fiber by Alkalinization Treatment: Effect of different concentration

Ismojo[1,2], K A Zahidah[1], E Yuanita[1], E Kustiyah[1,3] and M Chalid[1*]

[1]Department of Metallurgy and Materials Engineering, Universitas Indonesia, UI Depok New Campus, Depok 16424, West Java – Indonesia
[2]Department of Automotive Mechanical Engineering diploma III, Institut Teknologi Indonesia, ITI, South Tangerang, Serpong, Indonesia.
[3]Chemical Engineering, Universitas Bhayangkara Jakarta Raya 2nd Campus, Bekasi 17121, West Java – Indonesia

E-mail : chalid@metal.ui.ac.id

Abstract. The structure that plays an important role in the mechanical properties of natural fibers that will be used as reinforcement in polymer composites is crystalline cellulose. Chemical treatment is one method that is currently widely used for modifying the chemical composition of natural fibers to obtain fibers with high crystalline cellulose content. In this study, alkali treatment with variations in NaOH concentration was used to modify the surface of kenaf fibers. The effect of alkalinization treatment on the tensile properties of kenaf fibers has been carried out. The fiber is soaked in a sodium hydroxide solution with variations in concentrations of 4, 6, 10 and 15% by weight for 3 hours at room temperature and continued with the washing and drying process. Samples were then tested by tensile testing and characterized by using FTIR and FE-SEM instruments. Tensile testing of untreated and treated fiber is carried out 5 times for each variable. Based on the results of these experiments showed an increase in tensile strength of all treated fibers compared to untreated fibers and the highest maximum tensile strength was obtained from the treatment of alkalinization of fibers at a concentration of 10%. This result was also confirmed by the results of FTIR and FE-SEM characterization which have shown a decrease in amorphous content in kenaf fibers and fibrillated bundle fibers into fiber elements.

1. Introduction

At present, natural fiber is the center of attention as a substitute for reinforcing materials in polymer matrices, replacing synthetic fibers such as glass fiber, because it has several advantages, namely light, abundant, recyclable and environmentally friendly. Also, natural fiber is safe for process equipment, health and cheap [1,2]. Industry that considers the use of natural fiber as a reinforcing material replacing glass fiber and carbon is the transportation industry [2]. The use of natural fibers as reinforcing materials on synthetic polymer matrices has been applied to the automotive industry, but the application is still limited to the interior [3].

Among various types of natural fiber, Kenaf fiber (Hibiscus cannabis, L family Malvaceae) is one of the natural fibers that is a concern because of its rapid growth over a wide range of climatic conditions [4] and low cost [2]. In general, natural fiber consists of three main components, i.e. lignin, hemicellulose, and cellulose, which bind to each other in the form of bundles of fibers.

The mechanical properties of natural fibers are largely determined by the amount of cellulose content because cellulose has a semicrystalline structure while lignin and hemicellulose are amorphous. The higher cellulose content, mechanical properties of the fiber will be increased. Several options, including chemical treatments of natural fibers, are used to increase the crystalline portion of natural fibers and remove amorphous portions [4,5]

Alkali treatment, also known as mercerization, is one of the most widely used chemical treatments, to modify the surface of natural fibers because the process is simple and inexpensive. M. Suhairil Meon et al. Reported that the tensile strength of kenaf fiber tensile strength increases with increasing concentration of the NaOH solution [6]. B. Ahmed Amel et al. also researched kenaf fibers and reported that the tensile strength of kenaf fibers treated by 5wt% NaOH was higher than that of untreated fibers [7]. While Y Cao et al. shows the tensile strength of kenaf fibers with an alkaline treatment, 5% higher than without treatment and decreases at 10% and 15% [8].

The main objective of this research is thus to study the effect of alkali treatment on tensile properties and changes in chemical composition in kenaf fiber. Kenaf fiber obtained from the Bonorowo area, Jepara district, Central Java, Indonesia were treated with various concentration and tensile properties were investigated. To clarify changes in the resulting tensile properties, FT-IR and FE-SEM were used to investigate changes in compound composition and morphology of kenaf fibers from alkali treatments. This result is expected can be explored the potential of kenaf fibers to be used as a candidate reinforcing material in polymer composite.

2. Materials and experimental work

2.1. Materials

Kenaf fiber was obtained from the bast of the plant in Bonorowo, Jepara Regency, Central Java, Indonesia. The fibers were cleaned, washed and dried using a hairdryer. After dried, kenaf fibers were cut to 5 - 7 mm in length.

2.2. Chemical treatment of kenaf fibers

The kenaf fiber, which was prepared previously, was made in three different aqueous NaOH solutions (4, 6, 10 and 15 wt%) for 3 hours at ambient temperature. After being alkalized, kenaf fibers were washed several times until neutral with distilled water and dried using solar heat.

2.3. FT-IR measurement

Changes in the composition of compounds from treated fibers are compared with Fourier transform infrared spectra (Perkin-Elmer FTIR spectrometer)

2.4. Tensile strength of kenaf fibers.

The tensile testing of the alkaline treatment with different aqueous NaOH fibers was performed using UTM (Universal Tensile Machine) tensile testing machine by ASTM D1708-02 standard with a tensile speed of 100 mm/minute.

2.5. Morphology examination

The surface of kenaf fiber without treatment and alkali treatment results were examined by field scanning electron microscope (FE-SEM - FEI Inspect F50). All characterizations and testing were carried out in Metallurgy and Materials Engineering Department, Universitas Indonesia.

IOP Conf. Series: Materials Science and Engineering **703** (2019) 012030 doi:10.1088/1757-899X/703/1/012030

3. Results and discussion

3.1. Chemical composition

Fourier Transform Infrared spectroscopy was used to investigate changes in the composition of compounds in kenaf fibers after undergoing alkali treatment with different NaOH concentrations. FTIR spectra of untreated kenaf fibers and alkali treatment results are shown in Figure 1. From figure 1 (KV), it can be observed that untreated kenaf fibers show peaks at around 3335, 2894, 1731, 1594, 1423, 1160, dan 897 cm^{-1}. Peaks at 3335 cm^{-1} and 2894 cm^{-1} indicated O – H stretching vibrations and the hydrogen bond of the hydroxyl groups and C–H stretching of cellulose, hemicelluloses, and lignin, respectively [9,10]. The peak on 1731 cm^{-1} indicated the C=O stretching vibration of the acetyl groups in the hemicelluloses and lignin [8]. The peaks at 1594 dan 1423 cm^{-1} represents C=C stretching of the benzene ring of the lignin [10,11]. Another peak at 1160 cm^{-1} is a characteristic band for C – O stretching modes and C – O – C stretching vibration of cellulose, hemicelluloses, and lignin [10]. The final peak at 897 cm^{-1} is associated with the C – H of β-glycosidic linkages between the monosaccharides of Cellulose [9].

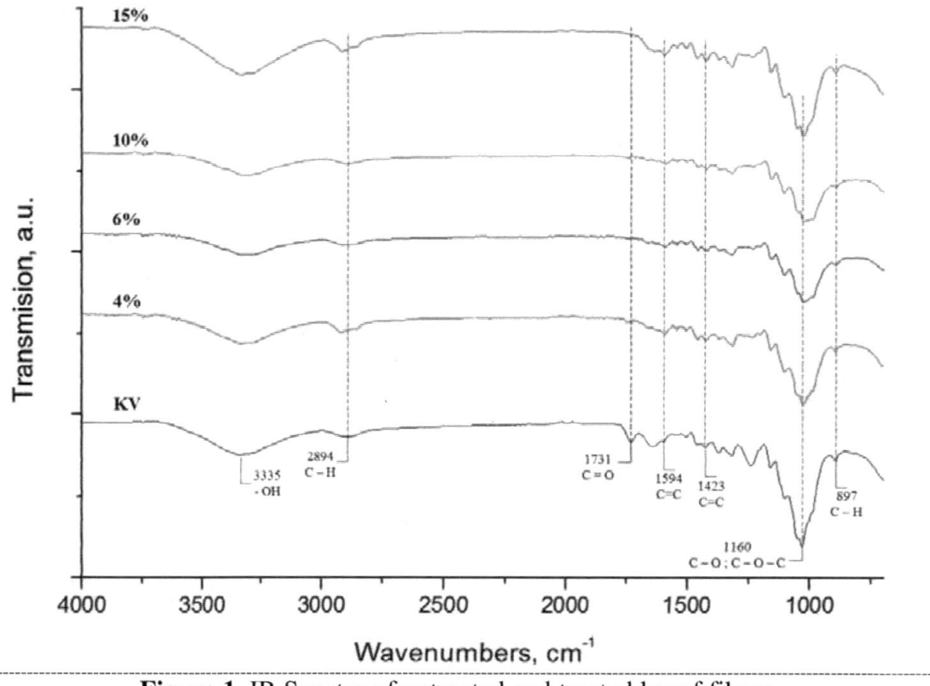

Figure 1. IR Spectra of untreated and treated kenaf fibers.

Fourier Transform Infrared spectroscopy was used to investigate Compared with the peak of kenaf fiber the result of the treatment of alkalinization at 10 wt.% NaOH, the peak at 1731 cm^{-1} is no longer visible. This indicates that part of lignin was removed from kenaf fibers after alkali treatment. While the peaks at 1594, 1423, 1160, and 897 cm^{-1} increase in transmission values which indicate a reduction in the molecules at the peak decreases. From these results, it can be concluded that alkaline treatment, causes loss of binding materials, such as hemicellulose, lignin, and amorphous cellulose, and increase the crystalline portion. While the loss of learning content in kenaf fibers due to the influence of the concentration of NaOH solution can be seen in Fig. 2.

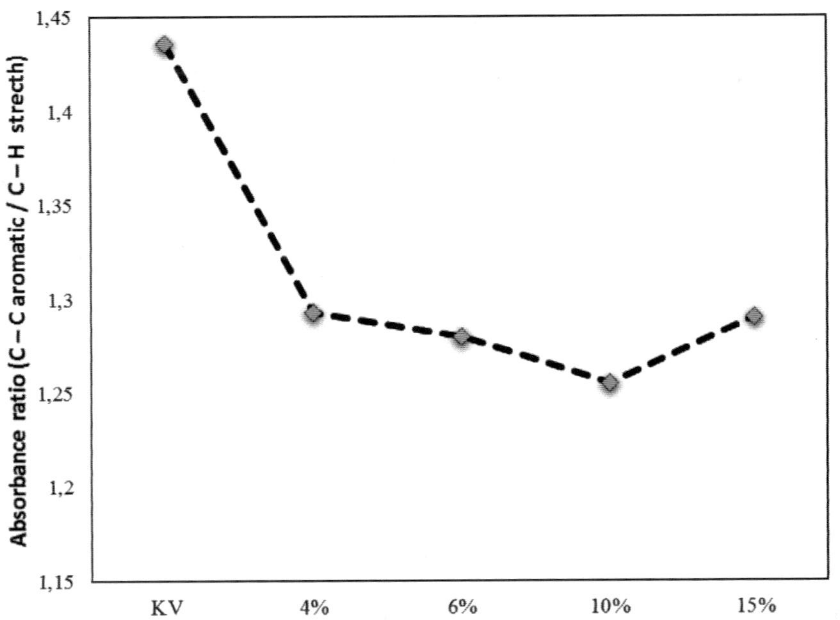

Figure 2. Effect of alkali treatment concentration on relative content of lignin

3.2. Tensile strength

One of the factors that influence the mechanical properties of natural fibers is the content of cellulose contained in and natural because cellulose has a semicrystalline structure while lignin and hemicellulose have an amorphous structure. The effect of the alkali treatment concentration is shown in Figure 3. From that figured, it was shown that the alkali treatment can increase the tensile strength of kenaf fibers when compared to untreated kenaf fibers. These phenomena occur because during the alkali treatment the amorphous content in kenaf fibers is reduced and Cellulose crystalline increased. The maximum value of the tensile strength of kenaf fibers was obtained in alkali treatment with a concentration of 10 wt.%. While on treatment alkalinization with a concentration of 15 wt.% NaOH, kenaf fiber tensile strength decreased. This phenomenon can be explained by the fact that during the treatment of alkali at that concentration a portion of the crystalline cellulose molecular chain of the kenaf fiber is broken and reduced. The reduced crystalline portion of the fiber causes the tensile strength decreased and this result is in line with the data shown in Figure 2 and the study has been reported by Y. Cao et.al. [8].

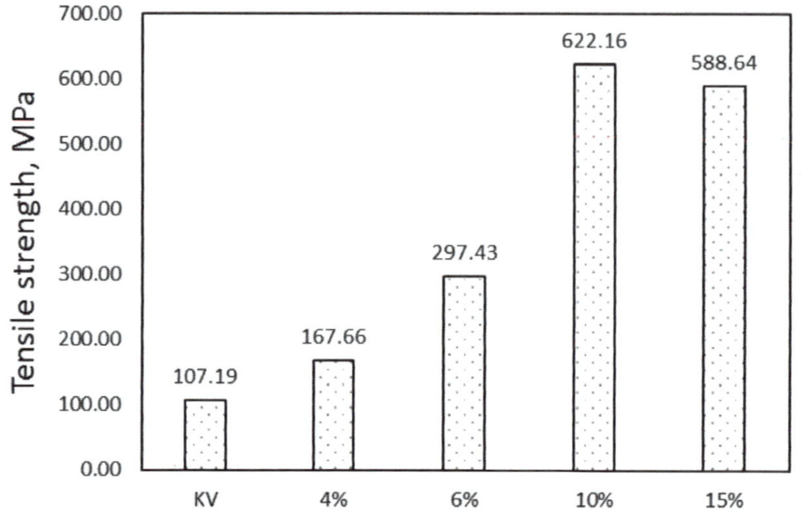

Figure 3. Tensile strength of untreated and treated kenaf fibers.

3.3. Morphology

SEM images in Fig. 4 & Fig 5 shows the untreated kenaf fiber and treated fiber through alkali treatment with 10 wt.% NaOH. From Fig. 4 it can be seen that kenaf fibers are still in a bundle or fused state and there are many impurities. While the fiber results in an alkalinization treatment (Fig. 5) fibrous kenaf fibers with a clean surface appearance. The fibrillated of kenaf fibers shows binding materials such as lignin and hemicellulose are removed after alkali treatment and bundle fibers break down into separate fiber elements. This result is under the results obtained from FTIR testing.

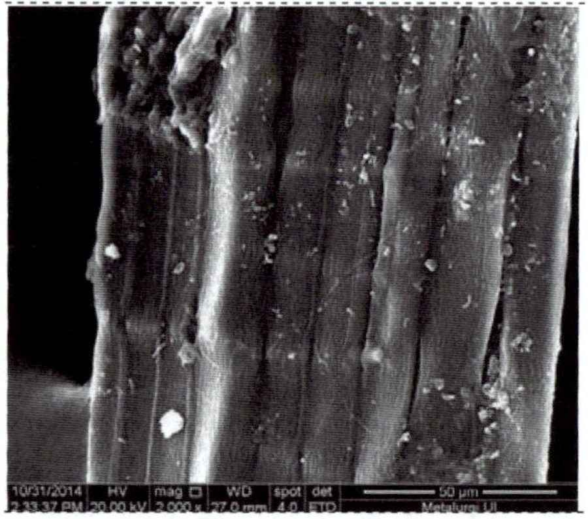

Figure 4. SEM imange of untreated kenaf fiber (kenaf virgin;KV)

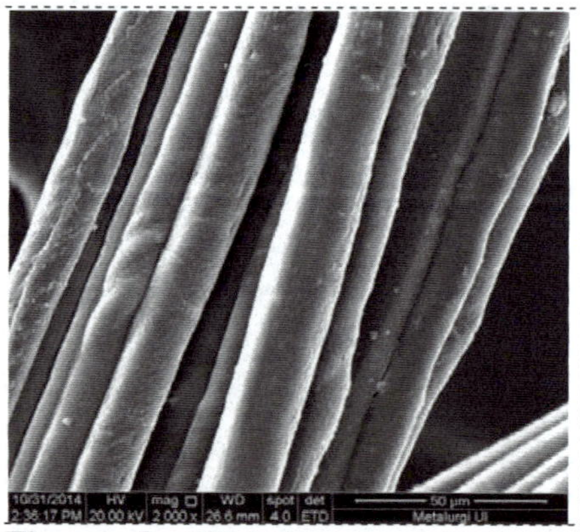

Figure 5. SEM imange of alkali-treated kenaf fiber with 10 wt.% NaOH

4. Conclusion

The effect of alkali concentration on the mechanical properties of kenaf fibers was discussed in this study. Alkali treatment, causes the kenaf fibers to become fibrillated and the amount of cellulose increases with the loss of amorphous content such as lignin and hemicellulose. Increasing cellulose content in fiber caused the tensile properties of kenaf fibers to increase and the maximum yield was obtained in alkali treatment with a concentration of 10 wt.% NaOH.

Acknowledgments
We appreciate and thank you for the support from DRPM UI through its program PIT9 Grant Scheme Which makes this research possible.

References

[1] Husnil Y A, Ismojo, Handayani A S, Setiaji D A and Chalid M 2019 The effect of alkalization and bleaching treatment of Sorghum fiber on the crystallinity index of PP composite *IOP Conf. Series: Mater. Sci. Eng.* **509** *012016*

[2] Holbery J and Houston D 2006 Natural-fiber-reinforced polymer composites in automotive applications *JOM* **58** 80-86.

[3] Ahmad F, Choi H S and Park M K 2015 *Mater. Eng.* **300** 10–24

[4] Pandey J K, Ahn S H, Lee C S, Mohanty A K and Misra M 2010 Recent advances in the application of natural fiber-based composites *Macromol. Mater. Eng.* **395** 975-89

[5] Handayani S, Husnil Y A, Handayani A S, Ismojo and Chalid M 2019 Application of waste sorghum stem (sorghum bicolour) as a raw material for microfibre cellulose *IOP Conf. Ser.: Mater. Sci. Eng.* **509** *012016*

[6] Meon M S, Othman M F, Husain H, Remeli M F dan Syawal M S M 2012 *Procedia Eng.* **41** 1587 – 92

[7] Amel B A, Paridah M T, Sudin R, Anwar U M and Ahmed 2013 *Ind. Crops Prod.* **46** 117– 23

[8] Cao Y, Sakamoto S and Goda K 2007 Effects of heat and alkali treatments on mechanical properties of kenaf fiber *Yamaguchi 16th International Conference On Composite Materials*

[9] Sahari J, Sapuan S M, Ismarrubie Z N and Rahman M Z A 2012 *Fibres Text. East. Eur.* **20** 2(91) 21-4.

[10] Rosa I M D, Kenny J M, Maniruzzaman M, Moniruzza M, Monti M, Puglia D, Santulli C and Sarasini F 2010 Effect of chemical treatments on the mechanical and thermal behaviour of okra (Abelmoschus Esculentus) fibres *Compos. Sci. Technol.* **71** (2) 246-254.

[11] Putera R D H *(Thesis) Extraction of Cellulose Fiber from Water Hyacinth Plants (Eichornia Crassipes) with Variation of Solvents* 2012 (Depok: Universitas Indonesia)

IOP Conf. Series: Materials Science and Engineering **703** (2019) 012031 doi:10.1088/1757-899X/703/1/012031

How do the Indonesian ecologically conscious millennials value upcycled clothing?

C A Parung

Faculty of Creative Industry, University of Surabaya, Raya Kalirungkut, Surabaya, 60293, Indonesia

E-mail : christabelannora@staff.ubaya.ac.id

Abstract. Upcycling is a form of green action in the area of product management and it is a new emerging trend nowadays. According to the findings of previous researches, Ecologically Conscious Consumer (ECC) tends to purchase more sustainable products, including environmentally-friendly fashion products. However, though upcycled products are classed as environmentally-friendly products, there has not been any research yet confirming that ECC in developing country likes Indonesia value and has interest in purchasing it. The aim of this research is to identify green consumer behaviour in Indonesia towards upcycled clothing. An online survey was conducted through online questionnaires to find the relationship between green consumer in Indonesia and environmentally-friendly fashion, also how the ECC value upcycled clothing as a form of eco-friendly fashion. The research shows that Indonesian millennials are mostly environmentally conscious and adopt green buying behaviour. It was found that many of green consumers in Indonesia are unfamiliar with the word 'upcycling', but some of them have already purchased it as form of sustainable deeds. Mostly the participants valued upcycled products more than recycle products for the ideas and they are valued as work of art. Profound statistical analysis was used to get the result of this research.
Keywords: upcycle, green consumer, ecologically conscious, sustainable clothing

1. Introduction

Fashion is one of the biggest and most successful business and industries in the world. It is stated by JEC Democratic, Bureau of Labour that almost 1.8 million people work in fashion industry and there are more than 200,000 textile and fashion products manufacturers in the United States (US). This phenomenon does not only happen in US, but also in other countries which shows that fashion industries are making very impeccable growth. Despite its success, it is nothing new for fashion industry to be considered as one of the biggest contributors of this world's pollution. The production process that turn raw materials into fashion products like garments, bags, shoes has negative impacts on environment, including water and air pollution, toxic chemicals and waste, etc. Kawamura stated that fast fashion as business model has made over-consumption among the consumers and leads to large numbers of waste [1]. This happens all around the world, including Indonesia which is ranked as top ten biggest textile producing countries in the world.

During the past decades, consumers' concern about environmental issues like excessive waste issue, global warming, water and air pollution have increased [2]. Consumers also started to show environmental awareness where they realize that their buying behaviour can directly affect environment. This type of consumers is called Ecologically Conscious Consumer or Ecologically Concerned Consumer (ECC), and it is defined as group of people who seek to buy / purchase products that does not have damaging impacts on the environment [3].

While fashion companies have to compete with each other in attracting consumers and make over consumption, consumers themselves have to 'make space' by throwing away or donating their clothes. This process is called post-purchase process in garment/clothing consumption which involves recycling, re-using, discarding and destroying clothing [4]. In this case, generally ECC tends to support recycling process more than destroying and discarding clothing since garments materials that are made from nylon and polyester is categorized as plastics contain toxic and non-biodegradable. Up-cycling is different from other forms of recycling because it is a conception of adding value to used products. Traditional recycling ways like down-cycle and recycle involve process like melting down and reconstructing process, and this process produces much CO2 which is not that friendly for the environment. In the other hands, up-cycling is considered less harmful.

Upcycling is now a trend and it has gained attention and popularity among the green fashion consumers. Some news suggests that it is popular on online marketplaces as well [5]. Upcycling should be studied further to place fashion in a more sustainable and environmentally friendly condition. However, like new other grounds, not many people are aware of this trend. The purpose of this study is to gain insights about the green consumers' behaviour of upcycled products in Indonesia since buying behaviour of fashion ECC is often considered complicated and unpredictable [6].

2. Literature Review

2.1 Upcycling as repurposing of clothing

Products repurposing (such as clothing, tools, etc.) has undeniably gained much attention these days, especially in literatures about manufacturing and production [7]. This phenomenon happened as an impact of increased social responsibility [8]. Innovators have developed extraordinary ways to reuse and repurpose materials waste. However, even though the word 'upcycling' is still unfamiliar for some people, it is not a new phenomenon. It has been done by people in 1930s and 1940s when people preferred to reuse their old stuffs than buying new ones.

Upcycling is a form of repurposing product in which the product itself is broken down into separate materials – which is called 'down cycling', and using those separated materials, make a new product out of it. Both recycling and upcycling are popular in developing countries that do not have many resources. Upcycling is often thought as creative way to reuse products. The main thing that differentiate upcycling from recycling, is that upcycling was meant to 'upgrade' the used products – nicer, better quality, etc. Therefore, upcycling demonstrates more effective and beneficial way in reusing the broken-down materials. For example, upcycling gives the designers to use textile waste to create new fashion, re-invent the style and adjust it with the current style. Many studies have confirmed that many discarded garments still have good quality at the disposal time, and with upcycling, people can optimize products' use and reduce waste materials. Researchers also found that upcycling process allows the consumer / users have some kinds of 'attachment' to the products. This product attachment is an emotional experience that users feel when experiencing the products [9]. Upcycled products are combination of used products with different end uses which probably have many stories behind. For those reasons, upcycled garments are usually more expensive than usual recycled garments. Even though the upcycling market can worth more than a hundred million dollars worldwide, not many countries in Asia support this practice. Korean upcycling market cannot reach the numbers that upcycling markets reached worldwide. This supports in their research that stated that many Asians do not really like wearing second-hand clothes [10]. However, even though it has been stated that upcycling markets in Asia do not expand as big as ones worldwide, the demand for the products are increasing.

2.2 Millennial ecologically conscious consumer (ECC)
In recent years, almost every consumer started to purchase products that do not cause environmental problems. This happens because the environment nowadays is way more important than it was before to them. This type of consumer is called ecologically conscious consumer nowadays. Generally, Ecologically Conscious Consumers are aware of environmental problems that lead to eco-friendly lifestyle, such as green purchase, goods disposal, etc. ECC knows that by purchasing products that are not harmful for environment, they can contribute to ecological preservation. There are several studies investigating the variables influencing ecologically-conscious behaviour of consumers. Anderson and Cunningham characterized that green consumers are usually 40 years old and above – with high level of education and higher social status. On the contrary, some researchers do not find any significant relations between age, sex and ecologically conscious behaviour [11]. However, Lu, Bock, and Joseph expressed that consumers in general are not driven to stop using certain products only to boycott the environmentally harmful products [12]. They think that price, design, and qualities are also important factors for them in choosing products. Generation Y, often termed as Millennial Generation, is a recent population, comprised of people in the age of 18 – 34 years old. According to Vermillion and Pearl, this group tends to be environmentally conscious than other age group [13]. Moreover, the educated millennials also tend to worry more about their health and community. However, they also seem to care more about their social status, which indicates that their intentions buying environmentally-friendly product might not always because they really care about the environment.

RQ1 Are millennials consumers ecologically conscious?

RQ2 How does millennials consumers' environmentally conscious behaviour affect their buying behaviour?

RQ3 How do the millennial consumers accept and value recycled and upcycled clothing as a form of supporting environmentally-friendly product?

3. Method
Quantitative study was chosen as the methodology of this research and the author picked survey as the data collection method. Survey method was conducted to get substantive outcome and ease the spread of questionnaires to the participants. Non-probability sample consisting of undergraduate students in several universities in Indonesia was recruited for this study. Purposive sampling technique was chosen because the questionnaires are needed to be given to specific group of people having certain characteristics: millennials, well-educated, and ecologically-conscious.

Firstly, faculty members and staff from each university were briefed and given the online questionnaires via email. Next, they emailed the link to the students' representatives and those representatives explained to their friends about the purpose of the survey. Interested students were suggested to give feedbacks and answers of the questionnaires in two weeks. Although non-probability sampling (purposive sampling) was used, the students that participated in the study have many interests and characteristics. The questionnaire itself was divided into several sections based on the issues being identified. It consisted of several sections with Likert scales as measurement that indicates the participants answer from 'never' to 'strongly agree'. The first 5 questions determined the demographics of the participants. The next few questions were meant to gauge their behaviour towards the environment. Data was analysed statistically by measuring frequencies, descriptive, and also using chi - square tests that allow the author to get the profound statistical result.

4. Discussion
The author received a total of 107 responses from graduated and non-graduated millennials, 54.2% of which were female and the rest (45.8 %) were male respondents. 88.8% had already graduated or were still studying in college and 11.2% did not go to college/ did not graduate. Most of these participants are from art and design background which are about 43% followed by social which are31.8%, STEM (Science, Technology, Engineering and Mathematics) which are 16.8%, and the rest are health and others. More importantly, 63.6 % of these participants are not familiar with the word 'upcycling'. One important and fundamental factor analysis was done on the questions measuring their environmental consciousness. Overall, this first analysis sought to distinguish millennials'

behaviour towards environment. Ahn and Park (1998) stated that to be considered as ecologically conscious consumers, they need to have eco-friendly lifestyle and have green purchase behaviour. These 4 items questioned stated that mostly these participants can be considered as ECC. Regarding to the first research question, it seemed that age, sex, and major background do not influence their environmental consciousness. Overall, all millennials are likely to be environmentally aware and conscious. Table 4.1. shows the percentage of Likert Scale answers towards the environmental related items.

Table 4.1. Principal component regarding to environment.

No	Items	Agree	Neutral	Disagree
1	I always look for rubbish bin when I am about to throw away rubbish	95.3	4.7	0
2	I always check the label of product to find out whether it's environmentally friendly	37.4	42.1	20.6
3	I am aware that garment industry has bad impact on environment	74.8	19.6	5.6
4	It is important for me to buy environmentally-friendly product	63.6	27.1	9.3

The following analysis sought to identify consumers' environmentally conscious behaviour and its effect on their buying behaviour. To answer this, crosstab analysis was done on the environmental concern segment and their 'green' fashion consumption. Chi-square test was done to see whether the independent variables were "fit". Green Fashion Consumption here was marked by green waste disposal habits, like clothes donation, and also green purchase like buying second-hand clothing, and fixing broken clothes. Those who tend to be environmentally conscious much likely to adopt green fashion consumption. There were several items regarding to these two segments, however there were three variables that crossed with asymp. Sig value <= 0.05. Table 4.2 indicates both dependent and independent variables which has valid correlations, and this answered the research question number two, the environmentally conscious millennials tend to have a green buying behaviour. There is a positive relationship between those two variables.

Table 4.2. Cross tabulation of Environmental Consciousness and Green Fashion Consumption Components

No	(Environmental Consciousness)	(Green Fashion Consumption)	As
1	I always look for rubbish bin when I am about to throw away rubbish	I always feel good donating my old clothes	.033
2	I always check the label of product to find out whether it's environmentally friendly	I love to buy second-hand clothing	.050
3	It is important for me to buy environmentally-friendly product	I tend to modify and fix my broken clothes than buy a new one	.004

Research question number 3 was the main question in this research. Frequency analysis was done using SPSS and overall, millennials tend to agree to swap clothes and buy second-hand clothes which indicate that they mainly have green purchase behaviour. However, the numbers are not significantly different for 'swap clothes' variables, the author had a chance to gain further information and this was because of pride. They do not mind swapping clothes with strangers, or even buying them, it is just their self-image they are worried about. Table 4.3 shows the frequency percentage of millennials recycled and upcycled clothing buying behaviour.

IOP Conf. Series: Materials Science and Engineering 703 (2019) 012031 doi:10.1088/1757-899X/703/1/012031

Table 4.3. Millennials Recycled and Upcycled Buying Behaviour

	Items	Agree	Neutral	Disagree
1	I like to swap clothes with my close friends and families	43.9%	25.2%	30.8%
2	I like buying secondhand clothes	61.7%	16.8%	21.5%
3	I buy secondhand clothes because it is cheap	50.5%	23.4%	26.2%
4	I buy secondhand clothes because it has unique style	55.1%	22.4%	22.4%
5	I prefer modified clothing than the unmodified ones	16.8%	43.9%	39.3%
6	I like buying upcycled clothing	55.1%	33.6%	11.2%

From the table above, it can be informed that millennials indeed love buying recycled and upcycled products. However, it is hard to know the buying intentions, whether it is as an act of environmental consciousness or they were only appreciating a work of 'art'. This might support Lu, Bock, and Joseph (2013) who indicates that even the most environmentally-conscious consumer do not always buy product over 'environmental concern'. The author took further analysis in the 21.5% who disagree in buying secondhand clothes. With the asymp. Sig of .000, it is confirmed that they dislike buying them over cleanliness reason.

Unlike the previous segments, here demographics play a consequential part. The cross-tab analysis between two variables (Sex and Modified-Unmodified Clothes), also Major background and Upcycled Clothes were undertaken which lead to an interesting result. Male participants tend to like unmodified ones (recycled clothes), while female participants have bigger number in 'Agree' and 'Neutral' than male.

Table 4.4. Sex and Modified Clothes Variable Counts

			Modified over the unmodified ones			Total
			D	N	A	
Sex	Male	Count	23	23	3	49
		Expected Count	19.2	21.5	8.2	49.0
	Female	Count	19	24	15	58
		Expected Count	22.8	25.5	9.8	58.0

Interesting results also came in the case study of 'Tesco bag'. On the questionnaires, the author gave an example of upcycled clothing that was made of an old shopping bag from a certain supermarket in England. Crosstab analysis was undertaken between major and their interest in certain upcycled products variables, and with the asymp.sig .002, the people who are interested the most in the upcycled products were from Art and Design background. In contrary, the people who are likely to disagree were from STEM background.

Table 4.5. Major and Upcycled Products Cross Tab

			Case 01 Tesco Bag Upcycled			Total
			D	N	A	
Major	Art and Design	Count	6	8	32	46
		Expected Count	11.6	10.7	23.6	46.0
	STEM	Count	11	4	3	18
		Expected Count	4.5	4.2	9.3	18.0
	Social	Count	7	11	16	34
		Expected Count	8.6	7.9	17.5	34.0

Health	Count	1	0	0	1
	Expected Count	.3	.2	.5	1.0
Others	Count	2	2	4	8
	Expected Count	2.0	1.9	4.1	8.0

5. Conclusions

In general, Indonesian millennials are considered to be ecologically conscious, and there are no correlations between gender, age and their behaviour. This behaviour is the one that leads to their green buying behaviour, which has been proven, is very much related. However, during their purchasing time, it is not clear yet whether the intention of their purchase is underpinned by their environmental consciousness. In some ways, millennials also take design, price, and style into consideration. Subsequently, their green behaviour also leads to their acceptance and how they value recycled and upcycled products. Even though they are not familiar with the word 'upcyling', unconsciously Indonesia millennials have taken part in such area Though both male and female participants fully agreed that upcycled products has to be more expensive than regular products because of the 'sense of art', not everyone choose to buy it more than the recycled ones. Female participants can accept upcycled products more than the male ones, but this does not mean that male participants reject it. From the questionnaires, the author can conclude that male participants prefer recycled ones because they do not want too many modifications (style-related). This study also proved that participants from art and design background tend to have more interest in upcycled clothing with its 'uncommon and peculiar style' than others from other background.

This research will be able to help brands and retails to target the environmentally-conscious fashion consumer and develop their awareness in green fashion consumption. Also, for educators and students, the findings of this research can be used for further observations in upcycling area.

References

[1] Kawamura Y 2005 *Fashionology: An introduction to fashion studies* (Oxford: Berg)
[2] Van Dam Y and Apeldoorn P 1996 Sustainable Marketing *J.Macromarketing* **16** pp 45-56
[3] Roberts J A 1996 Green Consumers in the 1990s: Profile and Implications for Advertising. *J Bus. Res.* **36** pp 217-31
[4] Ha-Brookshire J and Hodges N 2008 Socially Responsible Consumer Behavior? *Clothing Text Res J.***27** pp 179-96
[5] Goldsmith B 2009 Trash or treasure? Upcycling becomes growing green trend. *Retrieved* **9**
[6] Nordås H 2004 *The Global Textile and Clothing Industry post the Agreement on Textiles and Clothing* (Geneva : WTO Publications)
[7] Zink T, Maker F, Geyer R, Amirtharajah R and Akella V 2014 Comparative Liife Cycle Assessment of Smartphone Reuse: Repurposing vs Refurbishment *Int. J. Life Cycle Assess.***19** pp 1099-109
[8] Lee, Young & Halter, Holly &K P Johnson, Kim &Ju, and Haewon 2012 Investigating Fashion Disposition with Young Consumers. *Young Consumers*. **14** pp 67-78
[9] Schifferstein& P H and Zwartkruis-Pelgrim E P H 2008 Consumer-Product Attachment: Measurement and Design Implications. *Int. J. Des.***2**
[10] Xu Y, Chen Y, Burman, Ritika, Zhao and Hongshan 2014 Second-hand Clothing Consumption: A Cross-Cultural Comparison between American and Chinese Young Consumers *Int. J Consum. Stud.* **38** *Preprint* 10.1111/ijcs.12139
[11] McEvoy J 1972 The American concern with the Environment : Social Behavior *Nat. Resour. Environ.* pp 214-36
[12] Leslie Lu, Dora Bock, Mathew Joseph 2013 Green marketing: what the Millennials buy *J. Bus. Strategy* **34** pp 3-10 *Preprint* 10.1108/JBS-05-2013-0036
[13] Vermillin L and Pearl J 2010 Green marketing: making sense of the situation. *Proc. Acad. Marketing Stud.* **15** pp 68-72

IOP Conf. Series: Materials Science and Engineering 703 (2019) 012032 doi:10.1088/1757-899X/703/1/012032

Passive design implementation as sustainable development approach on vertical housing case study: Sentra Timur Residence

T Riotama[1] and H Herdiansyah[1]

[1]School of Environmental Science Universitas Indonesia, Jl. Salemba Raya No. 4 Central Jakarta 10430, Indonesia

E-mail : herdis@ui.ac.id

Abstract. Vertical housing development is a solution to solve the housing needs problem in Jakarta, which already had a land limitation issue. However, the development also contributes to environmental degradation. Building uses materials and energy in the development process, and when it operates. Most of energy use in Heating, Ventilating, and Air Conditioning (HVAC) to create physical comfort for user. Passive design strategies applied to maintain comfortable temperature in building such as building orientation, shading, and ventilation, so energy usage could be reduced. The purpose of this research was to analyse the implementation of passive design in vertical housing, as an approach to achieve sustainable building. Observation carried on Sentra Timur Residence, one of vertical housing in East Jakarta. This research used building data to create mass modelling to help analyse the passive design implementation. The results show that some passive design strategies are applied such as building orientation, shading, and ventilation, mainly in the housing unit. These strategies help to maintain thermal comfort inside the housing unit, so the users can feel comfort without using an air conditioner. It reduces energy usage in the building so negative impact on environment could be minimized.

1. Introduction

Human activities contribute to environmental issues, such as climate change and air pollution. Development carried out in order to fulfill human needs, sometimes does not consider the ecological impacts that cause environmental degradation. Therefore, the concept of sustainable development arises, in order to reduce environmental impacts and on the other hand improve the quality of human life.

The United Nations assigns 17 Sustainable Development Goals (SDGs), which one of them is sustainable cities and communities. Cities only occupy 3% of the land area, but their energy use reaches 60-80%. Building is one of the elements in the city as a place for people to live, work, and do various activities. However, on the other side, the building also contributes to environmental problems. Buildings produce 40% of existing carbon emissions and use resources such as material, water, and energy [1].

Jakarta is a high population density city, with about 10,3 million residents [2]. Besides humans, Jakarta is also filled with a lot of buildings, ranging from houses, offices, and skyscrapers. With the increasing number of residents every year, it will affect the needs of housing.

Housing backlog based on ownership in Indonesia amounted to 13,5 million units in 2014, and handling target was 6,8 million units in 2019 [3]. Jakarta, as a densely populated city in Indonesia also become a target of housing development. However, this city is already had land limitation issue, so the development directed to vertical housing.

The building development needs to consider the sustainability aspect, since the building give a negative impact to the environment such as consume a lot of material, water, and energy [4][5][6]. Building also consumes electricity to support the human activities inside, which is the source of electric power supply mostly come from fossil power. So, building also contributes to the production of emission [4]. Heating, Ventilating, and Air Conditioning (HVAC) is the most energy consumption in building [7] that needed to create thermal comfort for the user. Human thermal comfort can be achieved when the average temperature is around 23℃ -27℃ [8].

Jakarta has a high average temperature, around 28-31℃ [9], which means there are gaps to achieve thermal comfort. It needs some effort to reduce the temperature in the building. Air conditioner usually used to solve this problem, although it means there are more energy needed and emission produced.

In architecture, there are some strategies that can be applied, namely passive and active design. Different with active design that need mechanical device and high technology, such as air conditioner, passive design is applied through the planning process that makes building can adapt to the environment. Passive design strategies that commonly applied, such as building orientation, shading, and ventilation.

1) Building orientation
   Building orientation refers to the way building is placed on the site. The position of the building will affect the solar heat accepted by the building. Solar analysis could help to determine which direction the building should oriented. Usually, the building planner avoids north-west orientation because there is more heat exposure that accepted by the building, especially in the afternoon.
2) Shading
   Shading is used to control the amount of sunlight that expose the building. To apply shading strategy, there are some ways that can be used, first create façade that adaptive to sun movement, and the other way is using shading devices such as trellises, roller blind, and curtain.
3) Ventilation
   Ventilation is important element in building, it useful to air exchange in the building. Through ventilation, fresh air can enter the room and replace the air inside.

East Jakarta is part of Jakarta with the highest resident population among the other regencies. To fulfil the housing needs, there are a lot of residential developed in East Jakarta. Vertical housing development also increase, due to land limitation issue. One of the vertical housing in East Jakarta is Sentra Timur Residence.

Vertical building consists of some housing units, and every unit need to fulfil thermal comfort for its user. This paper aims to find the passive design implementation in Sentra Timur Residence and analyse its effect to the building. Passive design could help building to achieve thermal comfort, so energy usage will reduce, and make building more sustainable.

2. Methods

2.1. Study Area

This study held in Sentra Timur Residence, one of vertical housing in East Jakarta. This building located in a high-density housing area, and the surrounding area is filled with lot buildings, as shown in figure 1. Sentra Timur Residence consists of six towers, named Orange, Ruby, Green, Grey, Yellow, and Tosca Tower, as the color of each tower.

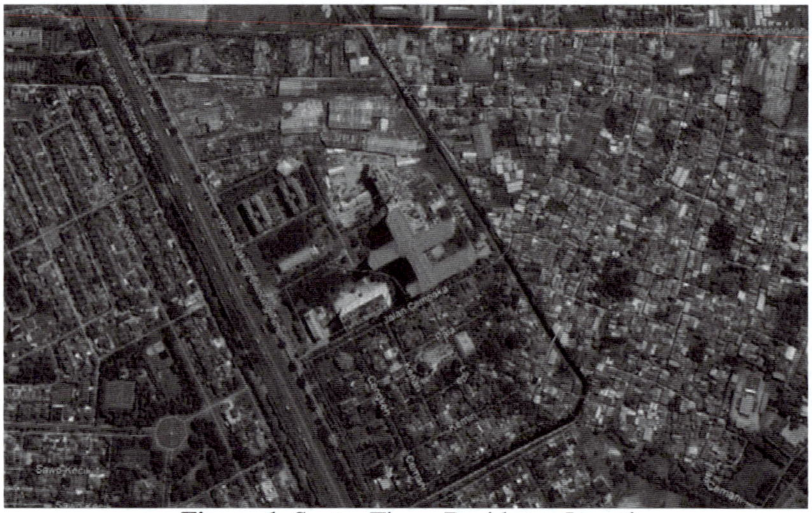

Figure 1. Sentra Timur Residence Location.

2.2. Data Collection

This study used building drawing such as plan, section, and elevation to create mass modelling. This model then analysed using solar path software. Other data such as solar movement, temperature, and map used to help to identify the issue. Observation also carried to identify the implementation of passive design in the building.

3. Results and Discussion

3.1. Building Orientation

Figure 2 showed the orientation of Sentra Timur Residence building and sun path in Jakarta. There are six towers, and as it is shown in Figure 2, none of the building exactly face east-west orientation. Two towers face southeast-northwest orientation, and four towers face southwest-northeast orientation. These orientations make the housing units not directly exposed to sunlight. Thus, the potential of heat entering the building through wall and openings can be reduced.

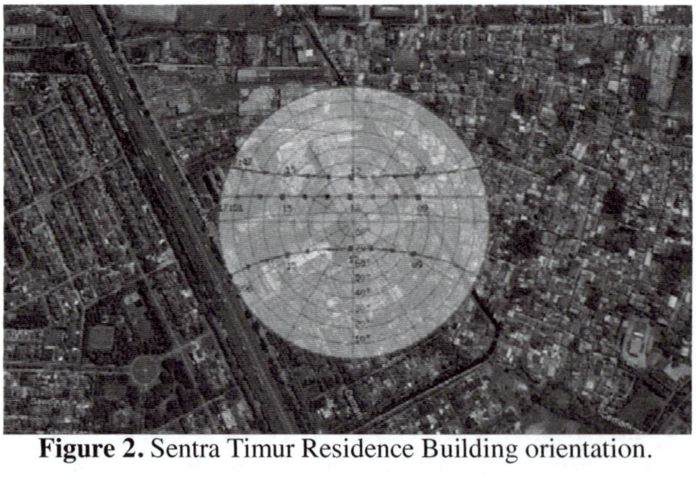

Figure 2. Sentra Timur Residence Building orientation.

3.2. Shading

Shading is useful for eliminating the heat of the sun entering the building. Shading can be achieved by façade design or using a shading device. In Sentra Timur residence buildings, shading is formed from an asymmetrical building mass. The towers position and height difference of the towers form shading for each other as can be seen in figure 3.

In figure 3a shows the condition when the morning when the sun rises, it can be seen that shading on the west side makes the units on that side cooler. Figure 3b shows the situation in the afternoon where the sun is on the west side. It can be seen that the shading formed covered many sides of the housing units.

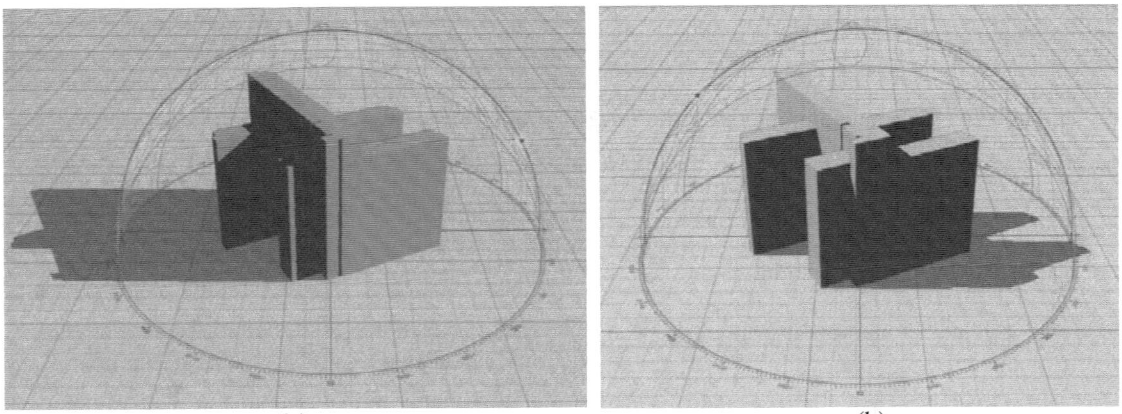

(a) (b)

Figure 3. Sentra Timur Residence Building Shading.

3.3. Ventilation

Figure 4 showed the unit plan of Sentra Timur Residence, which consists of a living room, two bedrooms and a toilet. Each unit is designed to have windows facing outwards. This window is a vent to put fresh air into the unit. The windows position in the room are in the bedroom 1 and living room. Considering the configuration of the room and windows, the room that needs air conditioner is bedroom 2 due to the absence of ventilation to flow fresh air inside. The use of air conditioners in the living room and bedroom 1 can be reduced by utilizing ventilation.

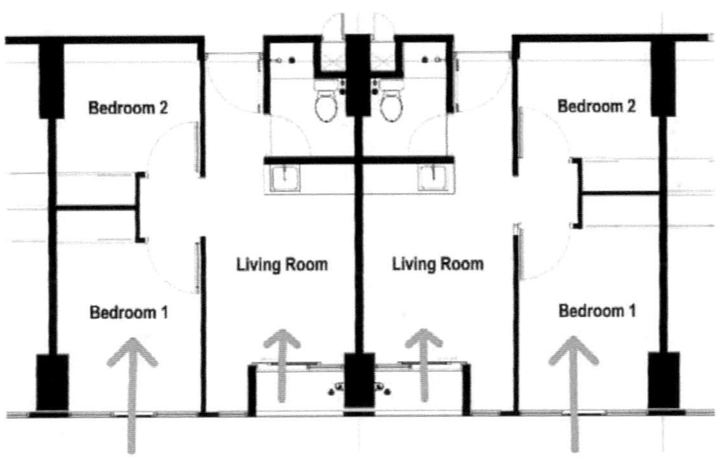

Figure 4. Housing Units Ventilation Illustration.

4. Conclusion

Sentra Timur Residence building applied passive design strategies, such as building orientation, shading, and ventilation. These strategies help to create thermal comfort inside the housing units, so air conditioner usage could be reduced. It means the energy usage also reduced and make the building more sustainable to contribute to achieving SDGs number 11, sustainable cities and communities.

Acknowledgments

This research is funded by Directorate General Higher Education (DIKTI), Ministry of Research, Technology and Higher Education of the Republic of Indonesia. [contract number NKB-1903/UN2.R3.1/HKP.05.00/2019].

References

[1] United Nations Development Programme 2019 *Goal 11 : Sustainable Cities and Communities* (Retrieved from https://www.undp.org/content/undp/en/home/sustainable-development-goals/goal-11-sustainable-cities-and-communities.html)
[2] Indonesia's Statistics Agency (BPS) of DKI Jakarta Province 2018 *DKI Jakarta Province in Figures* (Jakarta : Indonesia's Statistics Agency (BPS) of DKI Jakarta Province) p 53
[3] Directorate of Vertical Housing Ministry of Public Work and Housing 2015 *Strategic Planning of Vertical Housing Directorate*
[4] Priatman J 2002 DIMENSI *Jurnal Teknik Arsitektur* **30** p 167–175
[5] Hoffman A J and Henn R 2008 *Organ. Environ.* **21** p 390–419
[6] Sandi R. (2019). Environmental Impact On Digital Era By Architectural Production. *J. Environ. Sci. Sustainable Dev.* **2** (1) 26–47
[7] Kim N K, Shim M H and Won D 2018 *Energies* **11** 2690
[8] Sekatia A, Setyowati E, and Hardiman G 2018 *IOP Conf. Ser.: Earth Environ. Sci.* **213** 012044
[9] Farid V L and Wonorahardjo S 2018 *IOP Conf. Ser.: Earth Environ. Sci.* **152** 0–10

Development and usability evaluation of virtual guide using augmented reality for Candi Gunung Gangsir in East Java

I M Ronyastra, I Hapsari and F P Pani

Department of Industrial Engineering, University of Surabaya, Raya Kalirungkut, Surabaya, 60293, Indonesia

E-mail : imade.ronyastra@yahoo.com

Abstract. Despite being one of Indonesian cultural and historical heritage, Candi Gunung Gangsir failed to attract many visitors especially from the younger generations. Lack of attraction and supporting facilities in the vicinity is one of the reasons. This research aims to develop an interactive virtual guide enabled by augmented reality that will increase the attractiveness of the object. The methods used in this research are product design and development steps with consideration of human computer interaction, and usability. Primary data of customer needs are collected on and off site through questionnaires and interviews. Customer needs are then used to generate several product concepts to choose 1 final concept from through concept selection methods. Literature review and interviews are conducted to create an accurate story line for the virtual guide. The chosen concept is then developed to become a final application which also tested against the customer needs through a survey method. The result of this research is an offline Android-based augmented reality application to scan the 17 markers to show a3-D virtual guide character on the marker and inform the visitor about history of that particular spot. The usability testing was conducted to 20 respondents and revealed that the application can attract them to visit Candi Gunung Gangsir.

1. Introduction

Candi Gunung Gangsir, located in a housing complex in Beji, Pasuruan, Indonesia, is believed to be built at the end of the 11th century or at the beginning of the Majapahit era and was once used as a shrine to worship the God of Prosperity. Despite being a cultural and historical heritage, it failed to attract many visitors especially from younger generations. Our preliminary survey observed the visitor to this site on a weekend barely reach 20 persons due to lack of attraction and supporting facilities in the vicinity. Although it has undergone a restoration, some parts of Candi Gunung Gangsir is already destroyed which make it difficult for casual tourist to interpret the site [1]. To help improving the attractiveness of the site, this research aimed to develop a mobile augmented reality application that can serve as a virtual guide for the visitors.

The application can also be used to preserve important knowledge regarding the architecture and history of the site and it will help to transmit the knowledge in a more effective way [2] by overlaying the information onto point of interest [3]. The increasing adoption pace of augmented reality (AR) in the tourism sector [4] should make it easier to implement the application. Candi Gunung Gangsir is an outdoor site which provide a challenge in designing the application so that it must consider the ergonomic factors and product usability. Since there are no personal guides available at Candi Gunung Gangsir, we consider to create an application that can serve as a virtual guide for the visitors and bring us another challenge to create a good story line as it will be delivered in a virtual environments [5].

2. Related works

Using AR in to enhance engagement from user has gained more popularity as AR can help to create mixed reality by including virtual elements into an actual physical environments [6]. AR provide a means of improving users' experience and satisfaction. The use of AR in marketing is a promising and growing field [7] and it allows brand to enter consumer's domestic space with virtual offerings and has been practiced by leading brands such as IKEA, Wayfair, and Sephora [8] where the application enables customer to virtually try products. By providing additional information about the offering, AR can enhance a customer's real-world experience in interesting ways [9]. A personalized AR application should help customers in dealing with too much information provided by a common search engine [10].

The AR applications are also effectively implemented in the tourism business. Garau [3] used AR to improve the tourists' experience in Costello, Italy. AR has also been used for preservation purposes where knowledge of a heritage site is saved in the application. Haydar et.al [11] used AR to reconstruct underwater archaeological site and compared it with the virtual reality (VR). The historical information regarding Roman Theater at Byblos are used to reconstruct the site virtually [1]. Recently, Mah et.al [12] has generated a virtual tour to preserve the cultural heritage of Tampines Chinese Temple in Singapore. A mobile AR based on semantic web technology has been developed to provide contextual information about Injeongjeon Hall in South Korea [13].In Indonesia, Affan et.al [14] implemented AR as information and promotion media for Dieng tourism area.

Application design is not over once the application is ready to be released, it has to be reviewed for quality and improvement. A quality assessment helped the designer to identify shortcomings of the application's functionality [15] and use the information to improve the application. Usability is a popular metric for assessment where some design goals are used as the measurement dimensions [16]. Barnett et.al [17] used usability to evaluate mobile games users' experience in the view of first-time users. Yuan and Chee [18] evaluate their design of virtual tour guide using two dimension i.e. user satisfaction and agent's believability. Tahyudin and Saputra [19] used 4 dimensions i.e. text, graphic, animation, and interactivity to assess the user satisfaction of their application.

3. Development process methodology

The application development process is started with an exploratory data collection for users' insights through questionnaire targeting 100 respondents with age range of 17 – 25 years old, the number of respondents is adequate to represent the population based on Slovin's Formula with error tolerance of 10%. The data obtained from the survey are then used as considerations during the generation of application concepts.

The concept generation are done systematically by utilizing the concept combination table. Among the generated concepts, a selection process is done by comparing concepts against each other by using customer needs and its weight as selection criteria until one final concept is selected. The selected concept is then developed into a final product (Android application).

In order to get a better and more accurate information that will be conveyed through the application, an in-depth interview with historian, supported by literature review, was conducted so that the story line of the object's history, which is the main content of the application, can be plotted. Beside the story and timeline, some data regarding cultural aspects including how the people dress that is required to create the main character appearance in the application were also gathered

The augmented reality application is mainly developed using Unity 3D software while the assets (components) are developed using several other software such as:

- Corel Draw and Adobe Photoshop: designing markers, user interface, and 2D virtual guide character
- Cinema 4D, Zbrush, and Marvellous Designer: generating 3D character
- Adobe Audition: sound assets
- Android SDK: to create .apk file
- Vuforia SDK: to enable application reading the marker through android phone camera

After the application was developed, a usability testing was done by questioning 20 respondents fit to the target user group to ensure the human computer interaction is well considered by using 6 dimensions of usability [16] including: functionally correct, efficient to use, easy to learn, easy to remember, error tolerant, subjectively pleasing. We chose this metric not only because of its simplicity but it can also provide sufficient early feedback on the application without making it too hard for users to fill a more complex feedback questionnaire.

4. Results and discussion

4.1. Users insights, concepts generation, and concept selection

From the survey of 100 respondents, the frequency of visiting historical site is quite high as 67% claimed to visit at least once a year and the most visited site is Candi Borobudur. Only 11 respondents said that they already visited Candi Gunung Gangsir which confirms the lack of visitor to our object. 89% of respondents stated that they want to get more information on the history of the site and preferred the information to be conveyed through a video presentation and augmented reality instead of brochure, audio-only and in-site guide. The respondents preferred offline mobile applications (64%) while the file size should not exceed 300MB (74%). We identified 10 equally important needs that are expected to be delivered by the applications, as follows: ease of use, security, interesting interface, low file size, the use of augmented reality technology, the use of 3D objects, informative audio, and subtitles.

These insights helped us to generate 8 application product concepts by varying color scheme, type of marker, and the style of the virtual guide character as shown in Table 1. The evaluation of the concepts was done through concept screening that left us with 5 concepts which were then passed through the concept scoring which use the customer needs and its weight as the selection criteria where the score was given by comparing each concept with a reference concept.

Table 1. Concept Combination

Marker Type	Color Scheme	Character Style
Geographic	Basic	Modern character
Printed	Mixed	Classical Majapahit

The chosen concept is an augmented reality virtual guide application with basic color scheme, printed marker instead of geographical marker, and main characters designed to resemble the Majapahit era styles.

4.2. Content and assets generation

The main content of the application is information regarding the areas of interest within the main building of the site. We observe at least 14 areas of interest in form of relief carving on the body of the building and the photograph of the areas are unique enough to be used as markers to be scanned by the AR camera. Several examples of the markers are presented in figure 1. We also designed 3 additional markers as the place-holder for supporting information i.e. welcome greetings, the site architectural, and urban legend. The information required for creating the content is primarily gathered from in-depth interview with experts from Cultural Heritage Conservation Center (BPCB) Jawa Timur and supplemented by secondary data from the report on Candi Gunung Gangsir restoration in 2013. The data also helped us in designing the virtual guide character to resemble the persons from Majapahit

IOP Conf. Series: Materials Science and Engineering **703** (2019) 012033 doi:10.1088/1757-899X/703/1/012033

era. We designed 4 characters as shown in figure 2. The male characters are designed to wear sarong, while the female characters wrapped in seamless fabric. These characters are animated and dubbed during the presentation of each area of interest

Figure 1.AR markers using areas of interest

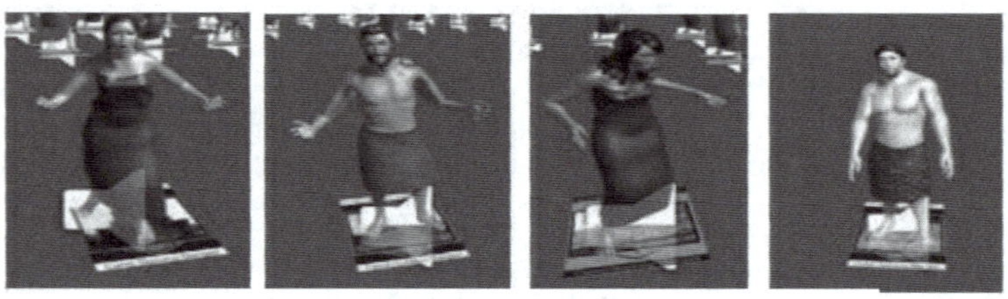

Figure 2. The virtual guide characters

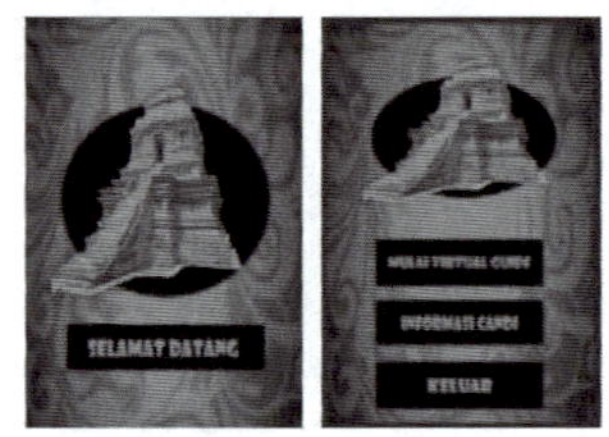

Figure 3. User interface

The next process is designing the application's user interface (UI) while ensuring the application must be easy to use. We chose a simple yet ergonomic user interface as shown in figure 3. The colour scheme is putting dark background and bright text to make it suitable with broad-daylight exposure. To align with the history of the site, we chose green background as it represents land fertility which can become a source of prosperity. The first button on the main page is designed to open the camera so that the user can scan the markers for virtual guide presentation. The second button will display text-based information of the site. Clicking the last button will close the application. The application file size is 146 Megabytes which is considered acceptable. Moreover, it can run in Android 4.0 version or newer, and smoothly runs on smartphones with 2GB of RAM.

4.3. Markers' placement

The markers will be placed near the relief carving so that the users can scan the marker while seeing the actual relief. The placement map is shown in figure 4. The marker is printed to fit into an A3 paper size (297 mm x 420 mm) where the virtual guide character will pop-up. The best viewing distance should be at 100 cm from the marker so that the virtual guide will be visible as a full adult person size as illustrated in figure 5.

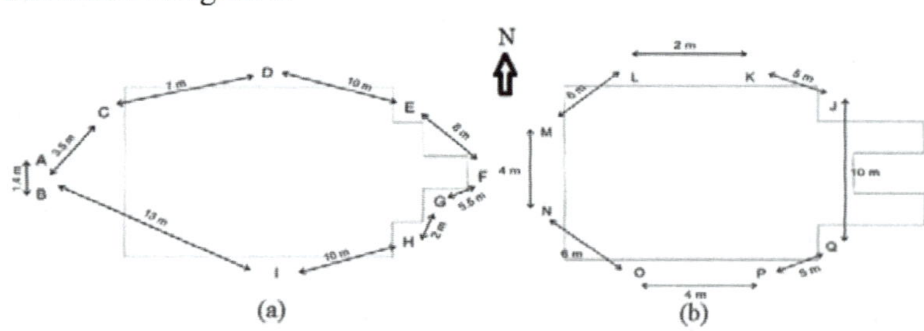

Figure 4. (a) markers placement on the ground floor of the building. (b) markers placement on the first floor of the building

4.4. Product and usability testing

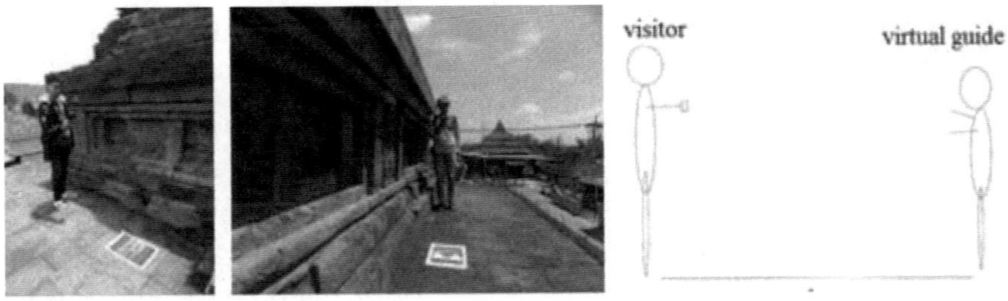

Figure 5. Suggested viewing distance and virtual guide size illustration

Product and usability testing were done by installing the application on a smartphone and have it tested by 20 respondents that belong to our target user group i.e. aged between 17 – 25, living in urban area. The survey was conducted in Surabaya instead of in Candi Gunung Gangsir premise as it will enable us to measure the interest to visit the site after testing our application. We used the purposive sampling techniques to choose our respondents.

The respondents were asked to fill out post-trial questionnaire consist of 6 usability questions using 5-point Likert scale and 1 binary question to assess their interest in visiting Candi Gunung Gangsir should the application is available on the site. The result of the usability questions is summarized in table 1 which indicates positive response for the application as the mean of each dimension is at least 4. The test also suggests that 90% of the respondent agrees that the application will increase their interest in visiting Candi Gunung Gangsir. We believe that the application would be able to help the site attract more visitors.

Table 2. Usability testing result

Dimension	Mean	Std. Deviation
Functionally Correct	4,85	0,36635
Easy to use	4,80	0,41039
Easy to learn	4,90	0,30779
Easy to remember	4,60	0,50262
Error tolerant	4,40	0,50262
Subjectively pleasing	4,85	0,36635

5. Conclusion and further research

This paper provides a systematic approach in designing a mobile augmented reality application that will serve as virtual guide in Candi Gunung Gangsir, East Java that should help the site to attract more visitors especially from the younger generations. Virtual guide should give visitors a better experience and better information regarding the history of the site. The application can as well preserve the site's history and architectural information and transfer it in a more engaging virtual environment to the visitors. The usability evaluation shown a positive feedback on the application and give early hint on the rising interest to visit Candi Gunung Gangsir.

As the application is only a part of a total solution to improve attractiveness of the site, further research area can be explored such as making the virtual guide to be interactive with the visitors by implementing artificial intelligence resembling a real personal guide who can take questions and give answers. Other research can also take the implementation as the focus and measure by how much actual visitors that the site can attract.

Acknowledgments
The authors wish to acknowledge LPPM Ubaya for financial support for this research and BPCB Jawa Timur for providing relevant literature regarding the history of the site.

References

[1] Younes G, Kahil R, Jallad M, Asmar D, Elhajj I, Turkiyyah G, and Al-Harity H 2017 Virtual and augmented reality for rich interaction with cultural heritage sites : A case study from the Roman Theater at Byblos *Digit. Appl. Archaeol. Cult. Herit.* **5** no. December 2016 pp 1–9
[2] Brondi R, Carrozzino M, Lorenzini C and Tecchia F 2016 *Informatica* **40** pp 311–6
[3] Garau C 2014 *Planning, Pract. Res.* **29** no. 3 pp 238–55
[4] M. C. Dieck M C and T. H. Jung T H 2017 Value of augmented reality at cultural heritage sites : A stakeholder approach *J. Destin. Mark. Manag.* **6** no. 2 pp 110–7
[5] Ibanez R, Aylett J and Ruiz-Rodarte R 2003 Storytelling in virtual environments from a virtual guide perspective *Virtual Real.* **7** pp 30–42
[6] Mota J M, Ruiz-rube I, Dodero J M and Arnedi 2018 Augmented reality mobile app development for all *Comput. Electr. Eng.* **65** pp 250–60
[7] Rauschnabel P A, Felix R, and Hinsch C 2019 Augmented reality marketing : How mobile AR-apps can improve brands through inspiration *J. Retail. Consum. Serv.* **49** no. March pp 43–53
[8] Scholz J and Du K 2018 We are at home : How augmented reality reshapes mobile marketing and *J. Retail. Consum. Serv.* **44** no. March pp 11–23
[9] Farshid M, Paschen J, Eriksson T and Kietzmann J 2018 Go boldly! Explore augmented reality (AR), virtual reality (VR), and mixed reality (MR) for business *Bus. Horiz.* **61** no. 5 pp 657–63
[10] Shi Z, Wang H, Wei W, Zheng X, Zhao M, Zhao J, and Wang Y 2016 Novel individual location recommendation with mobile based on augmented reality *Int. J. Distrib. Sens. Networks* **12** pp 1-13
[11] Haydar M, Roussel D, Maidi M, Otmane S and Mallem M 2011 Virtual and augmented reality for cultural computing and heritage: a case study of virtual exploration of underwater archaeological sites (preprint) *Virtual Real.* **15** pp 311–27
[12] Mah O B P, Yan Y, Tan J S Y, Tan Y X, Tay G Q Y, Chiam D J, Wang Y C, and Dean K 2019 Generating a virtual tour for the preservation of the (in)tangible cultural heritage of Tampines Chinese Temple in Singapore *J. Cult. Herit.* **39** pp 202-211.
[13] Kim H, Matuszka T, Kim J and Woo W 2017 Ontology-based mobile augmented reality in cultural heritage sites: information modeling and user study *Multimed. Tools Appl.* **76** pp 26001–29
[14] Affan B N, Suryanto A and Arfriandi A 2018 *Telkomnika* **16** no. 4 pp 1818–25
[15] Kourouthanassis P E, Boletsis C and Lekakos G 2013 Demystifying the design of mobile augmented reality applications *Multimed. Tools Appl.* **74** no. 3 pp 1045–1066
[16] Brinck T, Gergle D and Wood S D 2002 *Usability for the web* (San Fransisco: Mogran Kaufmann)
[17] Barnett L, Harvey C and Gatzidis C 2018 First Time User Experiences in mobile games: An evaluation of usability *Entertain. Comput.* **27** no. March pp 82–8
[18] Yuan B X and Chee Y S 2005 Design and evaluation of Elva : an embodied tour guide in an interactive virtual art gallery *Comput. Animat. Virtual Worlds* **16** pp 109–19
[19] Tahyudin I and Saputra D I S 2017 *Int. J. Electr. Comput. Eng.* **7** no. 6 pp 3500–6

IOP Conf. Series: Materials Science and Engineering **703** (2019) 012034 doi:10.1088/1757-899X/703/1/012034

The Role of Ergonomics in Suporting Supply Chain Performance in Manufacturing Companies: a Literature review

N Sampouw[1] and M Hartono[2]

[1] Postgraduate Program in Logistics and Supply Chain Engineering, Faculty of Engineering, University of Surabaya, Raya Kalirungkut, Surabaya, 60293, Indonesia
[2] Department of Industrial Engineering, Faculty of Engineering University of Surabaya, Raya Kalirungkut, Surabaya, 60293, Indonesia

E-mail : markus@staff.ubaya.ac.id

Abstract. Nowadays companies are required to work more productive yet maintain good quality. To increase productivity, company needs a conducive effort to optimize the work system. Supply chain management (SCM) is one form of competitive advantages that is applied to every industrial system. There are many factors which can influence the supply chain performance (SCF).Ergonomics is deemed to be one of significant drivers for SCF. This paper proposes and discusses the concepts related to ergonomics which is considered as macro ergonomics in supply chain system. SCM has been seen as macro ergonomics concerning on organization, people, technology, information and resources. This paper shows the importance of ergonomics at the inbound logistics (includes receiving, storing, inventory control and internal distribution management) and outbound logistics (the activites to get the finished product to customer) which are regarded as the critical supply chain activities. A scientific literature review on ergonomics as a critical determinant for SCF is discussed and provided as a theoretical contribution for further research related to the role of ergonomic to support supply chain performance.

1. Introduction

Supply chain management is a form of competitive excellence which is applied in every industrial system. The competitive strength of supply chain is achieved when the operation or act of adding value generates a maximum output. Supply chain is a network of companies that work in harmony to create and deliver product to the hands of end users [6]. In this era, every company is expected not only to work faster, but also to produce more high-quality products; however, there are still a number of businesses that do not consider the affecting factors and their risks. Ergonomics is the affecting factor which maintains the relationship between the company and the environmental aspects [2]. Ergonomics is divided into two sections which are micro- and macroergonomics. The former is the linkage of the workers, machines, methods and materials used, whereas the latter is a top-down sociotechnical system approach which is applied to an overall work system design of various interaction levels of microergonomics, such as human-job, human-machine, and human-software interface. Macroergonomics is something integrated because it covers knowledge, method, and equipment from sociotechnical system, industrial psychology, system design, physical ergonomics, and theoretical ergonomics. The use of macroergonomics in the supply chain performance evaluation is to analyze the structural work system by evaluating human, environmental, and equipment factors which are the support in production activities. The main purpose of macroergonomics is to harmonize the work system as a whole both in the micro and macro level, to increase the work satisfaction which is committed to the health and safety of the workers, and to increase productivity [12]. The company should build a macroergonomics concept that pays attention to the technical aspects (machines equipped with safety), organization/management, good work environment and competent operators/employees [1].

There are several studies done about the important role of ergonomics that could improve supply chain management performance. Pintelon&Azadeh (2016) stated that a systematic framework is by categorizing human factors in maintenance decision making which are human error, reliable calculation, workplace design/macroergonomics, and human resource management. Whereas Hasle & Jensen (2012) found that the ergonomic means and human factor to create international collaboration could affect global supply chain. Identifying five main challenges from global supply chain, especially the ones related to the sustainable social aspects, are the criteria for social sustainability, the role of key performance indicators in supply chain management, the constant changes in the supply chain, the challenges in building participation, and the agency development and mechanism regulations. The research done by Dewa, et al (2016) produces a model to conduct a sensitivity analysis to the implementation of corrective actions that is generated in the corrective models to reduce errors. According to Hartono (2008), ergonomics does not only focus on human and machine, but also on the interaction between culture and its surroundings. Ergonomic evaluation should also be understood and applied in vendors, suppliers, and all value chain components to achieve high level supply chain performance. This analysis uses Porter's value chain model. Supply chain is considered as value chain which covers supplier, the company itself, distribution and buyer. There are two value chains, which are primary activities and support activities. Primary activities include inbound logistic, operations, outbound logistic, marketing and sales, and services.

Based on several research results that have been conducted, it can be temporarily inferred that ergonomic factor in supply chain holds a vital role to the success of supply chain management itself. On previous studies, mostly ergonomics only focuses on operations, which is the product formation process. Due to this fact, this research discusses the importance of ergonomics on inbound and outbound logistics to improve supply chain performance. Inbound logistic is the role of all processes related to the reception storage, and distribution of internal input. Moreover, outbound logistic is the role of providing the products to the customers, such as inventory control and transportation.

The purpose of this research is to answer the question of how the role and contribution of ergonomic factor in the context of planning and supply chain operation is to inbound and outbound logistic. With many challenges and issues in operational management in supply chain, there is a need for an understanding and ergonomic management in planning and operating supply chain, which becomes a fundamental subject to be studied and managed.

2. Literature Review

2.1. Supply chain management

The term supply chain is defined in various ways. Pujawan (2010) stated that supply chain is the network of companies that work hand-in-hand to create and deliver product to the hands of end users. The purpose of every supply chain is to maximize the supply chain profitability, which is the difference between total revenue obtained from the consumers and the total of overall cost spent on a supply chain. According to Ravindran & Warsing (2013) customer is the single source of profit for every supply chain, which is met on a trade-off between responsiveness and efficiency. Supply chain responsiveness includes supply chain that is able to respond to the demand in a broad scope, have a short lead time, handle products with wide variation, and have a high service level.

2.2 Key performance indicator of supply chain

According to Miranda and Amin WT (2005), to develop a balanced performance measurement system, the company should consider financial and non-financial aspects. Even though financial performance measurement is vital to strategic decision-making and making external reports, control towards manufacturing operation and distribution is better handled with non-financial measurement. Supply chain performance measurement has an important role in knowing the company condition, whether it experiences decline or incline, and also the improvement as well as what kind of correction that should be conducted to enhance their performance. The measurement of supply chain performance includes company performance measurement of their internal and external process. The internal process happens within the company starting from the production planning to the product delivery to the customers. On the other hand, external process involves the relationship of the company with outside stages, which are supplier and customer. According to Rakhman(2006), performance measurement is important due to the following reasons: performance measurement can control direct and indirect performance;it will maintain the company's track to achieve the purpose of increasing supply chain;and it can be used to increase supply chain performance. Faulty measurement method can cause the decrease in supply chain performance. Supply chain can be directed after performance measurement is conducted.

2.3 Macroergonomics

Macroergonomics is a sub discipline of ergonomics which focuses on studying work system design [12]. A work system is comprised of a mutual interaction between personnel with hard and soft wares. A work system includes two or more individuals who cooperate to achieve a general purpose in an organization. Macroergonomics is known as an ergonomics sub discipline related to the correlation between humans, organizations and technology. Macroergonomics is an integrated discipline because it covers knowledge, methods, and equipment from sociotechnical system, industrial psychology, system engineering, physical ergonomics, and theoretical ergonomics. In its application, macroergonomics presents a valuable niche that none of these areas are overlooked. As science, macroergonomics directs to develop an understanding of a work system, behavior, or personnel that interact with hard or soft wares in an internal physical environment, external surroundings, and organizational structure as well as process to improve. Macroergonomics also studies profession, but it examines jobs and work system in a wider scope. Several points that are discussed in macroergonomics are organizational structure, interaction between people in an organization, and workers' motivation aspect. On the other hand, microergonomics only observes from occupational level, but macroergonomics views from both occupational and organizational levels. The table below states the differences between micro- and macroergonomics:

Table 1. Differences micro- and macroergonomics [12]

Characteristics	Microergonomics	Macroergonomics
Level of discussion	Micro	Macro
Work unit	Task, sub task	Work division, group
Purpose	Optimize work	Optimize work system
Focus	Detailing	Extensive review
Measurement tools	Generally measure physically, such as area, i-lumination, decibel, and time	Generally organizational and measure subjectively, such as the number of people, range of control, behavior, and moral.
Expertise application	Anatomy, psychology, psychology of perception	Organizational behavior, industrial psychology and organization

2.4 Porter's value chain

Value Chain Analysis is a strategy used to analyze the company's internal activities. It is a process where a company identifies the primary activity and support that increase the value of the product, then analyzes it to reduce cost or increase differentiation. Value Chain Analysis that is extensively used by companies is F e year 1985. The said model is

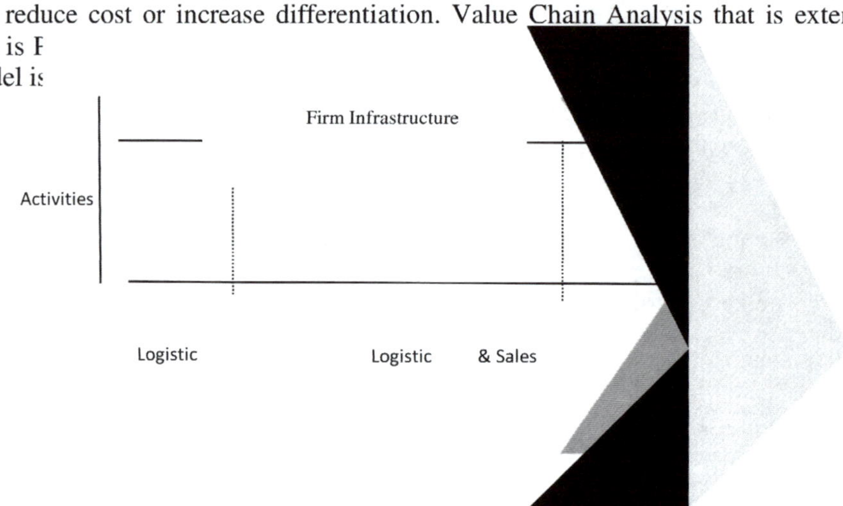

Figure 1. Porter's Value Chain [2]

Porter's Value Chain is divided into two categories, which are primary activities and support activities. The explanation of Porter's Value Chain category is as follows.

- Primary Activities
  The main activity is directly related to physical creation, sales, maintenance, and support from a product or service. Several components that support primary activities are:
  1. Inbound Logistic
     All process related to receiving, storing, and distributing internal input.
  2. Operations
     The action of transformation that changes the input to be the output, which will be sold to the customers.
  3. Outbound Logistic
     Give the customers their product or service. Outboung logistic is the activities to get finished product including warehousing, order fulfillment, transportation and distribution management.
  4. Marketing & Sales
     A process used to persuade the customers to buy the sold products.
  5. Service
     The action related to maintaining product value or services for the customers after the purchase.

2.5 The relation between supply chain performance and ergonomics

Not only does ergonomics focus on humans and machines, it also focuses on the interaction between culture and its surroundings. If a company is not supported by good ergonomics, supply chain performance will not run well. Ergonomics affects the accuracy and rate of a production process tremendously. The initial step to establish the right ergonomics to supply chain is by recognizing the ergonomic standard used. For instance, the product and design process standard that should be ergonomic and safe. Macroergonomics is a concerned with a broad view of ergonomics which is called technical system. Since supply chain management us regarded asa system of organization, there will be a closer relantionship between macroergonomics and supply chain management. The concern of how to maintain comfortable work station, flow of material, safety, team work and other. In today's competitive business world, employees are expected to work faster with more qualified result to meet customer needs. The biggest issue of ergonomic is how to maintain the employees or operators in appropriate way when they are working. All the workstation available should fits to the body dimension of operators, so they can work safely, comfortably and more productively [2]. Ergonomic can also be viewed in inbound and outbound logistic. Inbound logistic is from the suppliers to point of storage or use. The engineering ergonomist may have to work with supplier, providing specifications related to the material management. Other objectives for the procurement, transportation, warehousing must also be included in the ergonomics process. If the inbound logistic are not going well, it will hinder the production process such as inventory can not meet the demand. Outbound logistic is the activity to get finished product including distribution. If the distribution isnot going well, it will result in a late delivery product and thus, also failure to meet customer satisfaction.

3. Research Methodology

This research is theoretical; hence, the output produced is a study of literature, and data collection is yet to be done. This research focuses on the role of ergonomics in supply chain performance. The first step is collecting some journals which have a similar topic to this research. Material collection is an important stage. The database selection istaken through Emerald, Science Direct and Scholar.google.com. with keywords "macroergonomics", "supply chain". Criteria for journal selection is published from 2008-2018 (the last ten years). This stage is conducted by entering the title, abstract and conclusion which is similar to this research.

4. Analysis

The study related to the role of ergonomics in supply chain performance has been conducted by Hartono (2008). This research used Porter's Value Chain model with several additional variables to create the current complex field of ergonomics. Hartono (2008) stated that it is important to highlight ergonomics determinants such as employees' physical capability, safety, task requirements, equipment requirements, and policy or procedure in work system. These should be accounted for in every part of the company's supply chain. This research has not discussed further about the involvement of ergonomics in primary activities and support activities as one of the supporters for supply chain performance. Another study is by Hasle & Jensen (2012) which found the ergonomic means and human factor to create an international collaboration which could affect the global supply chain. Hasle & Jensen (2012) identify five main challenges of global supply chain, especially the ones related to the sustainable social aspects, which are the criteria for social sustainability, the role of key performance indicators in supply chain management, the constant changes in the supply chain, the challenges in building participation, and the agency development and mechanism regulations. The research done by Dewa, et al (2016) produces a model to conduct a sensitivity analysis to the implementation of corrective actions that is generated in the corrective models to reduce errors. Sensitivity analysis model is built by using Monte Carlo simulation. This model is equipped with human aspect that tends to be unclear, which uses the fuzzy concept. This model is capable of recommending cost effective corrective actions. However, this study has yet to consider the mental and inherent behavior aspects that are attached to human individuals. The summary of this study's differences with the previous one and the research position is as follows.

Table 2. Research position table

Characteristics	Hartono (2008)	Hasle & Jensen(2012)	Dewa, et al (2016)	This research (2019)
The scope of research	-	Operations (machine,	Operations (machine,	Inbound logistic (Inventory

Characteristics	Hartono (2008)	Hasle & Jensen(2012)	Dewa, et al (2016)	This research (2019)
		assembly)	assembly)	control & transportation) and outbound logistic distribution
Method	Porter's Value Chain	General ergonomics	Monte Carlo simulation	Porter's Value Chain
Performance measurement used	Did not measure performance	Occupational accident levels and attendance	Human error in carrying out the production process	Delivery time, number of complaints

Based on the table, this research focused on inbound and outbound logistic with the Porter's value chain method. Key performance indicator of this researh is delivery time and number of complainys. This research is expected to be a literature review for further reseacrh related to the role of ergonomics in supply chain performance.

5. Conclusion

After conducting reviews of several journals related, it can be concluded that macroergonomics has a vital role in a company's internal and external process. This will provide a considerable impact towards every process, supply chain included. Research results show that there are not many journals that discuss the effects of macroergonomics in supply chain on inbound and outbound logistics. To date, most of the discussions focus mainly on management from the company, and non-value-added activity on the production floor in the form of defects found. This becomes one of the considerations that could be used to conduct further research.

6. Acknowledgement

We would like to thank Directorate General of Learning and Student Affair, Ministry of Research Technology and Higher Education for giving us the financial support in this publication as implementation of grant program in the postgraduate improvement quality by the contract number (No.036/B4/PPK-SPK/VII/2019).

7. References

[1] Sukapto and Djojosubroto 2012 Penerapan ergonomi makro untuk meningkatkan keselamatan kerja dalam industri sepatu. *Indonesia J. Industrial Engineering* **21** 89-96

[2] Hartono 2008 Incorporating Ergonomic into Supply Chain Systems *J. Logist. Supply Chain Manage.* **1** 61-71

[3] Pintelon and Azadeh 2016 Human factors in maintenance: a review *J. Qual. Maint. Eng.* **22** 218-237

[4] Hasle and Jensen 2012 Ergonomics and sustainability – challenges from global supply chains *Work* **41** 3906-3913

[5] Shanmugam and Robert 2015 Human factors engineering in aircraft maintenance: a review *J Qual. Maint. Eng.* **21** 478-505

[6] Dewa P, Pujawan, and Vanancy 2017 Human errors in warehouse operations: an improvement model *International J. Logist. Syst. Manage.* **27** 299-314.

[7] Hancock 2012 Human factor *Encyclopedia of Human Behavior*

[8] Jan D, Ralph B, Peter B, Pascale C, Pierre F, William S M,John RW, and Bas van der D2012A strategy for human factors/ergonomics:developing the discipline and profession *Ergon.* **55** 377-95.

[9] Anika M and Ina F 2015 Collaborative information seeking environments benefiting

from holistic ergonomics *Library Hi Tech* **33** 439-59.
[10] Tarwaka, Solichul HA, Bakri,and SudiajengL 2004 Ergonomi untuk keselamatan, kesehatan kerja dan produktivitas (Surakarta: UNIBA PRESS)
[11] Santoso G 2004 Ergonomi Manusia, Peralatan dan Lingkungan (Jakarta: Prestasi Pustaka)
[12] Purnomo H 2012 Antropometri dan Aplikasinya (Yogyakarta: Graha Ilmu)
[13] Hendrick HW and Kleiner BM 2001Macroergonomics: an Introduction to Work System Design (Santa Monica: Human Factors and Ergonomics Society)

IOP Conf. Series: Materials Science and Engineering **703** (2019) 012035 doi:10.1088/1757-899X/703/1/012035

Fluazinam Potential as a Fungicide in Liquid Culture System for the Growth of *Haematococcus pluvialis* Microalgae

J R Witono, V Novianty, H Santoso, A Miryanti and A J Kumalaputri

Department of Chemical Engineering, Parahyangan Catholic University, Jl. Ciumbuleuit no. 94, Bandung, 40141, Indonesia.

E-mail: judy@unpar.ac.id

Abstract. All across biological system, fungal infection in freshwater systems is well known to obstruct the aquatic biomass population such as microalgae. However, there is a lack of extensive information on the suitable technique to control them. This research aims to assess the effects of the fungicide fluazinam as a single dose treatment on the *Haematococcus pluvialis* growth rate and the presence of chytrids fungus in a liquid culture system. The variables used in this experiment were the concentration of fluazinam given to the culture (i.e. 0.25; 0.5; 0.75; 1; 1.25 ppm) as a single dose treatment. The growth performance of *H. pluvialis* was observed in 6 days consecutively based on the cell density by a single direct absorbance measurement and the cell counting by haemocytometer. The best growth was achieved in a culture system which was given fluazinam at a concentration of 0.5 ppm (started from 1.0 x 10^5 cells/mL to 5.5 x 10^5 cells mL^{-1} on the 5th day) and no chytrids fungus are found in the culture system either. So, in conclusion, the use of fluazinam is a promising treatment for removing contaminants that inhibit *H. pluvialis* growth.

1. Introduction

More than 7000 species of green microalgae are found growing in a variety of habitats. One of them is *Haematococcus pluvialis* which becomes a commercial interest due to its ability to produce the highest level among all known astaxanthin producing organisms. It may reach up to 5 % dry weight in the encysted aplanospore state [1] and can reach 95% of the total carotenoids [2].

Astaxanthin (3,3′-dihydroxy-ß-carotene-4,4′-dione) is a bright red secondary carotenoid from the same family as lycopene, lutein, and ß-carotene, synthesized by some microalgae, plants, yeast [3]. Astaxanthin has a wide range of applications in the food, feed, cosmetic, aquaculture, nutraceutical, and pharmaceutical industries because of its free radical scavenging capacity [4]. The natural astaxanthin exhibited higher bioactivity as compared to the synthesized astaxanthin [5, 6]. In general, astaxanthin is produced by Haematococcusis through a two-stage culture: vegetative (green) and aplanospore (red) stages [7, 8]. At first stage cultivation, *H. pluvialis* cells are green and vegetative under favourable environmental conditions (i.e. nitrogen sources, light intensities, aeration rates, temperature and nutrients), and once subjected to stress conditions like high light, cells begin to synthesize and accumulate astaxanthin, transforming green vegetative stage to red aplanospores [9].

Due to the complex life-cycle of *H. pluvialis* it is important to maximize cell densities of alga at the "green stage" of cultivation by maximizing cell growth to obtain high biomass with high astaxanthin yield from the "red stage" [10]. However, vegetative cultivation of Haematococcus is the most problematic part due to the susceptibility to contamination which results in a slow growth rate or low cell concentration. Fungal contamination by chytrids has been recognized as one of the most serious hurdles for producing astaxanthin from the green algae *H. pluvialis* [11]. Moreover, the process of culturing a small volume *H. pluvialis* culture through the green swimmer stage to a large volume in the red cyst stage may take weeks due to the cultivation of Haematococcus involves growing, dividing, and accumulating carotenoids [12]. During this time the *H. pluvialis* cells are vulnerable, weaken the physical integrity of the cells (e.g., lysis) and easily attacked by contamination which reduces the chances of *H. pluvialis's* survival in both the green swimmer and red cyst stages [13].

Fluazinam is known as a fungicide with broad-spectrum activities which was commercialized since the early 1990s. The good characteristics of this fungicide control every stage of a fungal life cycle, multi-site action, no cross-resistance to other fungicides, safe and beneficial. Its chemical structure can be seen in figure 1 and its IUPAC name is 3-Chloro-N-(3-chloro-5-tri fluoromethyl-2-pyridyl)-α, α, α-tri fluoro-2,6-dinitro-p-toluidine.

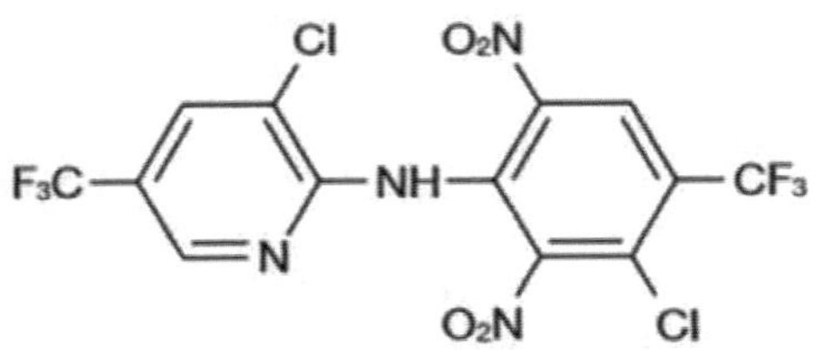

Figure 1. Fluazinam – phenyl pyridinamin

The species sensitivity of freshwater organisms such as invertebrates and algae to fluazinam has been assessed by Van Wyngarden et al. [14] and showed a good result. It has been tested as well in liquid culture systems for the growth of microalgae [15] combined with other fungicide. But until now it is hard to find research on the application of fluazinam as a single dose treatment. Therefore, the goal of this research is to study the application of fluazinam as a single dose treatment on the *Haematococcus pluvialis* growth rate and the presence of chytrids fungus in a liquid culture system.

2. Materials and Method

2.1. Materials and culture conditions

Haematococcus pluvialis UTEX Collection Strain 2505 was obtained from Algae Analytics, US, and has been cultivated for 1 (one year) in our laboratory. During that time, the inoculum was contaminated by fungus. Indeed, fluazinam (Nando 500 SC) with active ingredient 500g/L was purchased from Nufarm, Indonesia. The stock cultures were grown photoautotrophically in Modified Bolds 1N-3P medium without soil water, vitamin B12, and biotin (only thiamine 0.1 mg/L). The culture cell concentration was 1 x 10^5 cells/mL when 200mL *H. pluvialis* starter cultures were carried out into 250 mL erlenmeyer flask. The cultures were illuminated with white-blue LED lamps at light intensity of 3200 lux with dark: light cycle was 10:14 hours. The cells were grown in batch cultures and aerated at 0.2 L/min. under the temperature of the environment at 25.7 ± 1^0C without any additional CO_2 (see figure 2 below).

2.2. Investigating the effect of fluazinam

Five different concentration of fluazinam (i.e. 0.25; 0.5; 0.75; 1; and 1.25 ppm) were given to the contaminated culture (initial cell numbers 1 x 10^5 cells/mL) as a single dose. The contaminated culture grown in the absence of fluazinam was used as a comparison. Culture density and cells number were monitored in 6 days consecutively.

IOP Conf. Series: Materials Science and Engineering **703** (2019) 012035 doi:10.1088/1757-899X/703/1/012035

Figure 2. Set up of the experiments

2.3. The growth rate measurement

The optical density was measured using a spectrophotometer UV-6100PC Double Beam (Mapada, Shanghai, China). The absorption peak of green *H. pluvialis* cells was obtained at 680 nm. Cells number was quantified using an improved Neubauer haemocytometer (Marienfeld Superior, Lauda-Königshofen, Germany) at 400x magnification with a binocular microscope (Olympus, Tokyo, Japan).

The mean growth rate (μ) was calculated on the cell number basis according to the equation 1:

$\mu\,(day^{-1}) = \frac{\ln X_2 - \ln X_1}{t_2 - t_1}$ ..... (eq.1) ,where X_1 and X_2 in the equation represent the cell number at the times t_1 and t_2, respectively.

3. Results and Discussion

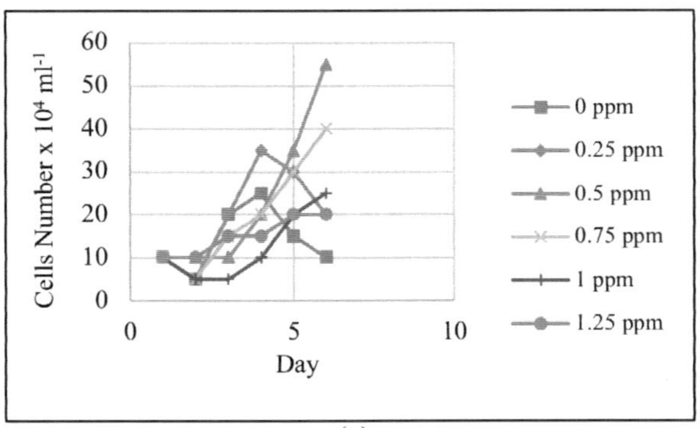

(a)

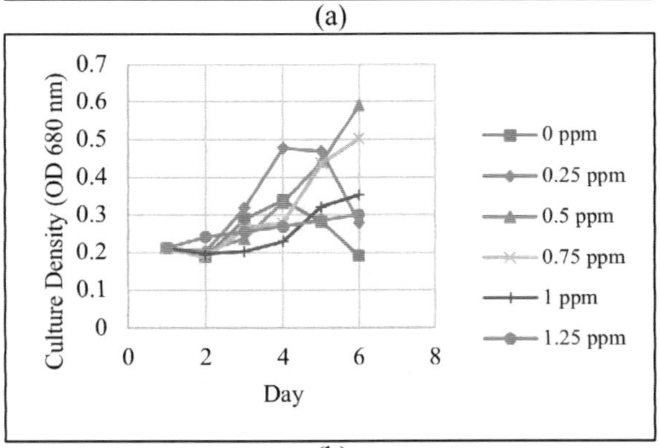

(b)

Figure 3. The Growth of *H. pluvialis* during treated by various concentrations of fluazinam (a) based on cell number (b) based on culture density

The presence of fungus in the liquid culture medium is not just a significant biological problem but furthermore is an economic challenge to the commercial cultivation of algae in industrial scale [16]. Han et.al [3] indicated that the fungal contamination by chytrids has been recognized as one of the most serious hurdles for producing astaxanthin from the green algae *Haematococcus pluvialis*. By applying fluazinam into the medium it was expected that the presence of fungus would be reduced. But as fungicide action has side effects on nonfungal microorganisms as well, therefore, it might affect the *H. pluvialis* growth rate [17]. It can be seen from figure 3, although the cell density (≈ optical density of the cell) represents the absorbance of both living and dead cells, whereas the cell number only measures the living cells, both of the graphs illustrate the same pattern.

3.1. The effect of fluazinam on the growth rate of H. pluvialis

From figure 3 can be seen that *H. pluvialis* in all cultures after being treated by fluazinam on the first day, did not grow, even some of the cell died especially at the culture with high concentration of fluazinam. The cells number of *H. pluvialis* in a contaminated culture in the absence of fluazinam and the presence of 0.25 ppm fluazinam just really collapsed on the 4th day. These were most likely due to the lack of nutrients source for the de novo protein biosynthesis of microalgae as filamentous fungus wrapped and bound from one to another alga. Besides, in the contaminated culture with 0.5 ppm or higher concentrations of fluazinam, more fungus could be reduced. Therefore, they were still increasing on the 4th day and so forth.

On the 6th day, the culture in the presence of 0.5 ppm of fluazinam resulted the highest cell number (5.5 $x10^5$ cells mL^{-1}), it was about 5 times from the initial number. The cell number in the culture with 0.75 and 1 ppm of fluazinam resulted in 4 $x10^5$ cells mL^{-1} and 2.5 $x10^5$ cells mL^{-1} respectively. They had a similar trend when viewed from the culture density side. The culture with fluazinam at a concentration of 0.5 ppm resulted in microalgae density at 0.59 ODU which was about 3 times the density of microalgae grown in the absence of fluazinam. It was also observed that the culture with fluazinam at concentrations 0.5ppm, 0.75 ppm and 1 ppm were still growing on the 6th day.

The culture grown in 1.25 ppm of fluazinam had the same number of cells ($2x10^5$ cells mL^{-1}) on the 4th and 5th day. However, there was a slight increase in culture density between 4 and 5 days of the culture grown in 1.25 ppm of fluazinam. That might happen because the cell density represents both living and dead cells as explained above. It means the rate of cell reproduced and collapsed were almost similar at that time.

This shows that the higher concentration of fluazinam might cause a decrease in microalgae growth. However, an effective concentration of fungicide may cause a decrease in microalgae growth but causes a greater reduction in the growth of a chytrid as well (can be seen in figure 5). Many of the fungicides used in agriculture inhibit respiration, and many are effective against a wide range of plant-pathogenic fungi. Fluazinam belongs to the secondary amine group, is cited as an uncoupler of oxidative phosphorylation, this compound inhibits reactions most likely on some proton-coupled uptake process necessary for fermentative growth within the cell [18]. Whereas some fungicides affect fungal respiration at the level of the enzyme complex system, other fungicides may impact respiration through other targets. Fluazinam sparks very unusual uncoupling activity in target cells. The metabolic state of their mitochondria was found to be inhibited after exposure to fluazinam. Consequently, ATP production is inhibited and downstream cellular metabolisms is interrupted [17]. Low concentrations of the uncoupler fluazinam inhibited conidial germination and germ tube elongation in fungus [19].

The culture with 0.5 ppm fluazinam gave the maximum growth rate (on cell basis) of *H. pluvialis* with the value 0.568 day^{-1} starting on the 4th day. Whereas before the 4th day, the culture without fluazinam and treated by 0.25 ppm of fluazinam supported the high growth rate of *H. pluvialis* (i.e. 0.805 day^{-1} and 0.973 day^{-1}). This condition is in line with the observation by some researchers, instead of killing the fungi, fluazinam also inhibits the growth rate of H. pluvialis by weakening cell walls [17].

After the 4th day, the cell number of *H. pluvialis* in the system with 0.25 ppm fluazinam and without fluazinam collapsed. This showed that the development of contaminants is more dominant than the cell resistance.

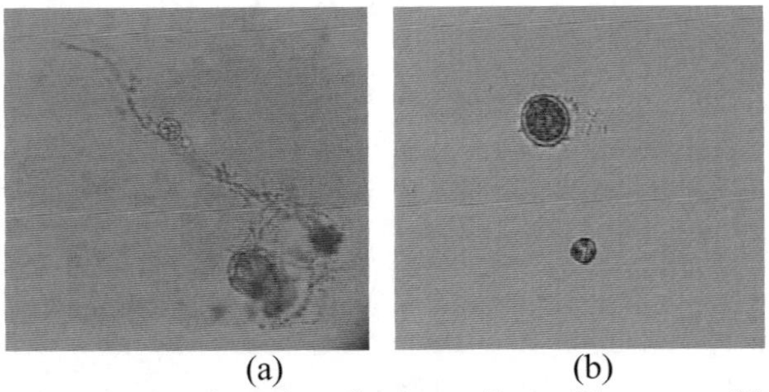

(a) (b)

Figure 4. *H. pluvialis* cells that collapsed because of fungus (a) in the absence of fluazinam and (b) in culture with 0.25 ppm of fluazinam at the 5th day

3.2. The effect of fluazinam on the fungus

Figure 4 (a) shows H. pluvialis cells that collapsed because of fungus in the absence of fluazinam on the 5th day whereas figure 4 (b) shows the dead and living cells of H. pluvialis in 0.25 ppm of fluazinam on the 5th day. Different microalgae may have different ranges of toxicity that determined by the growth of a microalga in the presence of certain concentrations of a fungicide. An effective concentration of a fungicide should not be toxic to the microalgae being cultured in the liquid system.

Figure 5 displays a macrozooid cell, green palmella cells, a sporocyst that was infected with chytrids, respectively (before the culture treated with fluazinam). As shown in figure 3, the growth of *H. pluvialis* treated by 0.5 ppm or higher concentration of fluazinam continued to increase from the first day being given a fungicide. This indicated that 0.5 ppm or higher of fluazinam in culture can inhibit the growth of chytrids (showed by no visible filament), even though the greater concentration of fluazinam has a different effect on the growth of microalgae. Figure 5 (b) displays healthy green motile cells, green palmella cell and a healthy cyst and sporocyst, respectively.

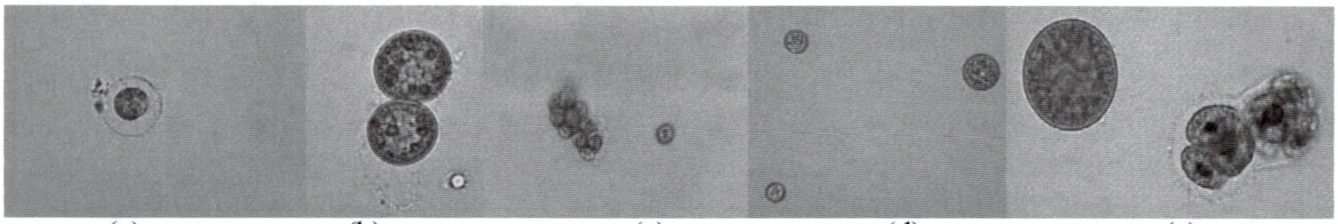

Figure 5. A macrozooid cell (a), green palmella cells (b), sporocyst (c) of *H. pluvialis* before treated with fluazinam and after treating with fluazinam (d, e)

4. Conclusion

Fluazinam as a single dose fungicide at a certain concentration (0.5 ppm) in the culture of the *H. pluvialis* can be an effective toxic to the growth of chytrids but do not turn off the growth of microalgae. It can be concluded that fluazinam at a certain concentration has a potential as a fungicide even as a single dose treatment.

Acknowledgment

The authors would like to express their thanks to the Directorate General of Research and Development Reinforcement (Ministry of Research, Technology and Higher Education) - Direktorat Riset dan Pengabdian Masyarakat, Direktorat Jenderal Penguatan Riset dan Pengembangan RISTEKDIKTI, which finances this research and also the Institute of Research and Community Service - Parahyangan Catholic University (LPPM – UNPAR) who assisted the authors in the preparation of this research. Special thanks are addressed to our research group team who made this research can be carried out.

References

[1] Wayama M, Ota S, Matsuura H, Nango N, Hirata A, and K S 2013 Three-Dimensional Ultrastructural Studyof Oil and Astaxanthin Accumulation during Encystment in the Green Alga *Haematococcus pluvialis PLOS One* **8** 1-9.

[2] Boussiba S, Bing W, Yuan J P, Zarka A, and Chen F 1999 Changes in Pigments Profile in the Green Alga *Haeamtococcus pluvialis* Exposed to Environmental Stresses *Biotechnol. Lett.* **21** 601-4.

[3] Han D, Li Y, and Hu Q 2013 Biology and Commercial Aspects of *Haematococcus pluvialis*, in *Handbook of Microalgal Culture: Applied Phycology and Biotechnology*, ed Richmond A and Hu Q (USA: John Wiley & Sons) 388-405.

[4] Shah M R, Liang Y, Cheng J Y, and Daroch M 2016 Astaxanthin-Producing Green Microalga *Haematococcus pluvialis:* From Single Cell to High Value Commercial Products *Frontiers in Plant Sci.* **7** 1-28.

[5] Ambati R R, Phang S M, Ravi S, and Aswathanarayana R G 2014 Astaxanthin: Sources, Extraction, Stability, Biological Activities and Its Commercial Applications—A Review *Marine Drugs* 12 128-52.

[6] Capelli B, Debasis B, and Gerald C 2013 Synthetic astaxanthin is significantly inferior to algal-based astaxanthin as an antioxidant and may not be suitable as a human nutraceutical supplement *Nutrafoods* **12**.

[7] Aflalo C, Meshulam Y, Zarka A, and Boussiba S 2007 On the relative efficiency of two- vs. one-stage production of astaxanthin by the green alga *Haematococcus pluvialis Biotechnol. and Bioeng.* **98** 300-5.

[8] Fabregas J, Domínguez A, Regueiro-Ramonde M, Maseda A, and Otero A 2000 Optimization of culture medium for the continuous cultivation of the microalga *Haematococcus pluvialis Appl. Microbial Biotechnol.* **53** 530-5.

[9] Göksan T, Ak I, and Kılıç C 2011 Growth Characteristics of the Alga *Haematococcus pluvialis* Flotow as Affected by Nitrogen Source, Vitamin, Light and Aeration *Turkish J. Fisheries Aquatic Sci.* **11** 377-83.

[10] Zhang Z, Wang B, Hu Q, Sommerfeld M, Li Y, and Han D 2016 A New Paradigm for Producing Astaxanthin from the Unicellular Green Alga *Haematococcus pluvialis Biotechnol. Bioeng.* **113** 2088-99.

[11] Hoffman Y, Aflalo C, Zarka A, Gutman J, James T Y, and Boussiba S 2008 Isolation and characterization of a novel chytrid species (phylum Blastocladiomycota), parasitic on the green alga *Haematococcus Mycological Res.* **112** 70-81.

[12] Masojídek J and Torzillo G 2008 Mass Cultivation of Freshwater Microalgae, in *Encyclopedia of Ecology,* ed. Jørgensen S E and Fath B D (Amsterdam: Elsevier) pp 2226-35.

[13] Carney L T and Sorensen K 2015 *Methods for Treating a Culture of Haematococcus pluvialis for Contamination using Hydrogen Peroxide*, ed A U Chandler (US: Heliae Development, LLC, Gilbert, AZ)

[14] Van Wijngaarden R P A, Arts G P H, Belgers J D M, Boonstra H, Roessink I, Schroer A F W, and Brock T C M 2010 The Species Sensitivity Distribution Approach Compared to a Microcosm Study: A Case Study with the Fungicide Fluazinam *Ecotoxicology Environ. Safety* **73** 109-22.

[15] McBride R C, Behnke C A, Botsch K M, Heaps N A, and Del Meenach C 2014 *Use of Fungicides in Liquid Systems* (San Diego: Shapphire Energy, Inc.).

[16] Carney L T and Lane T W 2014 Parasites in Algae Mass Culture *Frontiers in microbiol.* **5** 278-78.

[17] Yang C, Hamel C, Vujanovic V, and G. Y 2011 *Fungicide: Modes of Action and Possible Impact on Nontarget Microorganisms* (ISRN Ecology) pp 1-8.

[18] Allen J, Davey H M, Broadhurst D, Rowland J J, Oliver S G, and Kell D B 2004 Discrimination of Modes of Action of Antifungal Substances by Use of Metabolic Footprinting *Appl. Environ. Microbiol.* **70** 6157-65.

[19] Leroux P, Gredt M, Leroch M, and Walker A S Exploring Mechanisms of Resistance to Respiratory Inhibitors in Field Strains of *Botrytis cinerea,* the Causal Agent of Gray Mold. *Appl. Environ. Microbiol.* **76** 6615-30.

IOP Conf. Series: Materials Science and Engineering **703** (2019) 012036 doi:10.1088/1757-899X/703/1/012036

The use of pyrolusite to remove Pb and Cd in aqueous solutions: isotherm and thermodynamic

Y Fransiscus[1,2], M W B Kembie[1] and N M Tanusaputra[1]

[1]Department of Chemical Engineering, University of Surabaya, Raya Kalirungkut, Surabaya, 60293, Indonesia
[2]Centre for Environmental Studies, University of Surabaya, Raya Kalirungkut, Surabaya, 60293, Indonesia

E-mail : yunus@staff.ubaya.ac.id

Abstract. The utilization of natural material as an adsorbent to eliminate contaminants in the water is of great interest. Natural material can be directly used without further treatment that may lead to higher environmental impact. In this study, pyrolusite which is immensely available in Indonesia, especially in East Nusa Tenggara, was used to adsorb Cadmium (Cd) and Lead (Pb) in a single and simultaneous condition. Several operational parameters such as initial concentration, pH and temperature were applied to elaborate the sorption mechanism and to determine the adsorption capacity. The amount of absorbed Cd and Pb increased at a higher pH and reached the optimum condition at pH 7 for both. The adsorption isotherm for both adsorbates were best modeled following the Langmuir equation, giving an adsorption capacity as high as 108.70 mg/g and 123.45 mg/g for Cd and Pb respectively. Additionally, the thermodynamics data showed that the adsorption process in this study was endothermic process.

1. Introduction

The unexpected presence of Cd and Pb in the water system will cause serious effects on human health such as necrosis, proteinuria, gastrointestinal and neurology problem, even lead to malformation and cancer in the worst case [1-3]. Therefore, continual effort to find the most feasible technique to eliminate Cd and Pb from the contaminated water. Chemical precipitation, by adjusting the pH of the solution or after coagulation–flocculation process, is considered as the easiest method to remove heavy metals. However, this treatment produces excessive sludge that needs to be further treated [4]. Filtration, including membrane and ultra-membrane system; reverse osmosis and electro-dialysis are other systems, but less effective in term of operational and economic perspective [5,6]. Meanwhile, the application of the biological approach is still ongoing process as stability on the performance needs to be improved. Adsorption has been widely used on different scale because of its simplicity in operation and high efficiency over a wide range of concentrations [7]. Currently, there are many researches focus on the investigating and/or modifying adsorbent materials in order to find ones that have high adsorption capacity [8]. Assorted materials encompassing natural minerals (soils), industrial by-products (e,g. fly ash, steel slag), agricultural products (e.g. rice husk, sawdust, peanut shells), fishing waste (e.g. Shrimp and crab shells) have been reported to have high adsorption capacity [9]. Among those materials, natural minerals (soils) with oxides based content such as Iron and Manganese oxides are highly preferred. Oxides based minerals usually found with a poor crystalline structure and known as a potent scavenger of trace metals in soil, sediment, and water [1]. Manganese has been known as one of the most common minerals on earth [10]. One of the most important Manganese oxide minerals is pyrolusite, which has low economical value in the natural state. Pyrolusite is massively available in Indonesia especially in East Nusa Tenggara. For this particular mineral, single chains of edge-sharing Mn(IV)O_6octahedral share corner with other chains to build a framework structure accommodating tunnels of square cross sections that are one octahedron by one octahedron [11]. Pyrolusite is an essential material used as an oxidant in organic synthesis and for industrial battery, however information on its potential function in wastewater or water treatment area is still limited. The utilization of manganese (di)oxides as an adsorbent for water contaminants have been reported by several papers. Manganese oxide in synthetic form (AMO – amorphous manganese oxide and MOF - manganese dioxide formed in situ) was used to remove Cu, Cd, Pb and Zn [7,12]. The application of beta manganese dioxide (ß-MnO_2) to eliminate Pb(II) in a single condition was successfully presented by considering several parameters [1]. Meanwhile the use of natural manganese dioxide to eliminate Cu in aqueous solutions had been successfully presented by Demirkiran (2015), with the efficiency as high as 96.5%. In spite of the existing information, further investigation is still needed to have a better understanding on the performance of pyrolusite as an adsorbent for heavy metals. Based on that consideration, a set of batch experiments was conducted to evaluate the mechanism of pyrolusite to adsorb Cd and Pb in aqueous solutions.

2. Materials and Method

2.1. Materials

The entire chemicals used in this study, including $Pb(NO_3)_2$, $CdNO_3$, HCl, NaOH, $KMnO_4$, H_2SO_4 and Sodium oxalate were in analytical grade and no treatment needed prior to the utilization. Pyrolusite, an ore form of manganese dioxide was supplied by a private company in Surabaya. Characterization of pyrolusite was done by titration process to quantify the manganese oxides content, Fourier Transform Infra-red (FTIR, Shimadzu) to identify the surface functional groups and X-ray diffraction (XRD) to determine the structural condition. The content of manganese oxides in pyrolusite was as high as 84.76% (w/w) as described in previous work [4].

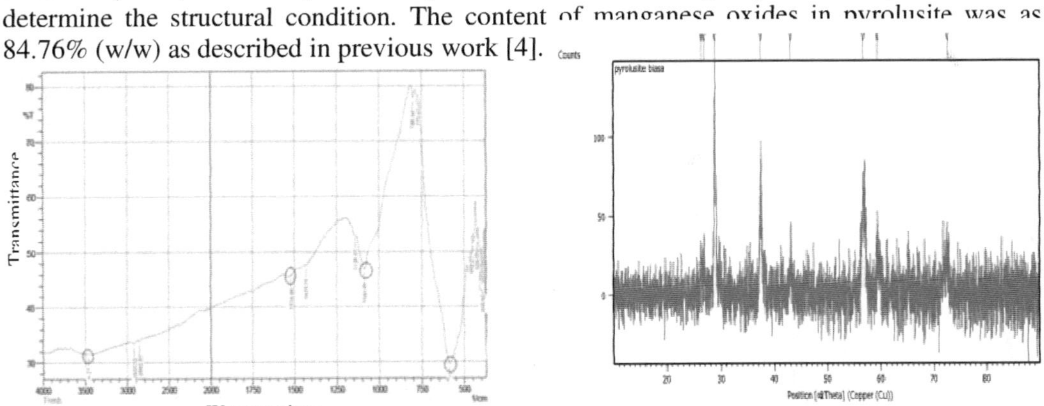

(a) (b)

Figure1.(a) FTIR spectra for pyrolusite; (b) XRD pattern for pyrolusite.

As described in figure 1(a), the result of FTIR analysis indicating a stretching vibration of Mn-O and Mn-O-Mn bonds at the peak of 590cm^{-1}. Two other peaks, at 1085cm^{-1}, expresses a vibration of hydroxyl groups that are joined to Mn atoms, and at 1519cm^{-1} reveals bending vibration of O-H and H_2O which insinuates that water molecule exists in the nanostructure. In addition, the rise at peak of 3464cm^{-1} indicating stretching vibration on O-H bond and bending vibration of the absorbed water molecule in the lattice. XRD analysis was conducted with Cu Kœ radiation (1.544À) and operated at the voltage of 40kV and tube current of 30mA. Figure 1(b) portrays the peaks that appear

(2Ø=28.8949°; 37.6337°; 43.0352°; 56.8375°; 59.4419°) are mostly reflections of manganese dioxide (except value at the peak of 2Ø=72.6963°), indicating a relatively high purity of the material.

2.2. Method

All batch experiments in this study were conducted at temperature T=298K and under ambient conditions. To evaluate the effect of pH on the adsorption process, 200 mL of heavy metal solutions (Cd or Pb) with the initial concentration of 10 mg/L were prepared into the conical flask, follow to that, 2.5 g/L pyrolusite was added. The flask was put on thermo–shaker for mixing at 160 rpm, and at certain interval sampling was done until reaching equilibrium condition. Sample solutions were filtered using membrane filter 0.45 µm (Sartorius) prior to analysis with UV – Spectrometer (HP series 8453). The desired pH level (4, 5, 6, 7) for both metals solutions was set using either 0.1 M HCl or NaOH. The reduction of heavy metals concentration in aqueous solutions expressed as removal efficiency that can be calculated with equation (1). The adsorption capacity (q_e) was defined by equation (2) and also used in the determination of isotherm models.

Removal efficiency (%) : $R = \frac{C_0 - C_e}{C_0} * 100\%$ (1)

Adsorption capacity (q_e, mg/g) : $q_e = \frac{(C_0 - C_e)}{m} * V$ (2)

Where C_0 is the initial concentration of Cd or Pb (mg/L), C_e is the concentration of heavy metals at equilibrium condition (mg/L), q_e(mg/g) is Cd or Pb adsorbed at given mass of adsorbent in the working volume (L). Adsorption isotherm models were developed by running batch experiments in different initial concentrations. The collected data of equilibrium concentrations for each metal were analyzed by using two adsorption models, Langmuir and Freundlich.

Langmuir : $q_e = \frac{b q_{max} C_e}{1 + b.C_e}$(3) Freundlich : $q_e = k_f C_e^n$(4)

where C_e is the heavy metals concentration in the solution at equilibrium (mg/L), b is Langmuir constant related to the affinity of binding sites (L/mg), q_{max} is the maximum adsorption capacity (mg/g), k_f is a Freundlich constant related to roughly adsorption capacity (mg/L), n is the constant related to adsorption density. In order to have an understanding on the effect of temperature, another series of batch experiments in three different temperatures (298K, 308K and 318K) were carried out. Thermodynamic data expressed as values of standard enthalpy change (ΔH°), standard entropy change $(\Delta S^{\circ})$and Gibb's free energy (ΔG^0). Those values can be determined by applying these equations:

$K_d = \frac{C_s}{C_e}$(5) ; $\ln K_d = \frac{\Delta S^{\circ}}{R} - \frac{\Delta H^{\circ}}{RT}$(6) ; $\Delta G^{\circ} = \Delta H^{\circ} - T\Delta S^{\circ}$(7)

where C_s and C_e are then removed and remaining concentrations of heavy metals, respectively. *R* (8.314 $Jmol^{-1}K^{-1}$) is the ideal gas constant, T is temperature in Kelvin. The values of ΔH^0 and ΔS^0 were defined from the slope and y-intercept on the plot between ln K_d versus $1/T$.

3. Results and Discussion

3.1. Effect of pH

Figure2(a) depicts the transition of Pb and Cd removal from the aqueous solutions at different pH. The trends were similar for both metals, initially steep at pH 4 to 6 then the change was relatively flat at the pH of 6 to 7. In all pH conditions, the removal of Pb was higher than Cd, with the best achievement at pH = 7 which were 84.06% and 81.82% respectively. At acid condition (pH < 7), the main species of Pb and Cd are Pb^{2+} and Cd^{2+}, thus the removal is mainly driven by the adsorption process. The surface charge of pyrolusite will then very crucial in determining the portion of the adsorbed metals. Several papers mentioned that pyrolusite (MnO_2) has a net charge of zero at the pH of 1.4 to 4.5, known as pH_{pzc} (point of zero charge) [13-15]. Lately from a deeper investigation, it was confirmed that the pH_{pzc} of pyrolusite is 4.2 [1]. This means that if the pH solution is above 4.2 the surface charge of pyrolusite will tend to be negative due to the deprotonation process, in which hydrogen ions leave the sites [16]. This phenomenon causes the decreasing of electrostatic repulsion between the adsorbent and metal ions, as a consequence, the interaction will be more intense since electrostatic attraction being dominant. Minor change in the portion of metals removed at pH 6 to 7 can be explained from the speciation of Pb and Cd, at pH 6.69 – 7 for Pb and 9.03 for Cd the solubility of both metals decrease significantly and tend to precipitate (as described in reaction 8 and 9), this instability diminishes the adsorption process.

$$Pb(H_2O)_3OH^+_{(aq)} + OH^-_{(aq)} \leftrightarrow Pb(H_2O)_2(OH)_{2(s)} + H_2O_{(l)} \quad (8)$$

$$Cd(H_2O)_3OH^+_{(aq)} + OH^-_{(aq)} \leftrightarrow Cd(H_2O)_2(OH)_{2(s)} + H_2O_{(l)} \quad (9)$$

3.2. Effect of Temperature

The removal of Pb and Cd through the adsorption process in three different temperatures is presented in figure 2(b). For both metals, the highest removal was achieved at 318K which were 88.07% and 84.60% for Pb and Cd respectively. The results indicate that the adsorption process of Pb and Cd onto pyrolusite was more favorable at higher temperature. A plausible elucidation for this is that dissolved metals need to destroy the hydration sheaths prior to the adsorption process. The destruction process needs energy, while the adsorption process of metals onto pyrolusite emits energy. The need for energy seems to be more than the available source, thus in higher temperature, the elimination of heavy metal ions is more effective. Adsorption mechanism of Pb and Cd on pyrolusite can be evaluated from thermodynamic data, as provided in Table I. The positive value of standard enthalpy (ΔH^o) relates to the information that the adsorption process is endothermic. The low value of ΔH^o in this experiment is similar to that found in [1], confirming that the adsorption process of Pb onto pyrolusite is weak, which also applies for Cd. However, since the value of ΔH^o for both metals is still bigger than 4200 J/mol, the weak adsorption process is still determined by chemical reaction [17]. The negative value of ΔG^o reflects the spontaneous process, at a higher temperature the process is more spontant. The lower ΔG^o value the lesser energy required for the attachment process of adsorbent – adsorbate [18]. In this case, the need of energy for adsorption process of Pb to pyrolusite is smaller compared to Cd. The positive value of ΔS^o suggests structural changes in adsorbate and adsorbent creating disorder-ness of the solid-liquid system [19].

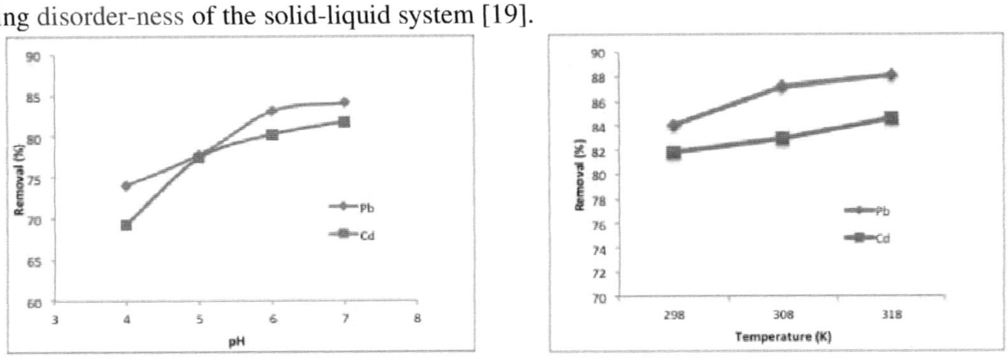

(a) (b)

Figure 2. (a) Removal of Pb and Cd in different pH value; (b) Removal of Pb and Cd in different temperature

Table 1. Thermodynamic data of the adsorption of Pb and Cd

T (K)	C_0 (mg/L)	ΔS^O (J/mol.K)	ΔH^o (J/mol)	ΔG^o (J/mol)
		Pb		
298	10	72.18	13335.66	-8173.98
308	10	72.18	13335.66	-8895.78
318	10	72.18	13335.66	-9617.58
		Cd		
298	10	52.15	7851.74	-7688.96
308	10	52.15	7851.74	-8210.46
318	10	52.15	7851.74	-8731.96

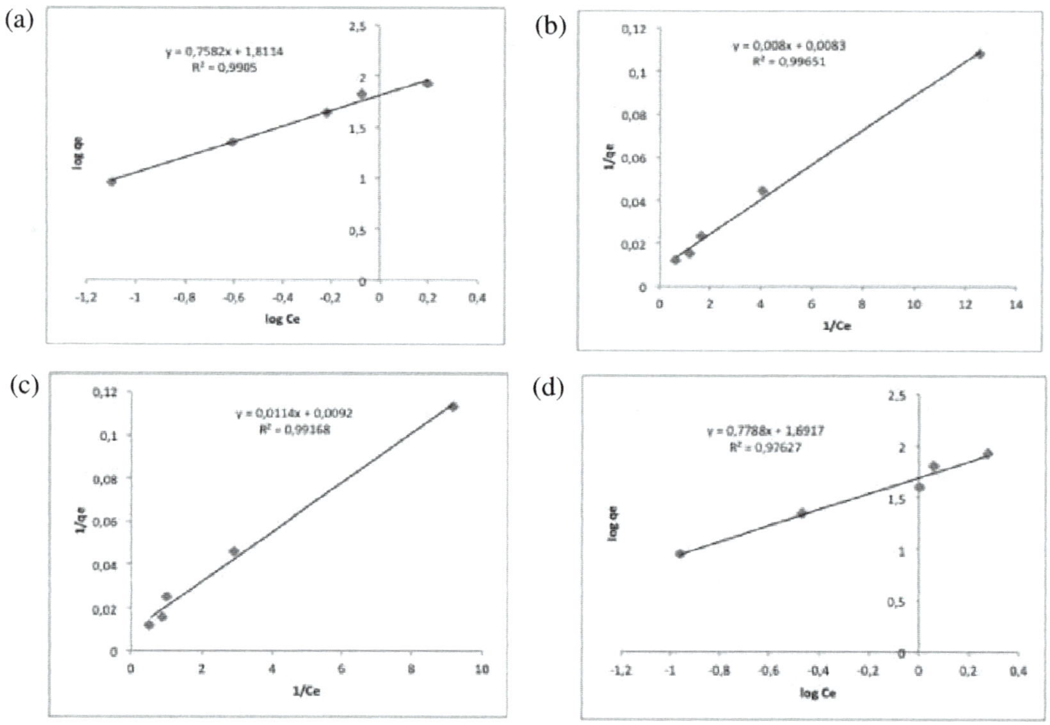

Figure 3. (a) Linear plot of Langmuir for Pb, (b) Linear plot of Freundlich for Pb, (c) Linear plot of Langmuir for Cd and (d) Linear plot of Freundlich for Cd

Table 2. Langmuir and Freundlich isotherm parameters for Pb and Cd

Model	Parameter	Pb	Cd
Langmuir	b (L/mg)	1.01	0.88
	q_{max}(mg/g)	123.45	108.70
Freundlich	K_f(mg/g)	64.71	49.09
	n	0.76	0.78

3.3. Adsorption Isotherms

In this study, Langmuir and Freundlich were applied to identify the isotherms adsorption that can be used to describe the distribution of adsorbate in the liquid and solid phases. The collected data of each metal concentration at equilibrium condition (C_e) were plotted versus the amount of adsorbed metal in pyrolusite (q_e) following the equation (3) and (4) for Langmuir and Freundlich model respectively.

The result of the analysis provides information that for both metals, Langmuir model fits better than Freundlich model to elaborate adsorption process onto pyrolusite (Figure 3), which is expressed by the higher value of R^2 (0.99651 and 0.99168 – Langmuir for Pb and Cd respectively compared to 0.99050 and 0.97627 – Freundlich for Pb and Cd). Langmuir model suggests that the adsorption mechanism is driven by the chemical process (chemisorption) through three stages, which are external diffusion, internal diffusion and chemical interaction. The finding that the Langmuir model is more suitable to the adsorption data also indicates that pyrolusite has a homogenous surface and as the result of creating identical adsorption activity. Pb and Cd are adsorbed by forming a complete monolayer all around the surface of adsorbate. By that, the assumption of interaction between adsorbate following the formation of monolayer as suggested by the Freundlich model is negligible. Following to that, the maximum adsorption capacity of pyrolusite for Pb and Cd determined by the preferred model was 123.45 mg/g and 108.70 mg/g respectively. The adsorption capacity of pyrolusite, a natural manganese dioxide, for Pb and Cd is superior if compared to other natural materials. For example, by using kaolin, fly ash, blast furnace slag and bentonite, it was reported the adsorption capacity (mg/g) for Pb were 4.50; 5.00; 5.50 and 7.60 respectively [20]. The use of Australian zeolite could reduce Pb and Cd as much as 9.97 mg/g and 6.72 mg/g, comparable to that the Korean zeolite that had the adsorption capacity of 8.90 mg/g and 7.10 mg/g respectively [11, 21]. Considering the finding, pyrolusite could be used as an effective and low cost adsorbent for heavy metals.

4. Conclusions

Pyrolusite was successfully tested as an alternative adsorbent for Cd and Pb in aqueous solutions. The thermodynamic data and isotherm model that the adsorption process for both metals is endothermic and spontaneous, indicating that chemical reaction drove the mechanism. The adsorption capacity of pyrolusite for Cd and Pb following the Langmuir model were 108.70 mg/g and 123.45 mg/g respectively.

References

[1] Zhao D, Yang X. Yang, Chen C, Wang X, 2010, *Chem. Eng. J.* **164** 49
[2] Carolin C F, Kumar P S, Saravanan A, Joshiba G J and Naushad M 2017 *J. Environ. Chem. Eng* **5** (3) 2782
[3] Asuquo E, Martin A, Nzerem P, Siperstein F and Fan X 2017 *J. Environ. Chem. Eng* **5** (1) 679
[4] Fransiscus Y, Widi R K, Aprilasti G O and Yuharma M D 2018 *Int. J. Adv. Sci. Eng. & Infor. Technol.* **8(3)** 818
[5] Lima J O, Ragassi M F, Gimenes M L, Vieira M G A and da Silva M G C 2017 *Chem. Eng. Trans.* **56** 1891
[6] Chuan F N, Wei F, Lin Z M and Yi G X 2018 *Trans. Nonferrous Met. Soc. China* **28** 794
[7] Qin Q, Wang Q, Fu D and Ma J 2011 *Chem. Eng. J.* **172** 68
[8] Kow K K, Kew P L, Yusoff R and Abdullah E C 2017 *Chem. Eng. Trans.* **56** 231
[9] Ahmed M J K and Ahmaruzzaman M, 2016 *J. Water Process Eng.* **10** 39
[10] Demirkiran N 2015 *Trans. Nonferrous Met. Soc. China* **25** 647
[11] Post J, 1999, *Proc. Natl. Acad. Sci. USA* **96** 3447
[12] Puppa L D, Komarek M, Bordas F, Bollinger J C and Joussein E 2013 *J. Colloid Interface Sci.* **399** 99
[13] Tripathy S S, Bersillon J L and Gopal K 2006 *Desalin.* **194** 11
[14] Tan W F, Lu S J, Liu F, Feng X H, He J Z and Koopal L 2008 *Soil Sci.* **173** (4) 277
[15] Gray M J,Malati M A, Rophael M W 1978 *Jour. Electro. Chem. Inter. Electrochem.* **89** (1) 135
[16] Chaudhry S A, Khan T A, Ali I, 2016 *Egypt. J. Basic Appl. Sci.* **3** 287
[17] Zou W, Han R, Chen Z, Jinghua Z, Shi J, 2006, *Colloids Surf., A* **279** 238
[18] Hu C, Zhu P, Cai M, Hu H and Fu Q 2017 *Appl. Clay Sci.* **143** 320
[19] Ajmal M, Rao R A K, Anwar S, Ahmad J and Ahmad R 2003 *Bioresour. Technol.* **86** 147
[20] Mishra P C and Patel R K 2009 *J. Hazard. Mater.* **168** 319
[21] Jeon C S, Park S W, Baek K, Yang J S, Park J G 2012 *Korean J. Chem. Eng.* **29** 1171

IOP Conf. Series: Materials Science and Engineering **703** (2019) 012037 doi:10.1088/1757-899X/703/1/012037

Power generation in a plant-microbial fuel cell assembly with graphite and stainless steel electrodes growing *Vigna Radiata*

K R S Pamintuan[1,2], K M Sanchez[1]

[1] School of Chemical, Biological, and Materials Engineering and Sciences, Mapua University, Intramuros, Manila, Philippines 1002
[2] Center for Renewable Bioenergy Research, Mapua University, Intramuros, Manila, Philippines 1002

E-mail: krspamintuan@mapua.edu.ph

Abstract. Plant-microbial fuel cells (PMFCs) are an interesting renewable energy technology that has the potential to generate clean electricity without competing with agriculture for land space. In this study, the electricity generation potential of mung beans (*Vigna radiata*) in a PMFC set-up with different electrode materials was explored. Three types of set-ups were prepared with five replicates each: PMFCs with stainless steel electrodes, PMFCs with graphite electrodes, and control pots without electrodes. This experimental set-up allowed for the evaluation of the better electrode material, and whether the PMFC environment harms or benefits the plants. The voltage gathered suggests that the potential difference generated in the PMFCs with differing types of electrodes were statistically the same ($\alpha = 0.05$). The same can be said for power and power density, although the system with stainless steel electrodes generated more power towards the end of the experiment. It was also evident that PMFCs with stainless steel electrodes experienced prolonged time lags due to the reduced biocompatibility of stainless steel. Polarization studies showed that a single PMFC is capable of generating power densities of 0.35 mW/m^2 and 0.12 mW/m^2 for stainless steel and graphite systems, respectively. The increased power density of PMFCs with stainless steel indicate the lowering of internal resistance brought by the stainless steel. Plants in the PMFCs set-ups were seen to grow faster, taller, and have higher pod output than those in the control set-up. These results indicate that the PMFC technology can be implemented in agricultural land for the continuous generation of passive electricity while growing food crops, eliminating the competition between energy generation and agriculture.

1. Introduction

The relentless energy crisis and environmental issues have prompted studies on diverse technologies as a source of alternative renewable energy [1]. An extended discipline of Microbial Fuel Cells (MFCs) that uses plants as the source of microbial organisms and converts chemical energy to electrical energy is the *Plant-Microbial Fuel Cell* (PMFC). This technology has gained attention and has been continuously studied for the last decade in which claims to produce non-destructible and sustainable green electricity via living plants [2]. Sustainable renewable energy is the next generation of energy transition and P-MFCs has great competency in reducing the dependence on the destructive use of fossil fuels. This bio-electrochemical system generates electricity concurrently without the need of disrupting the living plants thus defeating competition between arable lands for food production power generation plants that beguiles deforestation. It also eliminates the need of additional energy input from external sources as other biomass resources and renewable energy requires. With this alternative, the transport of biomass is no longer needed in which avoids the depletion of nutrients in the ecosystem [3]. PMFC technologies are based on the principle of rhizodeposition, which is the excretion of organic compounds from the roots that is the source of electrons and electricity generation by electrochemically active bacteria in a microbial fuel cell [4]. The PMFC is a version of a MFC where the system makes use of living plants as the source of microbial matter. Its roots are located within the anodic region assumed to be an anaerobic environment within the structured fuel cell allowing the system to act as a bioreactor that enables the conversion of chemical energy to green electrical energy. The requisite elements for Plant-Microbial Fuel Cell assembly are the electrodes, a membrane separator, an external circuit connecting the two electrodes and a resistor.

For Plant-Microbial Fuel Cells, present studies have demonstrated current production associated with plant rhizodeposits using plants from flower bearing species such as *Ipomoea aquatica, Typha latifolia* and *Canna indica* and plants from grass species such as *Oryza sativa, Glyceria maxima, Spartina anglica, Arundinella anomala, Lemna minuta* and *Arundo donax* [5]. The grass species in particular have shown to be useful for electrical power generation [6]. Previous researches focused on power production by the optimization of the reactor design and configurations, plant selection or nutrient media [6] and the development of bio-electrocatalytic interfaces [7].

Prior the launch of PMFCs to an industrial scale, numerous aspects of its study requires further undertaking. Several limiting factors present great challenge toward the engineered and commercialized application of a MFC Technology [8]. From poor electro-active bacteria kinetics, insufficient understanding of the electron transfer mechanism within the cell, the influence of complex environmental conditions and its effect on the microbial colonization and the knowledge of microbial community structures remains a large chunk of uncertainty in the study of PMFC systems [7]. For most PMFCs executed in laboratory conditions, energy sustainability and durability have yet to be achieved as the electrical power generated is considerably very low that lacks sufficiency to drive electronic devices available today [9]. The search for effective architecture and materials that will enable high generation of electrical energy in a PMFC system is constant as there are very few existing studies on the use varying electrode materials and their resulting performance in the power generation. In addition, the performance exhibited by the PMFCs are significantly affected not solely by the internal system but also by its external environment as well. This has not been considered thoroughly in the previous studies. Up to this date, a variety of selection of plant species and methods of creating PMFC assemblies and configuration are continuously studied in order to improve and industrialize this technology for green electricity generation.

The objective of this study is to design and construct a Plant-Microbial Fuel Cell assembly using graphite rods and stainless steel mesh as the electrode materials with Mung beans (*Vigna radiata Wilzeck*). Through the comparisons of the electrical potentials, the capability of electricity generation using the fuel cell chamber design utilizing different electrode materials was to be identified. This will discern which electrode material is more effective for the design system. Also, this study aims to evaluate the effect on PMFC elements in the growth and development of the model plants to evaluate whether a PMFC can be used to simultaneously generate electricity and produce biomass.

2. Materials and methods

2.1. PMFC assembly and set-up

Each single-chambered PMFC was composed of the electrodes, a resistor load, a semi-permeable membrane, the soil and the plant specimen. The stainless steel mesh and graphite rods were used as the electrode materials for the system. The resistor used had the load of 2000-Ω. A layer of cheesecloth was used as the membrane separator and copper wires were used as the external conductor that connected the electrodes with the resistor load.

The overall exterior that contained the assembly was made from polycarbonate plastic. Each fuel cell had the dimensions of 6-in x 6-in x 12-in (L x W x H). Two holes big enough for copper wires to pass through were made on one side of the customized pots with respect to the position of the electrodes. Additional holes were made on the bottom of the pot for drainage. From bottom to top of the container, the cell consists of 1-inch layer of garden soil, the cathode electrode, 5-inch layer of garden soil, the cheesecloth membrane cut in accordance with the dimensions of the pot, 4-inch garden soil, the anode electrode and finally the plant. The container was completely filled with soil which was patted thoroughly in every layer (Figure 1).

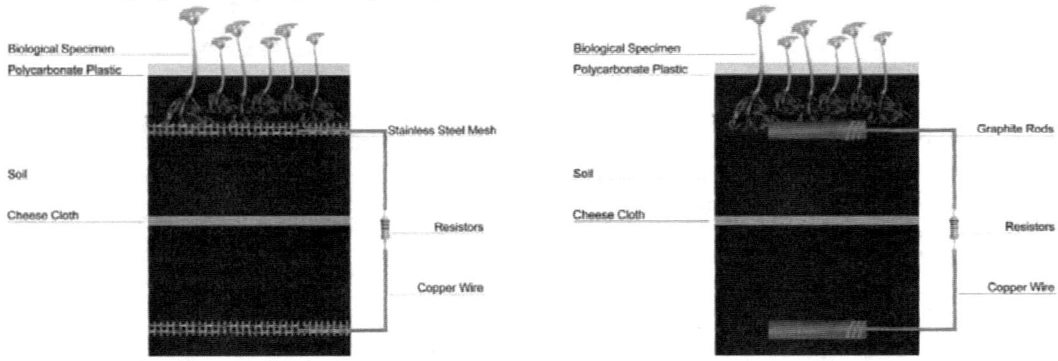

Figure 1. PMFC assembly with stainless steel (left) and graphite (right)

Green Mung Beans (*Vigna radiata Wilczek*) seeds that were sourced from the local market in Intramuros, Manila was the selected plant specimen for this experiment. The specimens were grown from seedlings in a soil medium composed of coconut coir and garden soil potting mixture sourced from Los Baños, Laguna. No additional nutrient media was used for planting. For outdoors planting, the seeds were required to be buried about 1-inch deep covered in soil as these particular seedlings require darkness to germinate. The seeds were planted on seed trays and were exposed to 8 to 10 hours of sunlight daily. The plant grow-out lasted for thirty days. During this period, plant care involved light watering to keep the seed bed moist to enhance the germination process. On the last day of plant grow-out period, the plants were transferred in the P-MFC set-up and were left for ten days to allow the microbial organisms to settle and consolidate with the soil. Continuous plant care was conducted. Upon plant transfer, the roots of the plant were fully covered with soil to avoid the plant from drying-out.

2.2. *Collection and treatment of data*

Voltage was recorded against an external resistor six times a day for 30 days straight using a digital multimeter (RS Pro RS14). Soil pH meter, hygrometer, and thermometer were used to monitor growing conditions. Current ($I = V/R$), power ($P = IV$), and power density ($P_D = P/A$) were computed from voltage readings taken against a 2000-ohm resistance.

2.3. *Polarization*

The polarization studies were conducted to indicate the power output of a certain fuel cell system under various resistor loads that are exhibited by polarization curves. This type of study is the most common method in testing a fuel cell's performance and determination of the optimum condition that it can achieve and is represented by a polarization curve that contains linear data that relates the power density and the current density. This allows comparisons with existing publications of fuel cell technology. The polarization studies conducted used sixteen different resistor loads varying from 5 Ω to 50,000 Ω. The voltage reading using each of these resistors were individually were recorded and compared to one another by producing a polarization curve showing the behavior of power generated

IOP Conf. Series: Materials Science and Engineering **703** (2019) 012037 doi:10.1088/1757-899X/703/1/012037

of a fuel cell chamber with respect to the step-wise incline of resistor loads. Current density was plotted against power density to complete the polarization curve.

3. Results and discussion

3.1. Voltage, power, and power density

The measured voltage and calculated power and power density are shown in Figures 2 and 3 below.

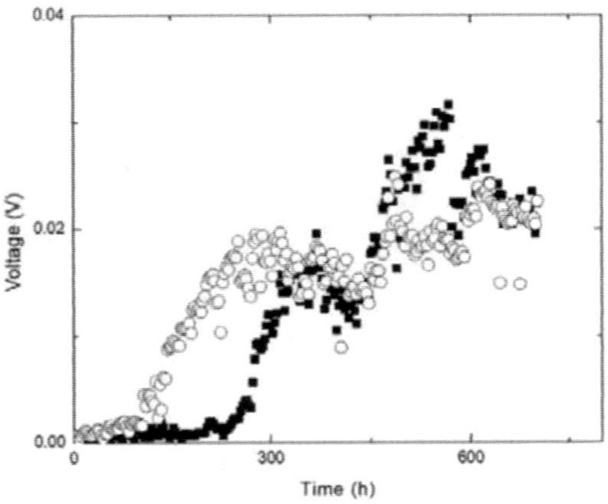

Figure 2. Voltage readings versus time (■: stainless steel, ○: graphite)

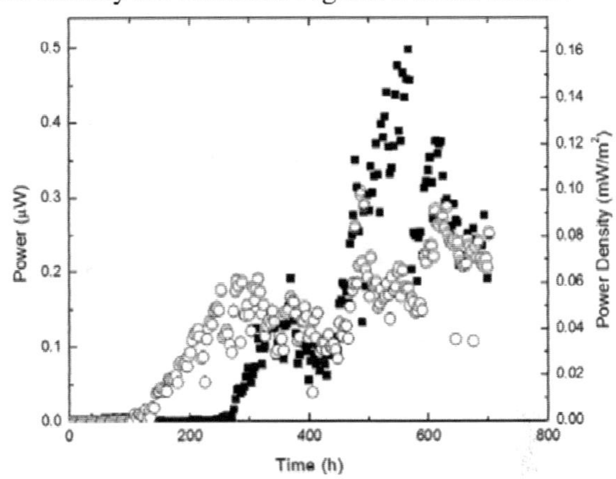

Figure 3. Power and power density versus time (■: stainless steel, ○: graphite)

There is no significant difference between the power and power density of both PMFC set-ups containing different electrodes ($\alpha = 0.05$). However, it is important to note that the PMFCs with stainless steel electrodes attained higher absolute power and power densities towards the end of the experimental period. The time lag for stainless steel systems was larger than that with graphite, presumably because of the inhospitable nature that stainless steel provides for bacteria (biocompatibility). The slow incline at the initial phase is due to the settlement of rhizodeposits where it is in the process of binding uniformly with the new soil. Right after, anaerobic reactions occur allowing the production of electron donors enabling the transfer of electrons that leads to the generation of electrical energy. The steady incline of power density generated means that the system is at a state of equilibrium and the environmental conditions are stable enough for the electricity generation.

3.2. Polarization

A polarization curve is made of the values of power densities under varying resistor loads. Through this fuel cell chamber's designed system will identify which resistor load will produce the optimum performance. The polarization studies conducted was done on the 30th day of data collection. At this point, the plant has reached maturity and is already bearing fruits.

By increasing the load stepwise, the polarization studies produced parabolic trends pinpointing the absolute maximum power generated found at the parabola's peak. This is the preferable resistance value to enable optimum conditions of the designed microbial fuel cells [10]. For the stainless steel systems, the maximum power density attained was 0.35 mW/m^2 under the resistance of 8200 Ω. This is 40% more than the power density using the default resistor of 2000 Ω which was 0.25 mW/m^2. This means that the applied external resistance was lower than that of the internal resistance, resulting to a lower yield. On the other hand, PMFCs with graphite electrodes obtained the maximum power density of 0.12 mW/m^2 under a resistance of 10000 Ω which is 50% larger than the power density under a resistance of 2000 Ω (0.08 mW/m^2). Overall, PMFCs with stainless steel electrodes obtained maximum power densities that are about 200% larger than graphite electrodes. Even though graphite is more biocompatible, stainless steel electrodes have lower resistances and thus contribute to the lower resistance of the PMFCs, thereby translating to higher power and power densities.

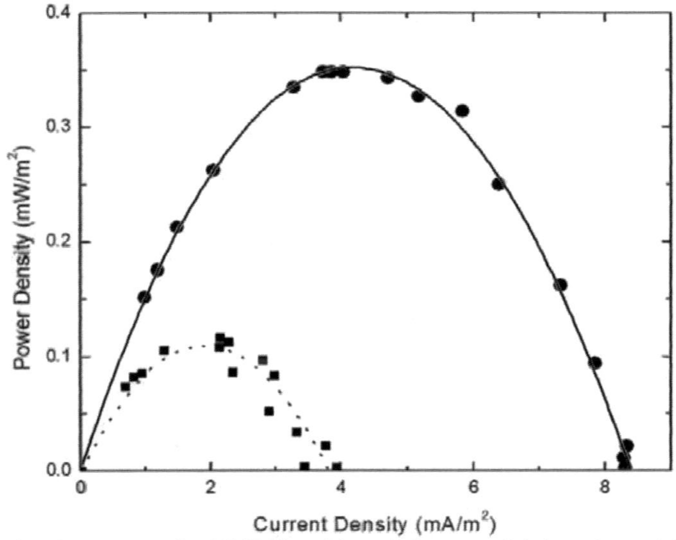

Figure 4. Polarization curves for PMFCs with stainless steel (●) and graphite (■) electrodes

3.3. Growth and development of plants

A separate set of trials (control) were made duplicating the entire PMFC set-up, only this time without electrodes. All replicates were exposed to the same conditions as those with electrodes. It is of interest whether the electricity generation of the PMFC has a positive or negative effect on the plants' growth. It was evident during the experimentation process that plants growing in the fuel cell assemblies with the electrode materials in their system showed more abundance and better growth as compared to the ordinary trial as the plants growing with the electrode elements appeared to be healthier and stronger. In the duration of the experiments, the trials with the electrode materials was observed to have greener leaves with taller, thicker and sturdier stems as compared with the control with brownish pale-green leaves and thinner stems. The maximum height of control was only about half the height of the plants with the electrode material. Upon monitoring, the highest average height measured for the plants in the PMFC trials reached 42 cm in height, while the height of those in control seemed stunted reaching to its highest average height of 22 cm. Fruit pods in the control were smaller compared to the pots with electrode material; the average length of the mung bean pods were about 8-cm in length for PMFC trials wherein the plants in a non-fuel cell environment was only about 4 to 5 cm long. In terms of maturity, the control plants matured later by two weeks than all the experimental trials conducted. The results of this study are in agreement with the results from a previous research [11] wherein the plants grown under polarized conditions (in PMFCs) have higher levels of intracellular nutrients compared to the control (no electrodes). One explanation could be the enhanced metabolic process of the plants stimulated by the electricity flow through the system,which caused the plant to accumulate more of its photosynthetic products [12]. This also leads to enhanced rhizodeposition, which explains the maintenance of power generation even with fluctuating sunlight levels.

3.4. Scale-up potential

The choice of the plant specimen was upon consideration of accessibility and availability in the area. Mung bean is included in the 21 major crops produced in the Philippines and according to the Philippine Statistics Authority's quarterly bulletin of the major vegetables and roots crops in the quarter of April to June 2018, the production of Mung bean had increased by 3.3% compared to the same quarter of the previous year, 2017. This translates to 24,800 metric tons of mung bean per annum. Approximately, for a yield of 1000 kg/ha, mung bean production would cover about 24,800 hectares of land which has the potential to generate about 760 MWh/y of passive electricity solely from the farming of mung beans. This can be made possible through stacking, as illustrated in a previous study [13].

4. Conclusions

The possibility of concurrent food and power generation is confirmed with the designed single-chambered PMFC for both systems utilizing graphite and stainless steel electrodes with *V. radiata*. However, the low levels of generated power density remains a challenge as it remains to be insufficient for applicable use. This may be due to the cell configuration and the used resistance within the system.

Polarization studies suggested that optimum levels of power density could be achieved by using resistor loads of 8200 Ω and 10000 Ω in stainless steel and graphite systems respectively. Executing the designed P-MFC under this resistor loads may produce power density up to 50% larger than the current power generated.

Through comparison between plants growing in the PMFC set up and normal plant set-up (control) without PMFC elements, the plants appear to be healthier and has more rigorous growth in contrast to the control with stunted growth and smaller pods. This is on the note that all plants are exposed to same environmental conditions.

With the land covered by the Mung bean production in the country, this study proves that it has high potential as a sustainable alternative green energy resource if only it can surpass the challenges of low power density production. Future studies must be focused on the amplification of power generated by the system to fuel the hopes on one day being able to generate clean electricity alongside agriculture.

References

[1] Zain S M, Ching N L, Jusoh S, and Yunus S Y 2015 Different Types of Microbial Fuel Cell (MFC) Systems for Simultaneous Electricity Generation and Pollutant Removal *J. Teknol.* **3** pp 13–9

[2] Strik D P B T B, Bert H V M H, Snel J F H and Buisman C J N 2008 Green electricity production with living plants and bacteria in a fuel cell *Int. J. Energy Res.* no. November 2007

[3] Helder M, Strik D P B T B, Hamelers H V M, Kuhn A J, Blok C, and Buisman C J N 2010 Concurrent bio-electricity and biomass production in three Plant-Microbial Fuel Cells using Spartina anglica, Arundinella anomala and Arundo donax *Bioresour. Technol.* **101** no. 10 pp 3541–7

[4] Pamintuan K R S, Gonzales A J S, Estefanio B M M and Bartolo B L S 2018 Simultaneous phytoremediation of Ni2+ and bioelectricity generation in a plant-microbial fuel cell assembly using water hyacinth (Eichhornia crassipes) *IOP Conf. Ser. Earth Environ. Sci.* **191** p 12093

[5] Lu L, Xing D and Ren Z J 2015 Microbial community structure accompanied with electricity production in a constructed wetland plant microbial fuel cell *Bioresour. Technol.* **195** pp 115–21

[6] Deng H, Chen Z and Zhao F 2012 Energy from plants and microorganisms: Progress in plant microbial fuel cells *Chem. Sus. Chem* **5** no. 6 pp 1006–11

[7] Santoro C, Arbizzani C, Erable B and Ieropoulos I 2017 Microbial fuel cells: From fundamentals to applications. A review *J. Power Sources* **356** pp 225–44

[8] Gong X B, You S J, Wang X H, Zhang J N, Gan Y and Ren N Q 2014 A novel stainless steel mesh/cobalt oxide hybrid electrode for efficient catalysis of oxygen reduction in a microbial

fuel cell *Biosens. Bioelectron.* **55** pp 237–41
[9] Qian F, He Z, Thelen M P and Li Y 2011 A microfluidic microbial fuel cell fabricated by soft lithography *Bioresour. Technol.* **102** no. 10 pp 5836–40
[10] Nikhil G N, Krishna Chaitanya D N S, Srikanth S, Swamy Y V and Venkata Mohan S 2018 Applied resistance for power generation and energy distribution in microbial fuel cells with rationale for maximum power point *Chem. Eng. J.* **335** no. May 2017 pp 267–74
[11] Hubenova Y and Mitov M 2015 Enhanced metabolic and redox activity of vascular aquatic plant Lemna valdiviana under polarization in Direct Photosynthetic Plant Fuel Cell *Bioelectrochemistry* **106** pp 226–31
[12] Nitisoravut R and Regmi R 2017 Plant microbial fuel cells: A promising biosystems engineering *Renew. Sustain. Energy Rev.* **76** no. March pp 81–9
[13] Pamintuan K R S, Clomera J A A, Garcia K V, Ravara G R and Salamat E J G 2018 Stacking of aquatic plant-microbial fuel cells growing water spinach (Ipomoea aquatica) and water lettuce (Pistia stratiotes) *IOP Conf. Ser. Earth Environ. Sci.* **191** p 12054

IOP Conf. Series: Materials Science and Engineering **703** (2019) 012038 doi:10.1088/1757-899X/703/1/012038

Gas sensitive properties of ZnO nanorods formed on silicon and glass substrates

V V Petrov[1], A P Starnikova[1], Y N Varzarev[1], K A Abdullin[2] and D P Makarenko[3]

[1] Southern Federal University, Research and Education and Centre "Microsystem technics and multisensor monitoring systems", Taganrog, 347922, Russia

[2] Al Farabi Kazakh National University, National Nanotechnology Laboratory of Open Type, Almaty, 050040, Kazakhstan

[3] JSC "VNIIHOLODMASH", Moskow 127410, Russia

E-mail : vvp2005@inbox.ru

Abstract. Nanostructured materials based on zinc oxide are being intensively studied, since such materials are used to create energy harvesters, gas sensors, and solar cells. Arrays of nanoscale ZnO nanorods are synthesized on silicon and glass substrates by the hydrothermal method. The formed ZnO nanorods with predominantly vertical orientation have an average transverse size of about 30–40 nm and a length of 500–600 nm. For electrical contact, V-Cu-Ni metallization was deposited over ZnO nanorods. The work investigated the gas-sensitive and electrophysical properties of the formed sensitive elements with respect to carbon monoxide (CO). It is shown that the sensor element based on ZnO nanorods on glass has a 30% higher sensitivity and lower (10 ppm) sensitivity limit than similar structures, but formed on silicon. Peculiarities of the response of gas sensors based on ZnO arrays of nanorods were determined, including the effect of temperature, the rate of gas flow, and the effect of the substrate. A method for measuring gas-sensitive properties has been developed. It is shown that sensors based on such sensor structures can be used as energy-efficient gas sensors, as well as detectors in early fire detection systems, robotic sensor systems, and other modern technologies.

1. Introduction

The properties of nanostructured materials based on oxide semiconductors are currently being intensively studied, since these materials have great potential for practical application. ZnO-based nanostructures are used to manufacture sensitive elements of gas sensors, piezo transducers and energy harvesters [1-4]. The principle of operation of gas sensors is based on the property of ZnO to change its electrical parameters during the adsorption of gas molecules.

The principle of operation of gas sensors is based on the property of ZnO to change its electrical parameters (for example, electrical conductivity) during adsorption of gas molecules. Resistive-type gas sensors based on ZnO nanorods formed by the method of pulsed laser deposition exhibit high sensitivity to carbon monoxide (II) [4]. However, research in this area is not enough. In particular, in [5, 6], the hydrothermal method formed arrays of ZnO nanorods on various substrates, but their gas sensitivity has not been sufficiently investigated. The purpose of this work was to study the gas sensitivity of gas sensors based on arrays of ZnO nanorods, to carbon monoxide (CO) at different temperatures.

2. Experiment

Arrays of ZnO nanorods were synthesized on p-type silicon and quartz glass by the hydrothermal method. A uniform distribution of the zinc acetate sol dissolved in ethanol on the substrate surface was achieved by centrifuging at a rotation speed of about 2000 rpm, further drying at 110-130 ° C and annealing at 350-450°C [5, 6]. The hydrothermal synthesis of ZnO nanorods was carried out in an aqueous solution of zinc nitrate and hexamethylenetetramine ($C_6H_{12}N_4$) in a glass beaker.
Hydrothermal treatment was carried out in the temperature range 90-97 ° C for 1-3 hours. The formed ZnO nanorods are predominantly vertical in orientation with a height of 590–660 nm and have an average transverse size of about 30–40 nm — figure 1 (a.) Then, contact metallization of V-Cu-Ni 0.2-0.3 μm thick was applied on them. The resistance of the obtained structures ranged from hundreds of kOhms during the formation of ZnO nanorods on silicon and up to tens of Mom during the formation of ZnO nanorods on glass.

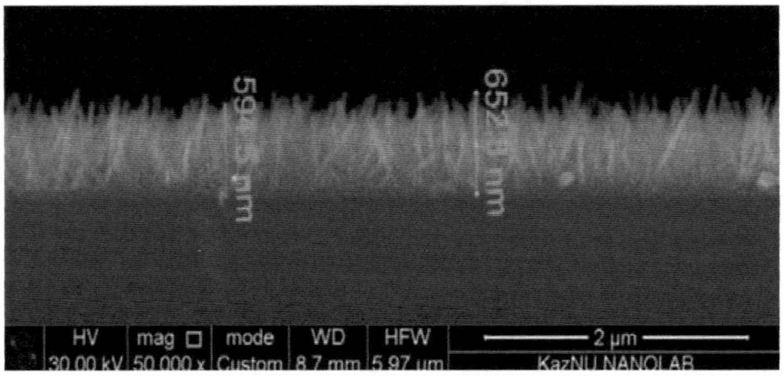

Figure 1. SEM images of the as-grown ZnO nanorod layer on a Si substrate.

Measurements of electrophysical and gas-sensitive properties were carried out on an automated bench for gas calibration of the collective use center "Microsystem Engineering and Integrated Sensors" — figure 2.

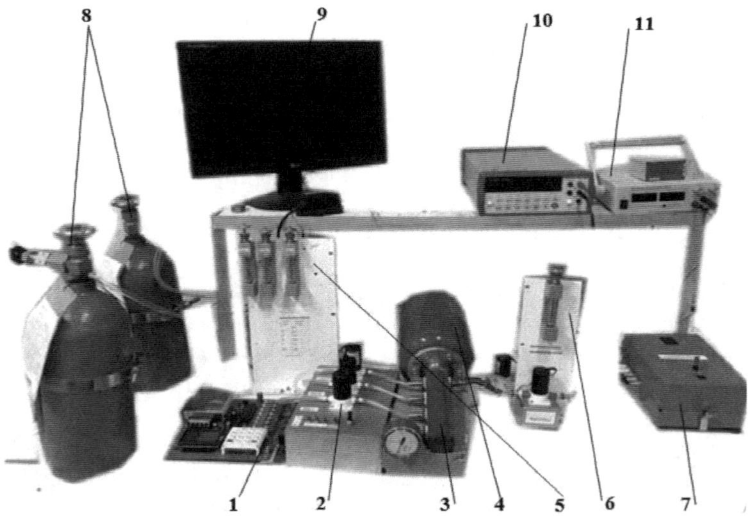

Figure 2. Automated gas sensor calibration stand: electronic control unit for the gas distribution system (1), solenoid valves (2), mixing chamber (3). receiver (4). unit for controlling the flow rate of the original gas components (5) unit for controlling the flow of gas

mixture (6); measuring chamber (7); cylinders with original gas components (8), personal computer (9), Keithly multimeter 2450 (10) heating control unit (11).

With the help of the stand, it is possible to automatically generate a mixture of gases up to three components of a given concentration with a lower concentration limit of 1 ppm or less, with a dilution factor of up to 100. The supply of the finished gas mixture is possible by controlling the speed of the gas flow is possible within 0.1–3.5 cm/min. The Study of the electrophysical and gas-sensitive properties of newly developed materials is possible by heating the samples to 400°C.

The measuring unit provides double shielding and allows high-accuracy measurement of parameters such as resistance of test samples of sensors and conductivity of samples of gas-sensitive materials in the range 0-10^{10} Ohm·cm (S^{-1}·cm).

For this case, the CO concentration was set in the range of 20-1000 ppm in air at operating temperatures of 100-270°C. The air and CO mixture were introduced at a flow rate of 0.3 dm^3/min.

The response of sensor elements based on ZnO nanorods on silicon was calculated using the formula $S=R/R_0$, and the response of sensor elements based on ZnO nanorods on glass was calculated using the formula $S=R_0/R$ (here R_0 is the resistance of the sensor element in the absence of gas; R is resistance sensor element when exposed to gas).

3. Results and discussion

It was experimentally determined that the minimum sensitivity of sensor elements based on ZnO nanorods on silicon was 100 ppm, and that of sensor elements based on ZnO nanorods on glass was 10 ppm - figure 3 (a, b). The operating temperature of sensor elements based on ZnO nanorods on silicon was near 200°C (figure 3, a). At the same time, the operating temperature of the sensor elements based on ZnO nanorods on glass was in the range of 250-270°C (figure 3, b (curves 2 and 3)). At 200°C, the response was unstable (figure 3, b (curve 1)).

(a)

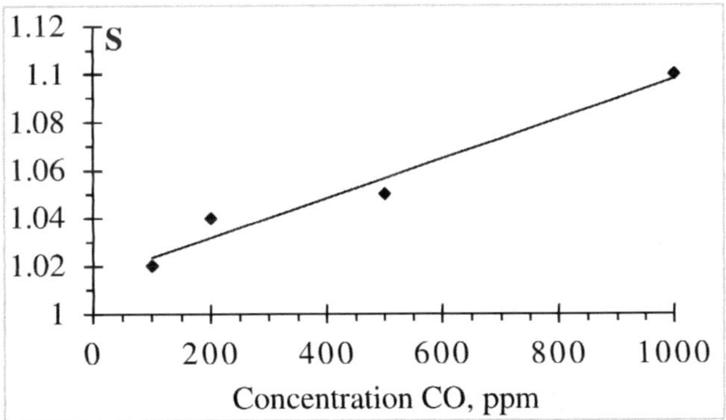

(b)

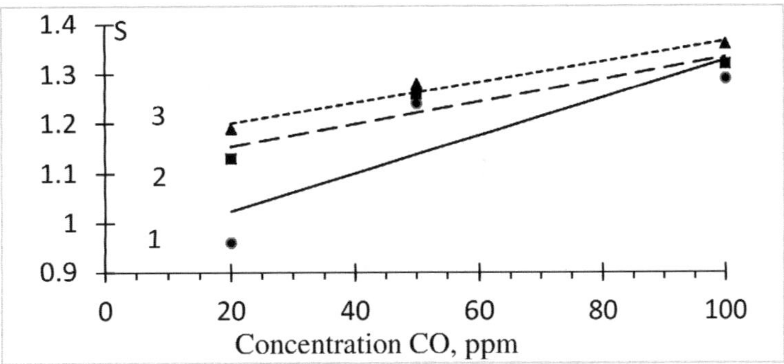

Figure 3. Dependence of the gas sensitivity coefficient of the investigated sensor elements based on ZnO nanorods on silicon on the CO concentration at 200 °C (a) and on glass (b) at 200 °C (1), 250°C (2) and 270°C (3).

Figure 4 shows typical responses of sensor structures based on ZnO nanorods grown on p-type silicon (figure 4, a) [7] and on glass (figure 4, b). The comparison shows that the response of the sensor formed on the glass became 1.3 times higher, and the sensitivity limit can reach units of ppm. Such parameters could not be achieved on sensor elements based on ZnO nanorods grown on p-type silicon.

In addition, the response to CO exposure for sensor structures based on ZnO nanorods grown on p-type silicon was inverse. When exposed to CO, the resistance grew (figure 4, a), but did not decrease, as it was when exposed to CO on the sensor structure based on ZnO nanorods grown on glass (figure 4, b). The result obtained is explained by the fact that zinc oxide is an n-type semiconductor. In the case when ZnO nanorods are grown on p-type silicon and there is contact metallization on top of them, two p-n-junctions are formed, which are connected towards each other. When measuring the resistance in such a structure (figure 5, a), the measuring current proceeds in two ways.

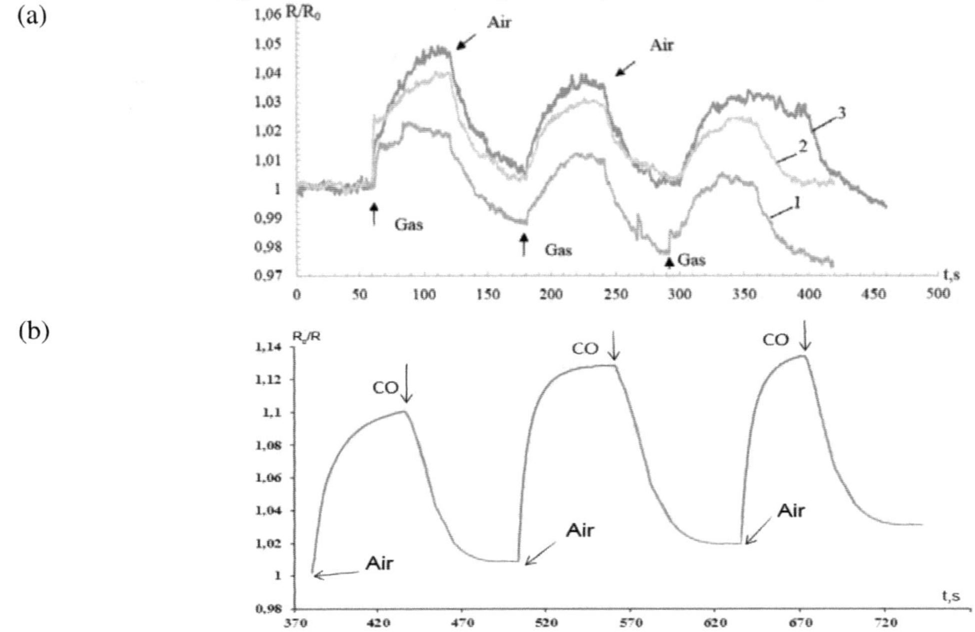

Figure 4. Normalized responses of sensor elements to CO 100 (1), 200 (2), 500 (3) ppm at 200 °C based on ZnO silicon nanorods (a) and response to CO (50 ppm) at 250 °C (ZnO on glass) (b).

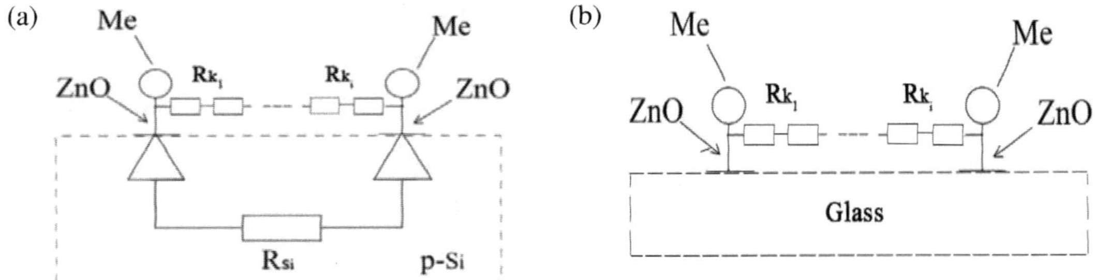

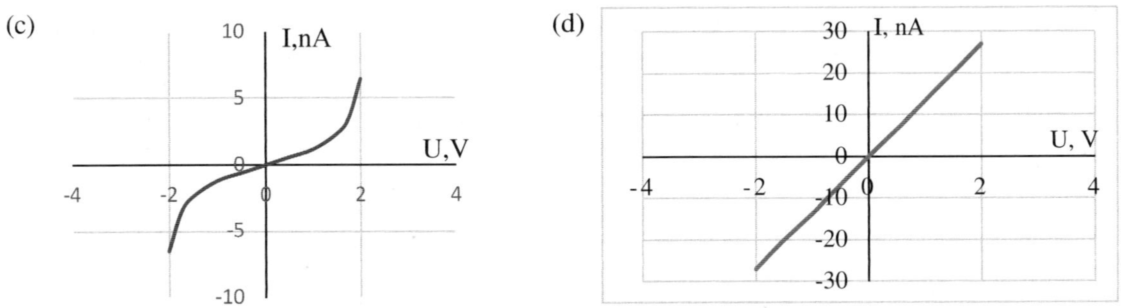

Figure 5. Electrical equivalent circuit (a, b) and volt-ampere characteristic (c, d) of the sensor structure based on ZnO nanorods on silicon (a, c) and glass (b, d).

The first path is from the first metal contact to the nanorods, on which it lies, and further, through numerous contacts of ZnO nanorods in contact with each other (Rk_i) to the second metal contact. The second way: from the first metal contact to ZnO nanorods, then through them to the silicon substrate, and through it to the second contact through ZnO nanorods. Since the resistance of the R_{Si} silicon substrate is less than the resistance of Rk_i nanorods in contact with each other ZnO nanorods, the current will mainly flow through the silicon substrate. As a result, two p-n-junctions are formed, connected towards each other - figure 5, a. In structure on the glass, it realizes only the first path of the passage of electric current - figure 5, b. These assumptions are confirmed by the current-voltage characteristics of the ZnO structures of nanorods grown on p-type silicon (figure 5, c), and on glass - (figure 5, d).

In the first case (figure 5, d), the current-voltage characteristic is non-linear and its form corresponds to the electrical equivalent circuit (figure 5, a). In the second case, the current-voltage characteristic is linear (figure 5, d), which corresponds to the electrical equivalent circuit (figure 5, b) and is consistent with the results presented in [8].

4. Conclusions

In this work, the sensitivity of sensory elements based on arrays of ZnO nanorods formed on p-type silicon and glass to carbon monoxide (CO) at temperatures of 200–270 ° C was investigated. Studies have shown that the shunting effect of two pn-junctions involved in opposite directions leads to a decrease in the gas sensitivity coefficient and the inverse response of the sensor structure based on ZnO nanorods grown on p-type silicon. The obtained response of the formed sensors based on the ZnO array of nanorods grown on glass was 30–40% higher than the response of a similar sensor structure based on the array of nanorods but grown on silicon. The maximum response of the sensory structure at 270 ° C was 1.36, when exposed to CO from concentrations of 100 ppm, and 1.28 and 1.19 at 50 and 20 ppm, respectively.

Thus, using gas sensors based on ZnO nanorods, one can measure the carbon monoxide concentration at a level of 10-50 ppm, which is necessary, for example, for the manufacture of gas fire detectors.

Acknowledgements
This work was financially supported by the Southern Federal University.

References

[1] Kołodziejczak-Radzimska A and Jesionowski T 2014 Zinc oxide-from synthesis to application: a review *Mater.* **7** 2833–81.
[2] Wang Z L and Song J 2006 Piezoelectric nanogenerators based on zinc oxide nanowire arrays *Sci.* **312** 242-6.
[3] Kumar R, Al-Dossary O, Kumar G, and Umar A 2015 Zinc oxide nanostructures for NO_2 gas–sensor applications: a review *Nano-Micro Lett.* **7** 97–120.

[4] Nikolaev A L, Kamencev A S, Lyanguzov N V, Petrov V V and Kaidashev E M 2017 Design and study of a high performance CO sensor based on ZnO nanowires *Proc. Int. Conf. on Phys. Mech. New Mater. Their Appl.* (Jabalpur: Nova Science Publishers, Inc) pp 441-6

[5] Abdullin Kh A, Bakranov N B, Ismailov D V, Kalkozova J K, Kumekov S E, Podrezova LV, and Cicero G 2014 Composite materials based on nanostructured zinc oxide *Semicondu* **48** 471-5.

[6] Gritsenko L V, Abdullin K A, Gabdullin M T, Kalkozova Z K, Kumekov S E, Mukash Zh O, Sazonov A Yu, and Terukov E I 2017 Effect of thermal annealing on properties of polycrystalline ZnO thin films *J. Cryst. Growth* **457** 164-70.

[7] Petrov V V, Starnikova A P, Abdullin K A, and Makarenko D P 2018 Features of the mechanism of gas sensitivity of the zinc oxide nanorods arrays to carbon monoxide *IOP Conf. Series: J. Physics* **1124**

[8] Park W J, Choi K J, Kim M H, Koo B H, Lee J L, and Baik J M 2013 Self-assembled and highly selective sensors based on air-bridge-structured nanowire junction arrays *ACS Appl. Mater. Interfaces* **5** 6802–7.

The study of the properties of lead zirconate-titanate films on silicon substrate after halogen lamps rapid thermal annealing

V V Petrov, A S Kamentsev, V V Polyakov, Y N Varzarev

Southern Federal University, Institute of Nanotechnologies, Electronics and Equipment Engineering, Taganrog 347928, Russia

E-mail : vvp2005@inbox.ru

Abstract. Ferroelectric thin films of lead zirconate titanate (PZT) are important is creation of ferroelectric MEMS, memory devices, sensors of physical quantities and converters energy. The thermal annealing is used to obtain satisfactory electro-physical parameters and optimal crystal structure PZT films. The rapid thermal annealing (RTA) in comparison with isothermal annealing has a possibility of selective annealing of individual layers of the multilayer PZT composition by selecting appropriate temperature and duration of the RTA. The purpose of this work is to study the phase-structural state of PZT films formed by high-frequency reactive plasma sputtering and subjected to rapid thermal annealing by halogen lamps. The PZT thin films with a thickness of 1.0 ± 0.1 μm were deposited by oxygen atmosphere high-frequency reactive plasma sputtering on silicon (100) substrates and silicon substrates with SiO_2 on surface. After applying, PZT film underwent RTA at temperatures of 600-800 °C and with speed 60 degrees/s. With X-ray phase analysis, the structure-phase composition of the PZT film is revealed. Besides that, the effect of RTA was investigated using electron microscopy. It is established that temperature change at RTA leads to a qualitative change in the phase-structural state of the PZT films as compared to their initial state. This gives a chance to use RTA in formation of the PZT films with given parameters.

1. Introduction

The use of materials in microelectronics in the form of nanostructured thin polycrystalline and epitaxial films (ferroelectrics, ferromagnets, multiferroics, etc.) is largely determined by the state of the art of creating these films. First of all, this is due to the size effects, the density of structural defects and the nature of interaction with the substrate. Thin ferroelectric films of lead zirconate titanate (PZT) are among the most important in the creation of ferroelectric memory devices and ferroelectric microelectromechanical systems (MEMS), as well as sensors of physical quantities and converters energy [1-4]. The use of nanoscale ferroelectric films requires solving a number of technological problems. First of all, these are the tasks of choosing materials, optimizing the composition and thickness of the film, the homogeneity of the film structure, etc. [5, 6]. Herewith, an important problem formation of PZT films is their thermal annealing in order to achieve the optimum crystalline structure and electrophysical parameters films at the minimum heat treatment temperature. The advantage of rapid thermal annealing (RTA) in comparison with isothermal treatment, is the possibility of selective annealing of the PZT lay by selecting the appropriate temperature, duration and speed of the RTA [7]. The positive effect of RTA on the structure and electrical properties of PZT thin

IOP Conf. Series: Materials Science and Engineering **703** (2019) 012039 doi:10.1088/1757-899X/703/1/012039

films was noted, for example, in [8–10]. The authors showed that the use of RTA of PZT films is an effective tool for changing the crystal structure and improving the electrophysical properties of PZT films. However, in these methods, the PZT structure cannot be formed without thermal annealing. At the same time, the formation of a perovskite crystal structure by high-frequency reactive plasma sputtering already occurs in the process of film growth. Long-term heat treatment for the formation in this method is not used. RTA, in this case, can be used to improve the structure of PZT films.

Therefore, the aim of this work is to study the phase-structural state of PZT films formed by high-frequency reactive plasma sputtering and subjected to rapid thermal annealing by halogen lamps in the temperature range 600–800 ° C.

2. Experimental

Ferroelectric PZT thin films of composition $Pb(Zr_{0,3},Ti_{0,7})O_3$ were deposited by oxygen atmosphere high-frequency reactive plasma sputtering on p-type silicon (100) substrates and on the similar silicon substrates with SiO_2 on surface. High-frequency reactive plasma sputtering was performed on the installation "Plasma 80 SE". PZT films were formed at a power of 400 W, a substrate temperature of 400 °C and an oxygen pressure in the chamber of 0.45 Torr for 30 minutes. The thickness of formed PZT film was 1.0±0.1 µm. After deposition the film, PZT was subjected to RTA by halogens lamps in air at temperatures of 600–800 ° C and speed 60 degrees/s. Rapid thermal annealing was carried out on the installation ITO-18M [11]. The surface of the initial samples and the samples after the RTA were examined using an electron microscope Nova NanoLab 600 - figure 1. The film thickness was measured using an MII-4 optical microscope. X-ray diffraction (XRD) analysis was performed on a DRON-3 unit at the CuK_{α} wavelength - figure 2. The results of the study of the structural-phase composition of the PZT films are presented in table 1.

3. Results and discussion

The surface morphology of PZT films before and after RTA at 700 °C is shown in figure 1.

a) b)

Figure 1. SEM images of the surface of the PZT films on silicon before annealing (a) and after RTA at 700 ° C (b).

XRD analysis was performed by comparing the form of diffractograms with the card 01-070-4055 from the PDF-2 database. To highlight the diffractogram peaks in the background of noise, a Savitzky–Golay smoothing filter with an interval width of 7 was used. The initial phase structure of the PZT film formed on both oxidized silicon and silicon plain substrates was identical and is shown in figure 2, diffractogram a.

Studies have shown that the influence of all RTA temperatures on the structure of PZT films formed on oxidized silicon substrates was catastrophic. The perovskite structure of PZT in all cases was transformed into a pyrochlore structure. Further investigation of such films became impractical.

At the same time, it was seen for the PZT films that the polycrystalline perovskite structure of the PZT is formed in the direction of crystal growth (110) and (001) with pronounced preferential directions (110) - figure 2, diffractograms b, c, d. The measured ratio of the intensities of the peaks I_{100}/I_{110} was 0.49-0.52 for the PZT samples that did not were exposed RTA and was expose the RTA at 500 °C and 600 °C. The ratio of peak intensities decreased to 0.43 for a sample annealed at an RTA temperature of 700 °C - figure 2. It can be seen that, at a temperature of RTA of 700 °C, a more pronounced formation of a crystalline perovskite (110) structure occurs. This trend is also noticeable in the growth of the size of the areas of coherent scattering of X-rays from PZT crystallites.

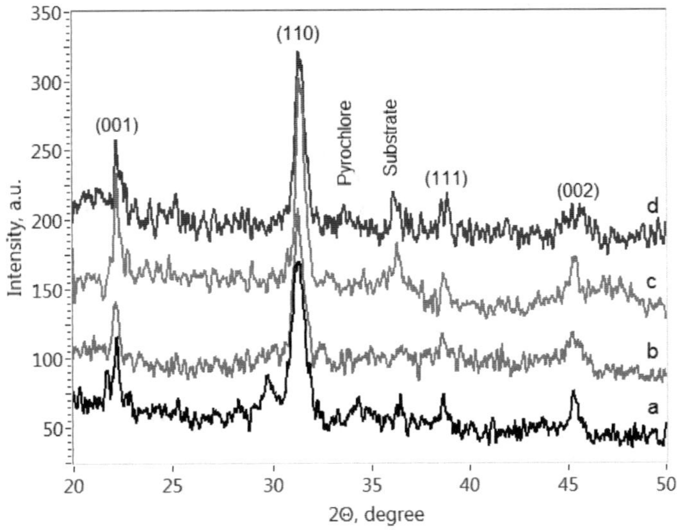

Figure 2. XRD patterns of PZT films before (a) and after RTA in the mode of 500 °C (b), 600 °C (c) and 700 °C (d).

The size of the coherent scattering regions was estimated using the Scherrer's equation [12]. The size of the coherent scattering regions of the perovskite (110) structure increased more than twice: from 13.0±0.3 nm for the PZT film without annealing to 30.0±4.9 nm for the sample annealed by RTA at temperature of 700°C – table 1. At the same time, the size of the regions of coherent scattering of crystallites of the perovskite (100) structure decreases from 51 to 18 nm. A similar behavior of the perovskite structure for isothermal annealing was observed in [13, 14].

Table 1. Dimensional factors PZT film

Sample	PZT block size (nm) (SEM image)	Size of coherent scattering regions from (001) crystallite (nm) (XRD analysis)	Size of coherent scattering regions from (110) crystallite (nm) (XRD analysis)	Film thickness (μm)	I_{100}/I_{110}
Without RTA	117	51	13	1,2	0.52
RTA 500	132	47	20	1,2	0.49
RTA 600	155	47	26	0,9	0.51
RTA 700	285	18	30	0,8	0.43

The growth of the crystallinity of the structure with increasing temperature RTA affects the increase in the sizes of the PZT blocks. The measurements were carried out on the basis of SEM images of the surface of the PZT films before and after the RTA – figure 1. From table 1 and figures 1 it can be seen that as the RTA temperature increases, the block size on the surface of the PZT film increases: from 117±3 nm, for the PZT film without annealing to 285±5 nm for the sample annealed by RTA temperature of 700°C. The latter suggests that as the temperature of the RTA increases, a PZT film is compacted, which is also confirmed by a decrease in the PZT film thickness with an increase RTA temperature.

Studies shown that RTA leads to a qualitative change in the phase-structural state of the PZT film compared to its initial state. RTA of PZT films on oxidized silicon substrates, the perovskite structure of PZT films is transformed into the pyrochlore structure. RTA of PZT films on non-oxidized silicon substrates, reverse processes occur, namely, their crystallinity improves. On one hand, a more pronounced formation of the crystalline perovskite (110) structure occurs. This is expressed in an increase size of the regions of coherent scattering of crystallites and a decrease in the intensity ratio of the I_{100}/I_{110} peaks. On the other hand, a seal of the PZT film is observed, which is expressed in an increase in the size of the PZT blocks and a decrease in the film thickness at higher RTA temperatures. Data analysis showed that PZT films annealed at 500 °C – 600 °C have the best uniformity. The structure of studied films is represented by nanoscale grains with an average size of 20–47 nm, some of which form columnar clusters that form in the direction of film growth, and is characterized by different grain sizes. The average grain size, the degree of granularity, and the number of columnar clusters in PZT films in range 500 °C – 600 °C RTA less than in other films studied.

4. Conclusions

The results obtained indicate that the observed differences in the structure and phase composition films of the initial state and after the RTA are mainly due to the existence and difference temperature gradient arising in the RTA process. Also, it should be noted that RTA leads to an increase in the crystallinity of PZT film and to the compaction of PZT film. Thus, the results of experimental studies have shown the possibility of using RTA to improve the structure of PZT films.

The results were obtained using the equipment of the centers for collective usage "Microsystem Engineering and Integrated Sensory" and "Nanotechnologies" of Southern Federal University.

Acknowledgments

The authors would like to express their gratitude to Pavlenko A V and Stryukov D V for carrying out XRD analysis.

This work was financially supported by Southern Federal University.

References

[1] Izyumskaya N, Alivov Y I, Cho S J, and Morkocet H 2007 *Crit. Rev. Solid State* **32** 111–202.
[2] Mukhortov V M and Yuzyuk Yu I 2008 *Heterostructures based on nanoscale ferroelectric films: production, properties and applications* (Rostov-on-Don: Southern Scientific Center of RAS)
[3] Kang M G, Jung W S, Kang C Y, and Yoon S J 2016 *Actuators* **5** 1–17.
[4] Petrov V V 2019 *Proceedings* **4** 40; doi:10.3390/ecsa-5-05870
[5] Kovalenko D A, Petrov V V, and Klindukhov V G 2014. *Izvestiya SFedU. Engineering Sciences* **4** 184.
[6] Kovalenko D A and Petrov V V 2015 *J. Nano- Electron. Phys.***7** 1–5.
[7] Polyakov V V 2014 Development and study of silicon dioxide nanostructured films forming on semiconductor structures surface *Advanced nano- and piezoelectric materials and their applications* ed Parinov I A (New York: Nova Science Publishers) **2** 19-60
[8] Zhu MW, Wang Z J, Chen YN, Kobayashi T, and Maeda R 2013 *Thin Solid Films* **540** 73–8.
[9] Lu J, Zhang Y, Ikehara T, Mihara T, and Maeda R 2007 *IEEE Trans. Ultrason. Ferroelectr. Freq. Control* **54** 2548-54.
[10] Yang J I, Welsh A, Sbrockey N M, Tompa G S, Polcawich R G, Potrepka D M, and Trolier-McKinstry S 2018 *J. Cryst. Growth* **493** 45–50.
[11] Svetlichny A M, Sechenov D A, Vorontsov L V, Burshtein V M, Braznik V A, and Polyakov V V 1990 *Russ. J. Electron. Ind.* **3** 5.

IOP Conf. Series: Materials Science and Engineering **703** (2019) 012039 doi:10.1088/1757-899X/703/1/012039

[12] Cullity B D and Stock R S 2001 *Elements of X-ray Diffraction* 3rd Ed. (New Jersey: Prentice Hall)
[13] Pronin V P, Senkevich S V, Kaptelov E Yu, and Pronin I P 2010 *J. Surf. Invest.-X-ray.* **9** 5–9.
[14] Pronin I P, Kaptelov E Yu, Senkevich S V, Shaplygina T A, Klimov V A, and Pronin V P 2010 *Proc. RAS. Phys. Ser.* **74** 1360–62.

IOP Conf. Series: Materials Science and Engineering **703** (2019) 012040 doi:10.1088/1757-899X/703/1/012040

Temperature Dependence of Electrical Properties of ZnO Nanorods Array

V V Petrov[1], Y N Varzarev[1] and K A Abdullin[2]

[1]Southern Federal University, Institute of Nanotechnologies, Electronics, and Equipment Engineering, Taganrog, 347922, Russia
[2]National Nanotechnology Laboratory of Open Type, al-Farabi Kazakh National University, Almaty, 050000, Republic of Kazakhstan

E-mail : vvp2005@inbox.ru

Abstract. Zinc oxide nanostructures (nanotubes, nanorods, nanowhiskers, etc.) are a promising material for various electronics devices for power system and green energy such as piezoelectric energy harvesters and solar cells. Electrical properties of ZnO nanorods grown on a glass substrate by the hydrothermal method in an aqueous solution of zinc nitrate were investigated in this work. The SEM micrograph shows that the grown nanorods are vertically oriented and the XRD pattern confirms its crystallinity with (002) preferred orientation. The temperature dependent resistivity of ZnO nanorods array measured from 300 to 523 K shows variation in resistance from $5\cdot10^9$ Ω to $3\cdot10^6$ Ω and temperature hysteresis and region with positive temperature coefficient of resistance at the heating that was not observed at cooling. The origin of this behavior is explained with the two competing processes - thermal generation of electrons and oxygen species adsorption on the ZnO surface. Oxygen adsorption leads to decrease in the number of oxygen vacancies that are electron donors, and consequently to a decrease in conductivity of ZnO nanorods. The values of activation energy of conductivity were calculated from the slope of linear regions on the Arrhenius plot which are 0.246, 1.466 and 0.248 eV for the heating curve and 0.650 and 0.315 eV for the cooling curve.

1. Introduction

Zinc oxide (ZnO) is a wide band-gap semiconductor and promising material for various electronics devices for power system and green energy such as piezoelectric energy harvesters and solar cells, gas sensors [1]. Recently, considerable attention has been paid to the formation and investigation of the ZnO nanostructures (nanotubes, nanorods, nanowhiskers, etc.) which are used in functional electronics devices [2, 3]. Therefore, an urgent task is to study electrical properties of such nanostructures and the effect of external influences, such as temperature, light, gas ambient on them. Temperature measurements of the electrical resistance of materials make it possible to obtain information about the mechanism of conductivity, in particular, to estimate the values of the activation energies of conductivity and to conclude that they are associated with levels of impurities or defects that are donors or acceptors of charge carriers. Previously, the temperature behavior of electrical conduction processes in ZnO have been studied by some groups [4-6]. In this works thermal activation of conductivity was observed over a wide temperature range and was associated with the influence of shallow donors - oxygen vacancies and impurities.

The study of the conductivity mechanisms in ZnO nanostructures is complicated by the fact that their developed surface adsorbs oxygen well, which has a significant effect on the conductivity mechanism. In this paper, the effect of oxygen adsorption on the temperature dependence of the resistance of an array of nanorods is shown.

2. Experimental

An array of ZnO nanorods was grown by hydrothermal technology in an aqueous solution of zinc nitrate $Zn(NO_3)_2$ $6H_2O$ and hexamethylenetetramine $C_6H_{12}N_4$ on a glass substrate during 1 hour at temperature 95°C [7, 8]. The surface morphology was studied using a scanning electron microscope (SEM) Nova NanoLab 600, and X-ray diffraction (XRD) analysis to study the crystal structure was done in a MiniFlex Rigaku diffractometer with CuKα radiation. The diffractogram of polycrystal ZnO (JCPDS card no. 00-001-1136) was used to characterize the phase composition of nanorods. To measure the electrical parameters of the material, metal contacts V-Cu-Ni with a thickness of 0.2-0.3 μm were applied to the sample surface by vacuum deposition. Resistance measurements in the temperature range of 300–523 K were carried out using a Keithley 2450 source meter.

3. Results and discussion

SEM studies have shown that the formed ZnO nanorods are mainly vertical, however, a significant number of nanorods are in contact with each other. Nanorods have a length of up to 0.5-0.6 μm [5] and an average transverse size of about 30-40 nm (figure 1). XRD analysis showed that the predominant growth direction of zinc oxide during the hydrothermal synthesis of ZnO nanorods is the direction (002) (Figure 2).

Figure 1. SEM micrograph of ZnO nanorods array

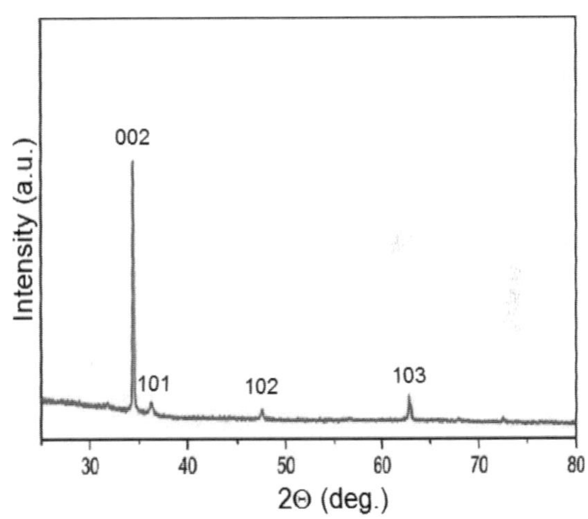

Figure 2. XRD pattern of ZnO nanorods

The temperature dependence of the resistance (figure 3) shows change in resistance by three orders of magnitude - from $5 \cdot 10^9$ Ω to $3 \cdot 10^6$ Ω. A distinctive feature of this dependence is that the heating and cooling curves at temperatures above 373 K do not coincide, that is, there is a temperature hysteresis.

On this dependence, we can distinguish regions that are well approximated by the Arrhenius equation [9]:

$$\rho = \rho_0 \exp\left(\frac{E_a}{kT}\right), \qquad (1)$$

where ρ_0 is preexponential factor, E_a is activation energy of conductivity and k is Boltzmann constant. Three such regions can be distinguished on the heating curve – I, II and III, and two on the cooling curve – IV and V (Figure 2). The values of the activation energy of the conductivity of a ZnO nanorods array calculated by equation (1) are given in Table 1.

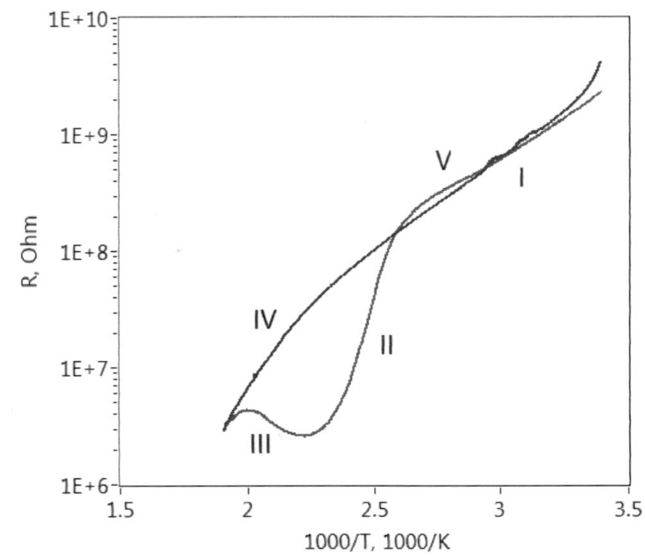

Figure 3. Temperature dependent resistivity of ZnO nanorods array

Table 1. Activation energy of the conductivity of a ZnO nanorods array

Region	Temperature band, K	Activation energy of conduction (E_a), eV
I	300-373	0.246
II	390-423	1.466
III	450-493	0.248
IV	523-463	0.650
V	400-300	0.315

It is known that the electrical conductivity of zinc oxide is determined by oxygen vacancies, which are electron donors [2], and, accordingly, the activation energy of conductivity is determined by donor levels formed by vacancies in the forbidden zone of ZnO. During heating, in regions I and II, the decrease in the resistance of the sample with increasing temperature is determined by the thermal generation of electrons. The different activation energy of conductivity for these regions indicates that two different donor levels are sources of electrons.

Region III is a region with a positive temperature coefficient of resistance; it is observed only during heating and is absent upon cooling. A similar behavior of the electrical resistance of zinc oxide with temperature was observed in [10], where sintered ZnO was studied, in [11] on ZnO pellets prepared by uni-axial pressing and in [12] on the ZnO nanostructures grown by sol-gel method. The increase in resistance with increasing temperature in this area can be explained by the influence of adsorbed oxygen. Oxygen molecules in the atmosphere at temperatures of 373–573 K are adsorbed on

IOP Conf. Series: Materials Science and Engineering **703** (2019) 012040 doi:10.1088/1757-899X/703/1/012040

the surface of zinc oxide and become negatively charged due to the capture of electrons from near-surface ZnO layers [13]:

$$O_2(gas) \leftrightarrow O_2(ads) \quad (2)$$

$$O_2(ads) + e^- \leftrightarrow O_2^-(ads) \quad (3)$$

$$O_2^-(ads) + e^- \leftrightarrow 2O^-(ads) \quad (4)$$

$$O^-(ads) + e^- \leftrightarrow O^{2-}(ads) \quad (5)$$

This leads to the formation of a depleted layer in the near-surface region of ZnO nanorods and a decrease in their conductivity [12]. The activation energy of conductivity on this site will correspond to the energy of oxygen adsorption. A further decrease in resistance after region III is probably caused by the predominance of thermal generation of electrons.

When the temperature decreases, extremums on the temperature dependence of the resistance is not observed, that may be due to the lack of oxygen species desorption. Sections IV and V on the cooling curve also correspond to two different values of the activation energy (Table 1).

Thus, the variations in conductivity during heating and cooling is due to two competing processes– thermal generation and recombination of electrons on one hand, and their capture by oxygen adsorbed on the surface on the other. Moreover, if in the first case, the equilibrium concentration is established almost instantly, then in the second it depends on the speed of the processes of adsorption and desorption, which is much lower and comparable with the rates of heating and cooling. Thus, the more dominant factor in the presence of hysteresis in the temperature dependence of the conductivity is oxygen species adsorption on the ZnO surface.

4. Conclusions

The temperature dependent resistivity of ZnO nanorods array obtained by hydrothermal method has been investigated. The values of activation energy of conductivity were calculated from the slope of linear regions on the Arrhenius plot. The origin of thermal hysteresis is explained with the two competing processes – thermal generation of electrons and oxygen species adsorption on the ZnO surface.

Acknowledgments

This work was financially supported by the Southern Federal University.

References

[1] Kołodziejczak-Radzimska A and Jesionowski T 2014 *Mater.* **7** 2833–81.
[2] Wang J, Chenb R, Xiangb L, and Komarnenic S 2018 *Ceram. Int.***44** 7357–77.
[3] Petrov V V, Starnikova A P, Abdullin Kh A, and Makarenko D P 2018 *J. Phys. Conf. Ser.***1124** 022017
[4] Li X, Qi J, Zhang Q, and Zhang Y 2012 *J. Appl. Phys.***112** 084313
[5] Lien C C, Wu C Y, Li Z Q, and Lin J J 2011 *J. Appl. Phys.***110** 063706
[6] Benton B T, Greenberg B L, Aydil E, Kortshagen U R, and Campbell S A 2018 *Nanotechnol.* **29** 415202
[7] Abdullin K A, Bakranov N B, Ismailov D V, Kalkozova J K, Kumekov S E, Podrezova L V, and Cicero G 2014 *Semicond.* **48** 471-5.
[8] Gritsenko L V, Abdullin Kh A, Gabdullin M T, Kalkozova Zh K, Kumekov S E, Mukash Zh O, Sazonov A Y, and Terukov E I 2017 *J. Cryst. Growth* **457** 164-70.
[9] Shalimova K V 1976 *Physics of Semiconductors* (Moscow: Energia)
[10] Takata M, Tsubone D, and Yanagida H 1976 *J. Am. Ceram. Soc.***59** 1–2, 4–8.
[11] Roy T K, Sanyal D, Bhowmick D, and Chakrabarti A 2013 *Mater. Sci. Semicond. Process.***16** 332–6.
[12] Santhosh Kumar A, Nagaraja K K, and Nagaraja H S 2013 *AIP Conf. Proc.***1536** 263–4.
[13] Rambu A P, Iftimie N, and Rusu G I 2012 *Mater. Sci. Eng. B-Adv.***177** 157– 63.

The kinetics oxidative degradation of chitosan in formic acid with the presence of hydrogen peroxide

E Purwanto[1,2], J Connor[1] and Y Ngothai[1]

[1]School of Chemical Engineering and Advanced Materials, The University of Adelaide, South Australia, 5005, Australia
[2]Department of Chemical Engineering, University of Surabaya, Raya Kalirungkut, Surabaya, 60293, Indonesia

E-mail: edy.purwanto@adelaide.edu.au

Abstract. The chitosan degradation in formic acid solution with the presence of hydrogen peroxide was investigated in this work. Temperature and reaction time had a significant effect on reducing the molecular weight of chitosan, and it is clear that it decreased the molecular weight faster at the first 15 minutes of reaction time. Therefore, the molar concentration of chitosan in the solution increased with temperature and reaction time. The reaction order was determined from the molar changes of chitosan on reaction time and found followed -0.1 reaction order to the molar concentration of chitosan. The Arrhenius equation was used to study the correspondence influence of reaction temperature on the degradation reaction rate constant. The results suggest that the values of degradation rate constant increased with higher reaction temperature. The value of activation energy was determined under the experimental operating conditions examined to be 99.8 kJ/mol. The FT-IR spectra demonstrated that there was no change in the chemical structure of chitosan before and after the degradation reaction.

1. Introduction

Chitosan is generally accepted as a product derivative of chitin that ordinarily acquired from crustacean shell through partial deacetylation reaction using base or enzyme [1]. Chitosan has a different structure compared to chitin concerning some acetyl groups present in chitin which have been replaced by atoms H. Hence, chitosan will have a varying degree of deacetylation (DD) with a chemical name as (1→4)-2-amino-2-deoxy-β-D-glucan.

As amino-polysaccharides, chitosan is extensively utilised for some purposes with great features regarding biocompatible, antibacterial, biodegradable, nontoxic, and environmentally friendly. Accordingly, chitosan is utilised in numerous applications, for example, in foods, biomaterials, medicine, cosmetics, and pharmaceutical products [2-6]. However, low solubility in water is the characteristic of original chitosan derived from the shell of shrimps as it has high molecular weight. These properties make the original chitosan have a limited utilisation in particular for medical applications. Therefore, it is necessary to obtain a chitosan product through the degradation process that accomplishes its higher solubility and preserves its biological properties. Oligomer chitosan with high solubility in solvents can be achieved through degradation reactions via several techniques. The degradation can be carried out by a chemical method using acids [7-10], high impact energy [11-13], and enzymatic [14-15].

Recently, acetic acid has been widely implemented in oxidative degradation using hydrogen peroxide as an oxygen carrier. However, there is little attention to the exploitation of formic acid as an acid solvent in the chitosan degradation with the utilisation of hydrogen peroxide. Formic acid may show better performance than acetic acid as a solvent when combined with hydrogen peroxide. The utilisation of hydrogen peroxide and synergised with formic acid has been widely employed, as shown in another type of oxidation reaction called epoxidation reaction [16-17]. Hence, this work aims to study the influence of reaction temperature and time on chitosan degradation and determine the kinetics of chitosan degradation in the formic acid solution with the presence of H_2O_2.

2. Materials and methods

2.1. Materials

The feedstock of chitosan with an average molecular weight of approximately 3300 kDa and a degree of deacetylation of 96.31% was used in this research. This chitosan was obtained from Xi' An Lukee Bio-Tech Co. Ltd. Hydrogen peroxide, formic acid, acetic acid, sodium hydroxide, and ethanol absolute were supplied by Chem-supply as laboratory reagent grade.

2.2. Feedstock preparation of chitosan solution

The starting of chitosan solutions was prepared in a glass reactor with dissolving chitosan to obtain 10 g/L chitosan in 1% formic acid solution. A calculated amount of formic acid solution was dissolved in the water to obtain 800 mL 1% formic acid solution. Following this, 8 grams chitosan dissolved in this acid solution and stirred with low rotation speed overnight to ensure chitosan has been soluble thoroughly.

2.3. Experimental setup and procedure

The chitosan degradation was carried out in a 1-L glass reactor equipped with a magnetic stirrer, condenser and thermocouple. The reactor was located on a heating plate and connected with temperature control, whose temperature could be maintained ±1°C. The reaction temperature was set by adjusting the temperature controller connected to the heating plate, and a calculated amount of hydrogen peroxide was added when the desired reaction temperature was reached with a final concentration 1% (w/v) in the mixture. The degradation reactions were run at several isothermal conditions and samples were taken at certain reaction times. After the degradation reaction, the samples were collected and cooled to terminate the reaction. It was then neutralised with 2 M NaOH solution to pH 8-9. The neutralised solution was then filtered off to separate solid phase (low molecular weight of chitosan) and liquid phase (oligomer chitosan soluble in the water). After that, the solid phase was washed with water until the remaining washing water was neutral. Next, the remaining impurities in the solid phase was removed using ethanol. Finally, the solid was vacuumed and then dried at 60°C to remove the water content. The solid product was analysed to determine its molecular weight using the viscometric method, and FT-IR was performed to monitor the formation and disappearance of functional groups of original and low molecular weight of chitosan.

2.4. Analytical methods

The viscosity-average molecular weight of chitosan (M_v) was measured using an Ubbelohde viscometer with a capillary diameter of 0.36 mm at 30°C. The several dilute solutions of chitosan in 1% acetic acid solution (C) were measured its flowing time (t_s) and the solvent acetic acid (t_0) in duplicate for each. The specific viscosity of the dilute chitosan solutions (η_{sp}) was calculated from relative viscosity (η_r) according to the following equation (1):

$$\eta_r = \frac{t_s}{t_0}, \quad \eta_{sp} = \eta_r - 1 \tag{1}$$

Huggins equation (2) was performed to determine the intrinsic viscosity (η) from the known value of specific viscosity (η_{sp}):

$$\frac{\eta_{sp}}{C} = [\eta] + k[\eta]^2 C \quad (2)$$

by plotting $\frac{\eta_{sp}}{C}$ vs C, the intercept at $C = 0$ was obtained as the intrinsic viscosity (η) [8]. The M_v value was determined through the Marck-Houwink equation (3):

$$[\eta] = K.M_v^a \quad (3)$$

The value of K and a for chitosan in 1% acetic acid solution are 0.0474 mL/g and 0.723, respectively [18]. Those values of K and a are constant and defined as the Mark-Houwink constants.

2.5. Kinetics model

Kinetic model of chitosan degradation has been developed according to the rate of chemical reaction definition. The degradation rate is expressed as the number of moles of cutting off the glycosidic linkage in the chitosan chain to yield the two pieces [19] according to the following equation (4):

$$R = \frac{d[M]}{dt} = kM^n \quad (4)$$

M is known as the molar concentration of chitosan (mol/L). The properties k and n represent the reaction rate constant and reaction order, respectively. In this case, the reaction order (n) typically has a negative value as with the increasing of chitosan solution concentration, the degradation rate will decrease. The molar concentration of chitosan can be calculated through equation (5):

$$M = \frac{C}{M_n} \quad (5)$$

Then, the solution of the differential equation (4) is described in equation (6):

$$M_t^{1-n} - M_0^{1-n} = (1-n)kt \quad (6)$$

where M_t is chitosan concentration at a certain reaction time (t) and M_0 is the initial concentration of chitosan. It is obvious that the value of C as the solution concentration remains constant (g/L), but the molar concentration (molarity in mol/L) each time increases with the reducing number of molecular weight through the degradation process. The M_n value is the number-average molecular weight can be calculated from the M_v value through equation (7) below:

$$M_v = \left[(1+a)\Gamma(1+a)\right]^{\frac{1}{a}} M_n \quad (7)$$

where $\Gamma(1+a) = \int_0^\infty e^{-t} t^a dt$ and the value of M_v is obtained from equation (3) above.

3. Results and discussions

3.1. The influence of reaction temperature and time on the molecular weight of chitosan

The reaction temperatures of degradation were varied at 40, 50 and 60°C for 120 min reaction time to study the influence of reaction temperature on the chitosan degradation. The corresponding of viscosity-average molecular weight (M_v) was plotted on a logarithmic scale in correlation with reaction time t at the different reaction temperatures, as depicted in figure 1. The results indicated that the M_v value decreased with the addition reaction time. The reaction temperature similarly demonstrated a substantial effect on chitosan degradation indicated by a significant reducing the molecular weight of chitosan. The corresponding rate of decrease for M_v value of chitosan in percentage was 98.26, 99.40, and 99.83% respectively for 120 reaction time from the initial M_v value of chitosan of 3300 kDa. The M_v value of chitosan dropped to 57.4, 19.8, 5.6 kDa for 120 min, respectively. Interestingly, the most significant reduction of M_v value was achieved at the first 15 minutes of reaction time with the corresponding rate of decline in percentage for M_v value was 94.33, 96.58, 98.47% respectively.

Figure 2 shows the correlation between the molar concentration of chitosan (M_t) vs reaction time (t) and the result indicated that the molar concentration of chitosan increased with the rise of reaction time. This finding implies that a random scission of glycosidic linkage occurred and results in a lower number-average molecular weight of chitosan. Therefore, it will increase the molar concentration of chitosan with additional reaction time.

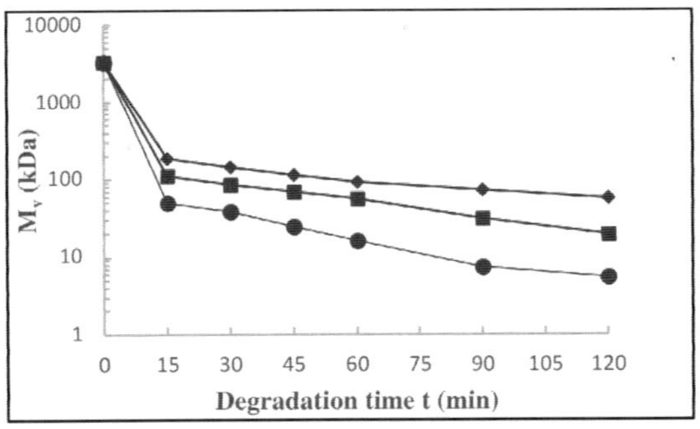

Figure 1. The influence of reaction temperature on the viscosity-average molecular weight of chitosan (M_v) for different temperatures: (♦) 40°C; (■) 50°C; (●) 60°C.

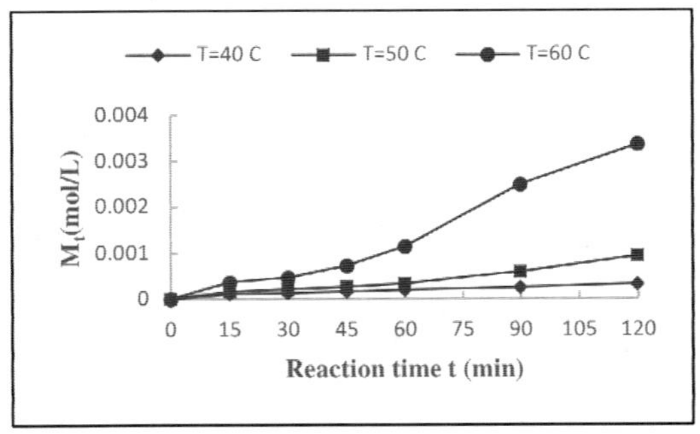

Figure 2. The molar concentration of chitosan vs time.

These results confirm a previous finding stated that reaction temperature plays an essential role for chitosan degradation [20]. The experimental results conclude that hydrogen peroxide in the formic acid solution may accelerate chitosan degradation during the degradation reaction. The reaction mechanism of chitosan degradation is mainly under the attack of powerful oxidant, namely, hydroxyl radical [21]. The reaction step is illustrated as follows, $R\text{-}NH_2$ first produces $R\text{-}NH_3^+$ in acid solution and decomposition of H_2O_2 also occurred as shown in equation (8) and (9) then summarised as a total reaction in equation (10). The reaction mechanism of chitosan degradation with hydrogen peroxide in the acid solution is as follows [21]:

$$R\text{-}NH_2 + H^+ \leftrightarrow R\text{-}NH_3^+ \tag{8}$$
$$H_2O_2 \leftrightarrow H^+ + HOO^- \tag{9}$$
$$H_2O_2 + R\text{-}NH_2 + H^+ \leftrightarrow R\text{-}NH_3^+ + HOO^- + H^+ \tag{10}$$

The hydroxyl radical (HO*) is easily produced from the decomposition of hydroperoxide anion as it is not stable, and the reactions in equation (11) and (12) take place.

$$HOO^- \rightarrow OH^- + O^* \tag{11}$$
$$H_2O_2 + HOO^- \rightarrow HO^* + O_2^*{}_- + H_2O \tag{12}$$

Eventually, the hydroxyl radical (OH*) attacks the glycosidic linkage in chitosan to produce oligomer chitosan according to the reactions in equation (13) and (14).

$$(GlcN)_m - (GlcN)_n + OH^* \rightarrow (GlcN^*)_m - (GlcN)_n + H_2O \tag{13}$$
$$(GlcN^*)_m - (GlcN)_n + H_2O \rightarrow (GlcN)_m + (GlcN)_n \tag{14}$$

3.2. Kinetics degradation of chitosan

Some different rate models have been suggested regarding polymer degradation [19, 22, 23]. However, in this paper, the reaction rate was modelled as a function of molar concentration with a negative order n according to equation (4) [19]. Utilising equation (6) as the solution of the differential equation (4), the reaction order (n) can be determined by performing trial and error of the n value on plotting $\left[M_t^{1-n} - M_0^{1-n}\right]$ vs t. The n value of -0.1 showed the best fit of the curve with the kinetic model. The finding of this study indicated that chitosan degradation followed -0.1 reaction order corresponding to the chitosan concentration.

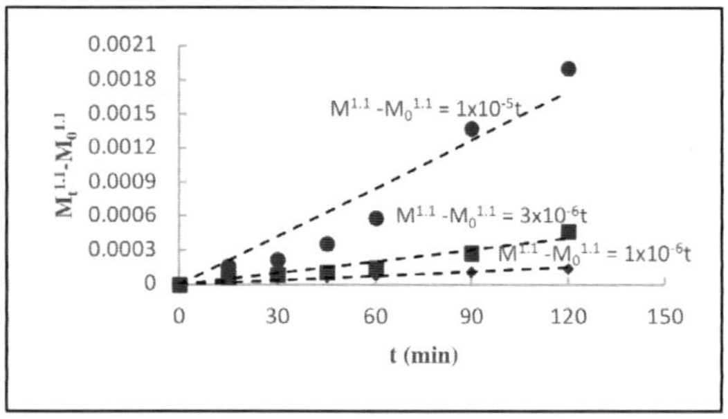

Figure 3. The changes of the molar concentration of chitosan fitted with -0.1 reaction order for different temperatures: (♦) 40°C; (■) 50°C; (●) 60°C.

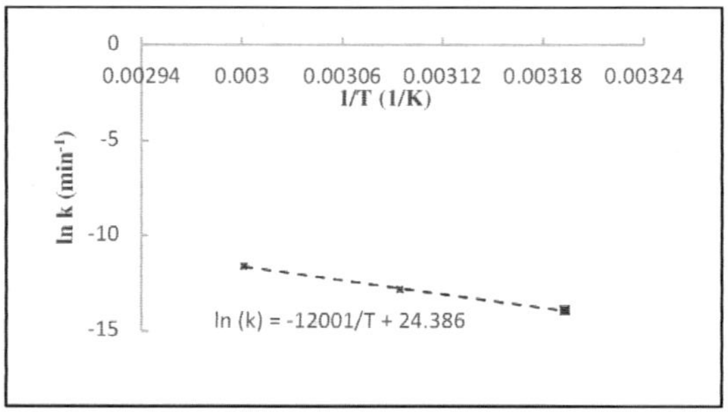

Figure 4. Plot Arrhenius to determine the rate constant

The changing for the molar concentration of chitosan was modelled according to -0.1 reaction order indicated a good fit with a coefficient of correlation $R^2 \geq 0.921$ (figure 3). The value of rate constants (k) for different temperatures: $9.1x10^{-7}$(40°C), $2.7x10^{-6}$(50°C), $9.1x10^{-6}$(60°C) $(mol/L)^{1.1}$/min. These rate constants are higher than compared to using sonolysis method with additional of Fe(III)/H_2O_2 in acetic acid solution with the value of k is in the range 1.538 – 4.923 x 10^{-9} $mol^{1.7}L^{-1.7}.min^{-1}$ for additional H_2O_2 between 0.0 – 1.2 mL [23]. They also reported that the reaction order was -0.7 using the same model used in this paper. This results showed that the chitosan degradation rate increased with reaction temperature indicated by the higher value of k as the temperature increase (figure 4). From the plot between ln(k) and 1/T, the value of activation energy (E_a) was obtained to be 99.8 kJ/mol. This activation energy value is considerably higher than that for degradation using electrochemical with electrode Ti/Sb-SnO_2 (E_a = 43.61 kJ/mol) and close to the value stated for acetic acid and H_2O_2 (Ea = 88.5 kJ/mol) [20, 24].

3.3. FT-IR spectra of degradation products

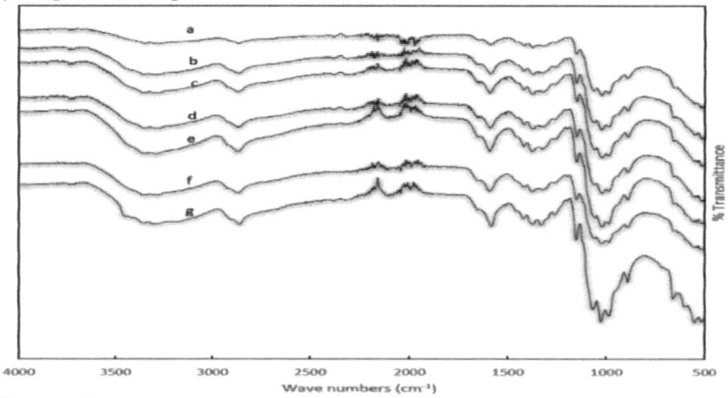

Figure 5. FT-IR spectra of chitosan for degradation at 60°C with reaction time 0 min (a), 15 min (b), 30 min (c), 45 min (d), 60 min (e), 90 min (f), 120 min (g).

The FT-IR spectra of chitosans for different reaction times at a reaction temperature of 60°C is described in figure 5. The peak which appeared at around 3368 cm^{-1} was referred to as the stretching vibration of N-H and O-H [25]. The peaks at 1648 cm^{-1} and 1324 cm^{-1} were assigned as amide I and amide III, respectively [26]. The peaks, as indicated at about 1597, 1156 and 1079 cm^{-1} were amide II, C-O-C, and C-O, respectively [25]. The β-D-(1→4) glycosidic bond shown at the peak in the range 1158-895 cm^{-1} [27]. The results of FT-IR spectra confirmed that there is no formation of a carboxylic group during the degradation reaction as there was no peak appeared at 1735 cm^{-1} attributed to the carboxylic group [28]. The results suggest that the chemical structure of initial chitosan and depolymerised chitosan remained the same and there was no significant difference with the increasing of the reaction time of degradation. Hence, it can be concluded that with the utilisation of H_2O_2 in 1% formic acid, there is a primary reaction occurred with the cutting off the β-glycosidic linkage, and no side reaction occurred with the formation of the carboxylic group during the chitosan degradation.

4. Conclusions

Oxidative degradation of chitosan in formic acid solution utilising hydrogen peroxide as an oxidising agent had been done and resulted in lower molecular weight of chitosan. The reaction time and temperature have a significant effect on the degradation reaction and the faster decrease occurring at the first 15 minutes. The degradation followed -0.1 reaction order to the molar concentration of chitosan and the activation energy is 99.8 kJ/mol. The chemical structure of chitosan remained identical during the degradation reaction and indicated only primary reaction occurring with the cutting off the β-glycosidic linkage to produce lower molecular weight of chitosan.

Acknowledgments
Mr Edy Purwanto would like to thank the Indonesia Endowment Fund for Education (LPDP), Ministry of Finance; collaborated with the Directorate General of Higher Education (DIKTI), Ministry of Research, Technology and Higher Education, the Republic of Indonesia for financial support.

References

[1] Struszczyk M H 2006 *12. Global Requirements for Medial Applications of Chitin and its Derivatives (Polish Chitin Society)* Monograph **11** 95 – 102.
[2] Shahidi F, Arachchi J K V and Jeon Y 1999 *Trends in Food Sci. & Technol.* **10** 37-51.
[3] Khor E and Lim L Y 2003 *Biomater.* **24** 2339-2349.
[4] Suh J K and Matthew H W 2000 *Biomater.* **21** 2589-2598.
[5] George M and Abraham T E 2006 *J. Controlled Release* **114** 1-14.
[6] Rinaudo M 2006 *Prog. Polym. Sci.* **31** 603-632.
[7] Hasegawa M, Isogai A and Onabe F 1993 *Carbohydr. Polym.* **20** 279-283.
[8] Jia Z and Shen D 2002 *Carbohydr. Polym.* **49** 393-396.
[9] Zamani A and Taherzadeh M J 2010 *BioResour.* **5** 1554-1564.
[10] Tsao C T, Chang C H, Lin Y Y, Wu M F, Han J L and Hsieh K H 2011 *Carbohydr. Res.* **346** 94-102.
[11] Choi W S, Ahn K J, Lee D W, Byun M W and Park H J 2002 *Polym. Degrad. Stab.* **78** 533-538.
[12] Li K, Xing R, Liu S, Qin Y, Meng X and Li P 2012 *Int. J. Biol. Macromol.* **51** 767-773.
[13] Hai L, Diep T B, Nagasawa N, Yoshii F and Kume T 2003 *Nucl. Instrum. Methods Phys. Res. B* **208** 466-470.
[14] Lin H, Wang H, Xue C and Ye M 2002 *Enzyme Microb. Technol.* **31** 588-592.
[15] Lee D X, Xia W S and Zhang J L 2008 *Food Chem.* **111** 291-295.
[16] Goud V V, Patwardhan A V, Dinda S and Pradhan NC 2007 *Chem. Eng. Sci.* **62** 4065-4076.
[17] Campanella A, Fontanini C and Baltanas M A 2008 *Chem. Eng. J.* **144** 466-475.
[18] Knaul J Z, Kasaai M R, Bui V T and Creber K A M 1998 *Chem. Eng. J.* **76** 1699-1706.
[19] Taghizadeh M T and Abdollahi R 2011 *Ultrason. Sonochem.* **18** 149-157.
[20] Chang K L B, Tai M C and Cheng F H 2001 *J. Agric. Food. Chem.* **49** 4845-4851.
[21] Tian F, Liu Y, Hu K and Zhao B 2004 *Carbohydr. Polym.* **57** 31-37.
[22] Cai Q, Gu Z, Fu T, Liu Y, Song H and Li F 2011 *Polym. Bull.* **67** 571-582.
[23] Taghizadeh M T, Abdollahi R and Orang N S 2012 *J. Polym. Environ.* **20** 208-216.
[24] Gu Z, Cai Q, Liu Y and Li F 2013 *J. Polym. Environ.* **21** 479-486.
[25] Huang Y, Wu Y, Huang W, Yang F and Ren X 2013 *Polym. Degrad. Stab.* **98** 37-43.
[26] Wang W, Du Y, Qiu Y, Wang X, Hu Y, Yang J, Cai J and Kennedy J F 2008 *Carbohydr. Polym.* **74** 127-132.
[27] Ma F, Wang Z, Zhao H and Tian S 2012 *Int. J. Mol. Sci.* **13** 7788-7797.
[28] Li K, Xing R, Liu S, Qin Y, Meng X and Li P 2012 *Int. J. Biol. Macromol.* **51** 767-773

Drying of celery leaves *(Apium graveolens L.)* using a PV/T solar dryer

L Sapei[1], E Tarigan[2,3], D N Sugiarto[1] and D Gianluca[1]

[1]Department of Chemical Engineering, Faculty of Engineering, University of Surabaya, Raya Kalirungkut, Surabaya 60293 East Java, Indonesia
[2]Department of Electrical Engineering, Faculty of Engineering, University of Surabaya, Raya Kalirungkut, Surabaya 60293 East Java, Indonesia
[3]Center for Renewable Energy Studies, University of Surabaya, Raya Kalirungkut, Surabaya 60293 East Java, Indonesia

Email: lanny.sapei@staff.ubaya.ac.id

Abstract. Celery (*Avium graveolens L.*) has been naturally used to treat hypertension and cancer. Fresh celery contains a high water content of approximately ±86% which may lower the quality upon storage for a long time. Drying was applied to reduce moisture content in order to prolong the shelf-life. Drying of celery was carried out using a self-designed photovoltaic-thermal (PV/T) solar dryer as well as conventional dryers such as oven and vacuum oven for the comparisons at an approximate temperature of 49°C. The PV/T solar dryer was able to simultaneously convert the solar energy into thermal and electrical energy. The electrical energy was in turn used to drive the blowers to create a forced convective flow of heat transfer. Drying in the PV/T solar dryer was carried out in 3 different weather conditions while 3 different pressures (10; 40; 70 kPa) were tested during drying in the vacuum oven. The PV/T solar dryers demonstrated the best results in terms of constant drying rate (0.986 kg H_2O/min.m^2) and drying time (87 minutes). Total chlorophyll content retained was 5.42 mg/L. The PV/T solar dryer seemed promising to be further developed for an efficient drying process since it required low energy consumption while using the costless renewable solar energy.

1. Introduction

Indonesia is amongst the world's richest nations in terms of its biodiversity including plants and herbs. There are more than 30,000 herbs present in Indonesia, however only 25% of them of which medicine efficacy have been known and moreover only about 1,200 species have already been processed as raw materials for medicinal traditional herbs or so called *jamu* [1]. Celery (*Avium graveolens L.*) contains micronutrients, vitamins and phenolic compounds and widely used as herbs for lowering hypertension and for cancer prevention [2, 3]. Herbs as well as natural plants generally contain high moisture content which could accelerate the microbial growth leading to physical deterioration. An effective drying process is required to reduce the water content thus improving the shelf life of celery.

Several drying methods were conducted in order to reduce moisture content in celery [4, 5]. The most suitable drying techniques involved low temperature of less than 70°C to preserve the nutrient contents [5]. However, there has been no investigation yet related to drying of celery in a solar dryer. Solar dryer has become promising as a drying alternative for drying herbs using the renewable energy with preserved texture of dried samples [6]. The solar energy is able to be converted into thermal energy by means of a solar panel. Moreover, a hybrid solar dryer using an amorphous Si type PV/T system was capable to convert solar energy into thermal as well as electrical energy. Therefore, this system could produce high energy efficiency besides a much higher performance stability compared to the individual solar dryer [7]. This hybrid system has been increasingly applied in diverse applications due to the high demand of the use of renewable energy, especially in Europe [8, 9].

This research objective was to reduce the moisture content of celery by using a hybrid PV/T solar dryer. This self-designed PV/T solar dryer was expected to be able to convert a higher solar energy per drying surface area since the electrical energy would in turn drive the blower inducing active drying facilitated by force convection mechanism. The drying characteristics were studied and compared with those obtained by using the conventional dryers such as oven and vacuum oven drying. The bioactive content particularly chlorophyll content retained in the dried samples were also determined. The use of hybrid solar dryer could be promoted as a dryer means alternative which was pollution free while using a cost free and renewable solar energy for supporting many sustainable green processes.

2. Materials and methods

2.1 Materials

Celery (*Apium graveolens L.*) was purchased from Wonokromo local market, Surabaya, Indonesia; technical grade acetone 98% (Brataco, Indonesia); demineralized water

2.2 Drying of celery leaves

Celery leaves were separated from the rods and other parts and then cleaned to remove soils and other dirt. The leaves were subsequently cut into 2 parts whereby each part had the dimension of about 3 x 4 cm. Sample amounts of about 15 grams were weighed with the balance (Mettler, Toledo) and closely loaded into the aluminium drying pan (29 x 25 cm) in a single layer.

A self-designed photovoltaic/ thermal (PV/T) solar dryer was shown in details in Figure 1. The solar dryer was mainly made of aluminium for the body, glass roof at the top, and a drying cabinet consisting of 4 trays supported by hooks made of plywood boards. A PV/T collector connected to the solar dryer simultaneously converted the solar energy into thermal and electrical energy. It consisted of amorphous Si solar cell panel covered by glass accompanied with battery and controller system. The battery stored electrical energy that was used by the blowers to improve the air circulation. The drying pan containing samples was only put onto the first tray in the middle during the experiment. Drying of celery was conducted in 3 different days between 9.00 am until 16.00 pm with ambient air temperature ranged between 30-35°C resulting in different averaged temperatures within the drying cabinet. The air temperatures were monitored by reading the thermometer (Figure 1D) at the same time with sampling i.e. every 1 minute for the first 30 minutes, followed by every 2 minutes until 1 hour and then every 3 minutes until constant sample weight was attained. The air temperatures were then averaged.

Drying was also conducted using oven (Venticell 111 Model, Medcenter, Germany) and vacuum oven (BOV-30V Model, Biobase, China) at controlled temperature of 49°C corresponding to the optimum averaged temperature obtained in the solar dryer. Vacuum pressures of 10, 40, and 70 KPa were varied during drying at 49°C in the vacuum oven. Drying was carried out until constant weight of sample was attained within the tolerance of ± 0.005 g.

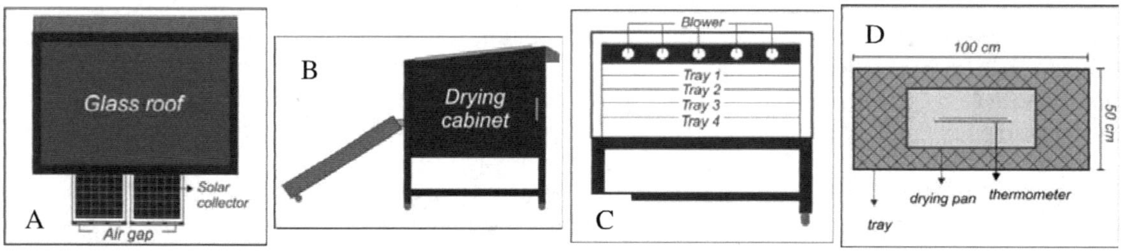

Figure 1. The self-designed PV/T solar dryer. A) Top view; B) Side view; C) Inside the drying cabinet; D) Air temperature measurement around the drying pan.

Results were presented as drying curves whereby free moisture content (X) were monitored versus the drying time [10]. Free moisture content were gained from the moisture content at certain time (X_t) deducted by equilibrium moisture content (X^*).

$$X_t = \frac{W_t - W_d}{W_d} \quad (1)$$

where X_t = moisture content at certain time (kg H_2O/kg dry weight); W_t = sample weight at certain time (kg); W_d = sample dry weight (kg) obtained after sample weight was constant after a certain time.

$$X = X_t - X^* \quad (2)$$

where X = free moisture content (kg H_2O/kg dry weight); X_t = moisture content at certain time (kg H_2O/kg dry weight); X^* = equilibrium moisture content (kg H_2O/kg dry weight).

The drying curve characteristic was also represented by the drying rate (R) versus X.

$$R = \frac{-L_S}{A}\frac{dX}{dt} \quad (3)$$

Where R= drying rate in kg H_2O/m^2.min; Ls = sample dry weight (kg); A = sample surface area (m^2);dX/dt is the slope of linier trend line of X *vs.* t for the constant rate period, where as dX/dt for the falling rate period was calculated using the equation (4) which was the slope of exponential function model as shown in equation (5).

$$\frac{dX}{dt} = \frac{a}{t} \quad (4)$$

$$X_t = a \times \ln t + b \quad (5)$$

Finally, the loss of moisture content in %wet basis was calculated as follows.

$$\text{Moisture Loss (\% wet basis)} = \frac{W_i - W_f}{W_i} \text{x } 100\% \quad (6)$$

where W_i= initial sample weight (g); W_f= final sample weight (g).

2.3 Determination of chlorophyll content

Celery leaves were pounded in a mortar with pestle. Afterwards 3 ml acetone solution was added into 0.02 g of celery leaves powder followed by water addition up to 25 ml in a volumetric flask. An aliquot of 8 ml was taken 3 times into different centrifuge tubes. Centrifugation was conducted for 20 seconds at 3200 rpm in a centrifuge (Shanghai, China) to separate the sediment with supernatant. The supernatant was decanted into a filter to remove out the suspended solid particles. Clear supernatant liquid of 2 ml was mixed with 3 ml acetone prior to absorbance test in a spectrophotometer (Lambda EZ 150, Perkin Elmer, USA). The absorbance values were measured at λ= 665 nm and 649 nm. Both chlorophyll A and chlorophyll B contents were calculated using the equation proposed by Wintermans and De Mots [11, 12]. Total chlorophyll content in mg/L was the sum of chlorophyll A and chlorophyll B. The absorbance measurements were conducted in triplicates at each wavelength.

3. Results and discussion

The drying characteristics of celery leaves in the PV/T solar dryers measured at 3 different average temperatures were depicted in Figure 2. It was obvious that the higher the temperature the shorter the time required to reach the equilibrium moisture content indicated by the constant sample weight. The higher drying temperature also resulted in the higher drying rate. This was in line with the previous drying study on rhizomes [6, 11]. The average temperatures measured in the vicinity of tray 1 where the drying pan was located during the experiment was within the range of 43–55°C which were relatively high compared to the ambient air temperature. The increased temperature was due to the generated thermal energy derived from the solar energy. The heat transfer mechanisms involving radiation, conduction, and forced convection influenced the overall transfer of thermal energy into the drying cabinet. The first tray received the highest thermal energy via solar energy radiation [6] through the glass roof while heat conduction occurred along the dryer walls. The solar radiation impinged into the solar panel was converted into thermal and electrical energy. The force convective mechanisms occurred as the results of self-driven blower was induced by electrical energy produced by the PV/T solar dryer. The blower induced an external force which accelerated the convective heat transfer. The simultaneous thermal and electrical energy derived from solar energy seemed quite efficient in improving the drying rate inside the solar dryer.

The drying performance in the PV/T dryer was compared with those in conventional oven and vacuum oven dryers as could be seen in Figure 3. It turned out that the highest drying rate (Figure 3B) and thus the shortest drying time (Figure 3A) was obtained upon drying using the PV/T solar dryer at the approximately similar drying temperature of ~49°C. The constant drying rate of ~16 times and ~4 times higher compared to those of oven and vacuum oven respectively as seen in Table 1. The constant drying rate obtained from this PV/T solar dryer was much higher with the order of about 2 compared with the previous developed solar dryer without self-driven blower [6]. The lower constant drying rate obtained by using the conventional dryers (< 0.3 kg H_2O/min.m^2) was due to the lower rate of convective heat transfer in contrast to the PV/T solar dryer (~1 kg H_2O/min.m^2). This was also reflected by the highest moisture loss of about 82% and shortest drying time of about 87 min obtained upon celery drying in PV/T solar dryer.

The celery drying in the oven required a shorter time compared to that required in the vacuum oven though the constant drying rate obtained in the vacuum oven was higher than that obtained in the oven. The oven was equipped with a small blower enabling the air circulation and forced convective heat transfer mechanism which accelerated the drying process. In the vacuum oven, the vacuum pressure played a significant role in lowering the boiling point of water in spite of quite low convective heat flow inside the chamber. The lower pressures tended to reduce the drying time although only fair differences were noticed.

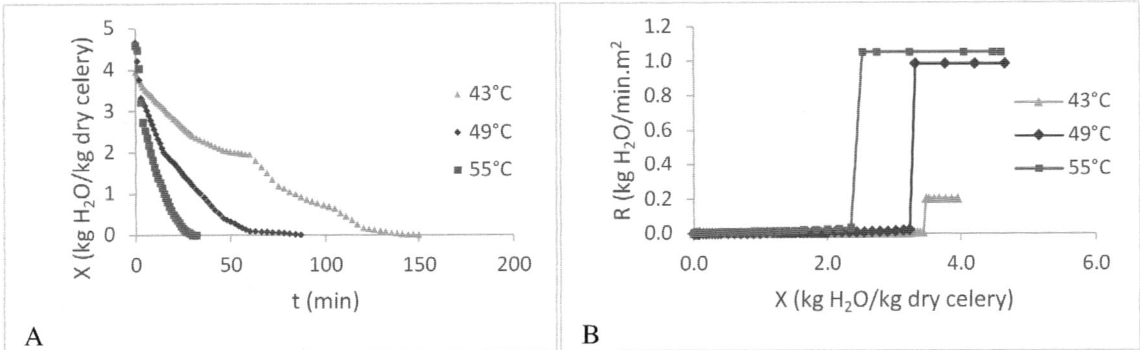

Figure 2. Drying curves characteristics of celery leaves at different averaged drying temperatures in the PV/T solar dryer. A) Free moisture content *vs.* time; B) Constant drying rate of celery leaves *vs.* free moisture content

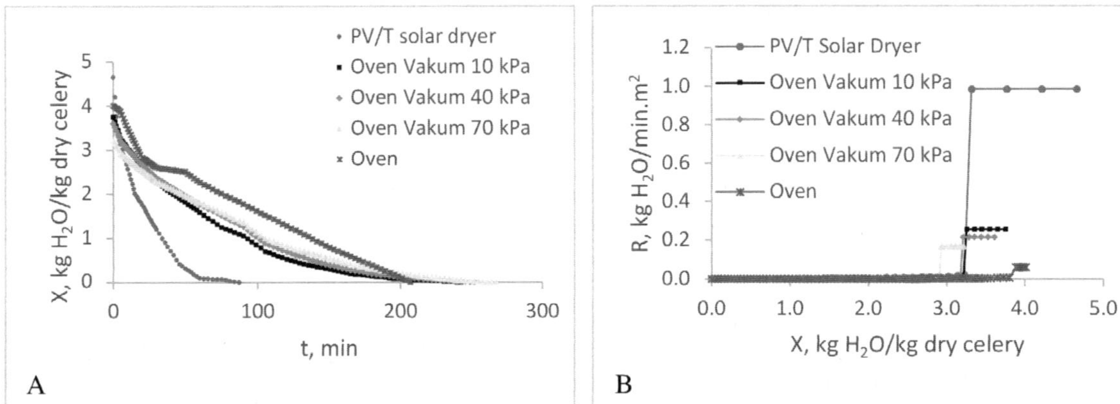

Figure 3. Drying curves characteristics of celery leaves at drying temperature of approximately 49°C in various dryers. A) Free moisture content *vs.* time; B) Constant drying rate of celery leaves *vs.* free moisture content

Table 1. Comparisons of drying time, constant drying rate, moisture loss and chlorophyll content during drying of celery leaves in various dryers at various drying conditions.

Types of dryers	Drying condition		Drying time (min)	Constant drying rate, (kg H_2O/min.m^2)	Moisture loss (% wet basis)	Total chlorophyll content (mg/L)
	P (kPa), Boiling point(°C)	T (°C)				
PV/T solar dryer	101.35 (100°C)	43	150	0.204	79.76	5.71 ± 0.027
	101.35 (100°C)	49	87	0.986	82.1	5.42 ± 0.040
	101.35 (100°C)	55	32	1.054	82.07	5.37 ± 0.020
Oven	101.35 (100°C)	49	207	0.060	79.3	5.48 ± 0.064
Vacuum oven	70 (89.64°C)	49	267	0.171	76.2	7.51 ± 0.086
	40 (72.45°C)	49	252	0.217	78.2	7.74 ± 0.022
	10 (45.81°C)	49	243	0.259	78.9	7.85 ± 0.110
Fresh celery	-	-	-	-	-	9.45 ± 0.015

Finally, the chlorophyll content in the dried celery was determined to study the retention of a sort of bioactive content after the drying process. The total chlorophyll content decreased of about 17-40% after the drying process as depicted in Table 1. This huge decrease in chlorophyll content could be caused by the milling process generating higher heat effect for the dried samples compared to fresh samples. The highest total chlorophyll content was retained in the samples dried using vacuum oven whereas chlorophyll content of celery dried in PV/T solar dryer was similar with that dried in the oven regardless the length of drying time. Slow drying process accompanied by the water evaporation at a much lower temperature seemed slow down the chlorophyll deterioration. Chlorophyll seemed to be sensitive towards drying temperatures as well as drying rates. The experimental results implied the great potential of PV/T solar dryer in substituting the use of energy intensive conventional dryers such as oven and vacuum oven. The renewable and cheap solar energy should be more intensively and optimally used to enhance the sustainable process particularly in drying process of herbs or plants in the tropical country.

4. Conclusions

Drying of celery leaves in the self-designed PV/T solar dryer seemed superior in comparison with the other conventional drying methods such as oven and vacuum oven dryer in terms of constant drying rate, drying time, and chlorophyll content of final dried celery though operational temperature was hardly controlled. The drying temperature was highly dependent on the sunlight intensity. The higher the drying temperature was, the higher the constant drying rate and thus the shorter the drying time was. The constant drying rate of celery leaves in the PV/T solar dryer was approximately16 times higher and 4 times higher compared to those obtained from oven and vacuum oven at drying temperature of about 49°C. The relatively high constant drying rate combined with short drying time was due to the predominant forced convection facilitated heat transfer mechanism created by solar driven blowers besides heat conduction and heat radiation mechanisms. The celery weight reached a constant weight within 87 minutes which was almost half of the required time within oven and vacuum oven without deteriorating the chlorophyll content. The chlorophyll content in PV/T solar dried celery leaves ranged from 4.1–5.7 mg/L which was a bit higher compared with those obtained using oven and vacuum oven. The abundant and renewable solar energy captured by a PV/T solar dryer seemed quite promising for various drying purposes in an efficient and cheap way.

Acknowledgments

The authors wish to thank Ir. Rudy Agustriyanto, S.T., M.Sc., Ph.D., IPM and Ir. Natalia Suseno, M.Si. for some fruitful discussion. The research was funded by Ministry of Research, Technology and Higher Education of the Republic of Indonesia under the research grant scheme of PTUPT (*Penelitian Terapan Unggulan Perguruan Tinggi*) 2019 under the contract number: 030/ST-Lit/LPPM-01/DRPM/Mono/FT/III/2019.

References

[1] PT. Sido Muncul 2015 *Delivering The Vision-Annual Report PT. Sido Muncul, Tbk for the year 2015* (Jakarta: PT. Sido Muncul)
[2] Saputra O and Fitria T 2016 *Jurnal Majority* **5** 120-5
[3] Sukohar A and Arisandi R 2016 *Jurnal Majority* **5** 95-100
[4] Alibas I 2014 *Food Sci. Technol.* **34** 394-401
[5] Arief U M, Suryanto A and Suryanto S 2014 *Sainteknol: Jurnal Sains dan Teknologi* **12** 171-81
[6] Soebiantoro F A, *Tarigan* E, Hwa L, Halim V P and Sapei L 2018 *Pertanika J. Sci.Technol.* **26**1 265-74
[7] Tyagi V V, Kaushik S C and Tyagi S K 2012 *Renewable Sustainable Energy Rev.* **16** 1383–98
[8] Noro M, Lazzarin R and Bagarella G 2016 *Energy Procedia* **101** 496-503
[9] Ellabban O, Abu-Rub H, and Blaabjerg F 2014 *Renewable Sustainable Energy Rev.* **39** 748-64
[10] Geankoplis C J 2003 *Transport Process and Separation Process Principles 4th ed* (New Jersey: Prentice Hall) pp 559-80
[11] Wintermans J F G M and De Mots A S 1965 *Biochim. Biophys.Acta,Biophys. Incl. Photosynth* **109** 448-453
[12] Novichonok E V, Novichonok A O, Kurbatova J A and Markovskaya E F 2016 *Photosynthetica* **54** 130-7
[13] Sapei L, Tarigan E, Lie H, Halim V P and Soebiantoro F A 2017 *KnE Life Sciences* **3** 110-17 Knowledge E.

IOP Conf. Series: Materials Science and Engineering **703** (2019) 012043 doi:10.1088/1757-899X/703/1/012043

Mass transfer kinetic model and removal capacity of acid blue 29 adsorptions onto activated carbon

P Setyopratomo, H R Priyantini, and R Agustriyanto

Department of Chemical Engineering, University of Surabaya, Raya Kalirungkut, Surabaya, 60293, Indonesia

E-mail: puguh@staff.ubaya.ac.id

Abstract. So far, the handling of dye pollution remains a challenge. One of the promising methods to reduce the dyes content from such waste is adsorption. The purpose of this study is to examine the suitability of kinetic models and to determine the removal capacity of acid blue 29 adsorptions onto activated carbon. Two mathematical models were developed to evaluate the adsorption kinetics. The first model only took into account the convective mass transfer of the dyestuff to the surface of the adsorbent, while the second model also took into account the diffusion of the dyes in the adsorbent pore. The result showed that the adsorption process followed the Langmuir equilibrium and the kinetic fits to the second model in which the adsorption rate is controlled by convection mass transfer in the liquid phase as well as the intra-particle molecular diffusion that occurs in the pore. The dye removal capacity inversely proportional to the dye concentration and reaches 86.6% when the initial dye concentration is 75 ppm.

1. Introduction

Up to recent decades, the contamination of industrial waste streams by synthetic dyes still becomes a problem which needs special attention [1]. Dyes are widely used in industries such as cosmetics, leather, printing, paper, food, and textiles. During the coloring process, approximately 10-15% of the dyes will enter the waste stream [2]. Dyestuff molecules are stable, so that when these chemicals enter the human or animal body or plant tissue. These substances will be stable in these tissues for a very long time[3].

The adverse effect of these chemicals on an aquatic ecosystem is to reduce the oxygen content and penetration of light in water. Besides, in the ecosystem, the pollutant inhibits photosynthesis and growth of biota [4]. Moreover, it was reported that exposure to a small amount of dye (<1 ppm) can cause several adverse effects such as skin irritation, dermatitis, allergy, cancer and even mutagenic effects on aquatic organisms and human [5].

Many methods and technologies were developed to remove these pollutants such as coagulation, precipitation, filtration, ion exchange, membrane technology, and adsorption. Among these methods, adsorption is seen as a very prospective choice because it is technically simple, high efficiency and economically competitive [6,7].

Studies of acid blue 29 adsorption using different adsorbents have been carried out by several researchers, among others: observation of the removal efficiency using activated carbon-clay composite [8], evaluation of the effect of dyes' structure on the efficiency of the adsorption [9] and investigation of the adsorptive potential of chemically activated olive pomace boiler ash [10]. In this works, acid blue 29 adsorptions onto activated carbon was carried out in a batch system. Two kinetic models were developed. In the first model, the transport of adsorbate to be assumed only occurs from the bulk solution to the particle surface while the second model takes into account the intra-particle diffusion of adsorbate molecules to the active sites through the pore of the particles. The study was focused on getting a suitable kinetic model applied to this adsorption system.

2. Theory and experiment

2.1 Adsorption equilibrium correlation

Adsorption equilibrium is a condition where the solute concentration in the liquid phase at the interface is in equilibrium with the concentration of the solute at the sorbent at the interface. Three mathematical relationships are generally used to represent adsorption equilibrium, i.e: Langmuir (Eq. 1), Freundlich (Eq. 2) and Linear (Eq. 3).

$$\frac{1}{C_{\mu}} = \frac{1}{C_{\mu,max}} + \frac{1}{C_{\mu,max}k}\frac{1}{C_{eq}} \tag{1}$$

$$C_{\mu} = K\,C_{ceq}^{1/n} \tag{2}$$

$$C_{eq} = HC_{\mu} \tag{3}$$

where, C_{μ}: the amount of adsorbate adsorbed per mass of the adsorbent at equilibrium, $C_{\mu,max}$: the theoretical monolayer saturation capacity, C_{eq}: solute concentration in solution at equilibrium, k, K, n: constant, H: a distribution constant.

2.2 Modeling

In this study, two adsorption models were developed and observed. The first model is called model I, while the second model is called model II. The model I assumes that adsorption only occurs on the surface of the adsorbent and there is no intra-particle concentration gradient. Then model I adsorption mechanism consists of (1) mass transfer of the adsorbate from the solution bulk to the surface of the adsorbent through the film layer on the outer surface of the adsorbent; (2) adsorption of the adsorbate at the surface of the adsorbent.

The first model implies that, in the liquid phase, the rate of decrease in adsorbate concentration is equal to the rate of mass transfer of the solute from the solution to the adsorbent. Meanwhile, the solute mass transfer rate from the solution to the adsorbent will be equal to the rate of accumulation of the adsorbate on the adsorbent. Therefore, the following two equations apply to the first model.

$$\frac{dC_L}{dt} = -\frac{k_c A}{V}\left(C_L - C_L^*\right) \tag{4}$$

$$\frac{dC}{dt} = \frac{k_c A}{m}\left(C_L - C_L^*\right) \tag{5}$$

where, C_L: the solute concentration in liquid phase, k_c : the convective mass transfer coefficient, A: the outer surface area of the adsorbent, C_L*: the concentration of the adsorbate at the liquid phase interface, V : the volume of solution, t : time, C : the concentration of adsorbate on the adsorbent and m : mass of adsorbent. The initial condition are: at t=0, $C_L = C_{Lo}$, C = 0. The boundary condition is $C_N = H\,C_L^*$, where C_N: concentration of solute at the solid surface.

The second model assumes that the adsorbent is porous media, so that the mass transfer of the adsorbate occurs with three mechanisms, namely: (1) mass transfer of the solute from the solution bulk to the surface of the adsorbent through the film layer of the liquid phase, (2) intraparticle diffusion of solute from the adsorbent surface to the center of the adsorbent through the pore, (3) adsorption of the solute at the active sites in the pore. Assuming the adsorbent particles are spherical and have uniform diffusivity and porosity, the following differential equations apply to the adsorption of the second model.
At the middle of the solid:

$$\left(-\mathrm{De}\, 4\pi r^2 \left.\frac{\partial C}{\partial r}\right|_r\right)-\left(-\mathrm{De}\, 4\pi(r+\Delta r)^2 \left.\frac{\partial C}{\partial r}\right|_{r+\Delta r}\right)=4\pi r^2 \Delta r \frac{\partial}{\partial t}\left(\varepsilon C + C_\mu \rho_{ads}\right) \quad (6)$$

At the center of the solid:

$$\mathrm{De}\, 4\pi(\frac{\Delta r}{2})^2 (\frac{\partial C}{\partial r}) = \frac{4}{3}\pi(\frac{\Delta r}{2})^3 (\varepsilon + \frac{\rho_{ads}}{H})(\frac{\partial C}{\partial t}) \quad (7)$$

On the surface of the adsorbent:

$$k_c\, 4\pi\pi^2 (C_L - C_L{*}) = \mathrm{De}\, 4\pi\pi^2 (\frac{\partial C_N}{\partial r})_{surface} \quad (8)$$

Where, C: concentration of solute at solid, ρ_{ads}: density of adsorbent, ε: ratio of the solution volume to the total volume, ρ_{ads}: adsorbent density, R: adsorbent radius, r: position in adsorbent in the radial system, D_e: effective diffusivity.

2.3. Experiment
The acid blue 29 dye was purchased from a local chemical supplier. Meanwhile, activated carbon was obtained from Merck Millipore. The activated carbon has a particle size <100 μm and bulk density: 150 - 440 kg/m^3. Batch adsorption experiments were carried out using a 250 ml beaker with constant speed stirring using a magnetic bar. The volume of acid blue 29 solutions used in each batch is 100 ml with the initial concentration was varied by 75, 125, 185, 225 and 275 ppm. While the mass of activated carbon used was varied as much as 0.1, 0.1125, 0.125, 0.1375 and 0.15 g. Thus the ratio of the mass of activated carbon to the volume of the solution varies by 0.001, 0.001125, 0.00125, 0.001375 and 0.0015 g/ml. The concentration of acid blue 29 in solution at certain times and at equilibrium was measured using a UV-Vis spectrophotometer - Agilent 8453.

3. Results and discussions

3.1 Equilibrium analysis
In this experiment, to ensure that adsorption equilibrium has been reached, the batch adsorption experiment was carried out overnight. The acid blue 29 concentration in activated carbon and in solution at equilibrium was obtained from the experiment, which was marked as dots, and plot of Langmuir, Freundlich and Linear equilibrium correlations is shown in figure 1. While the values of the parameters of the equilibrium correlation are presented in table 1. The results showed that the Langmuir and Freundlich equilibrium correlations are better suited to the experimental data than the Linear correlation. Linear correlation generally only applies to low concentrations. Based on the value of the coefficient of determination, the correlation of Langmuir is the most suitable.

Since the Langmuir equilibrium correlation is developed based on the assumption that the adsorbent surface can only adsorb one layer of the adsorbate molecules, then this result confirmed that the acid blue 29 molecules absorbed on the surface of activated carbon form a single layer (monolayer) with an adsorption capacity of 156.25 mg/g. Adsorption of several types of acid dyes with activated carbon followed the Langmuir correlation [11]. The adsorption capacity obtained in this study is higher than that obtained by Marrakchi et al. which is 104.83 mg/g using activated carbon – clay composite [8] and lower than that obtained by Auta et al. which is 453.12 mg/g using waste tea activated carbon [12].

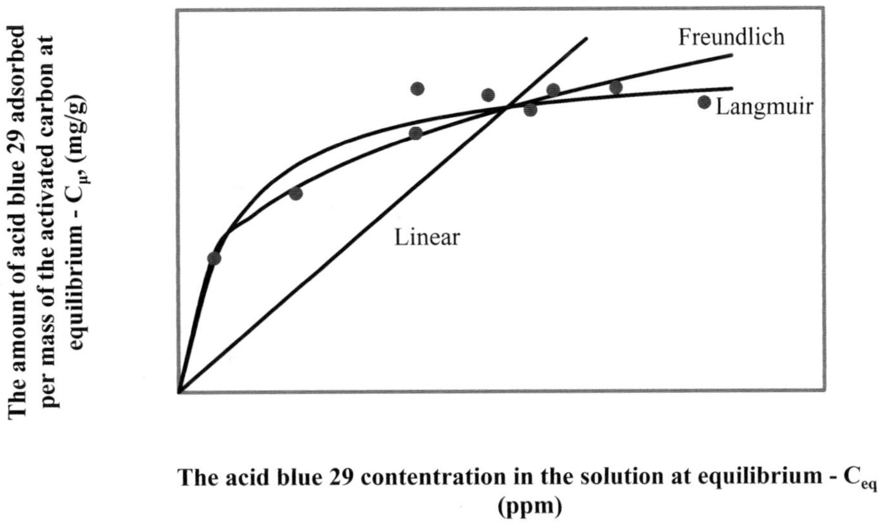

Figure 1. The acid blue 29 adsorptions equilibrium curve on activated carbon

Table 1. Equilibrium parameters of acid blue 29 adsorption on the activated carbon.

Equilibrium adsorption model	Parameters		The coefficient of determination (R^2)
Linear	H	0.063 g/L	0.523
Freundlich	K	31.728	0.910
	1/n	0.320	
Langmuir	k	0.067 L/mg	0.956
	$C_{\mu,max}$	156.25 mg/g	

3.2 Suitability of the model with experimental data

To solve the first and the second model, the finite difference method was applied. Then, the numerical computation was run to simultaneously solve a set of the differential equations using the Matlab program. In the computation, the adsorption equilibrium parameters obtained from the experiment were used. The numerical computational results, in which the acid blue 29 concentration change concerning to time during the adsorption process, are presented in figure 2 as a continuous line. Meanwhile, the related real values which were obtained from an experiment taken at several certain times were displayed as discrete data in the form of dots.

The results clearly show that the second model is more suitable than the first model. This finding has confirmed that acid blue 29 adsorptions onto activated carbon occurs with a mechanism that involves intra-particle diffusion through the pores of the activated carbon. Both external mass transfer and intra-particle diffusion play a role in dye adsorption. At the initial stages, the external mass transfer is more significant, while at the final stages the mass transport is mainly controlled by intra-particle diffusion [13].

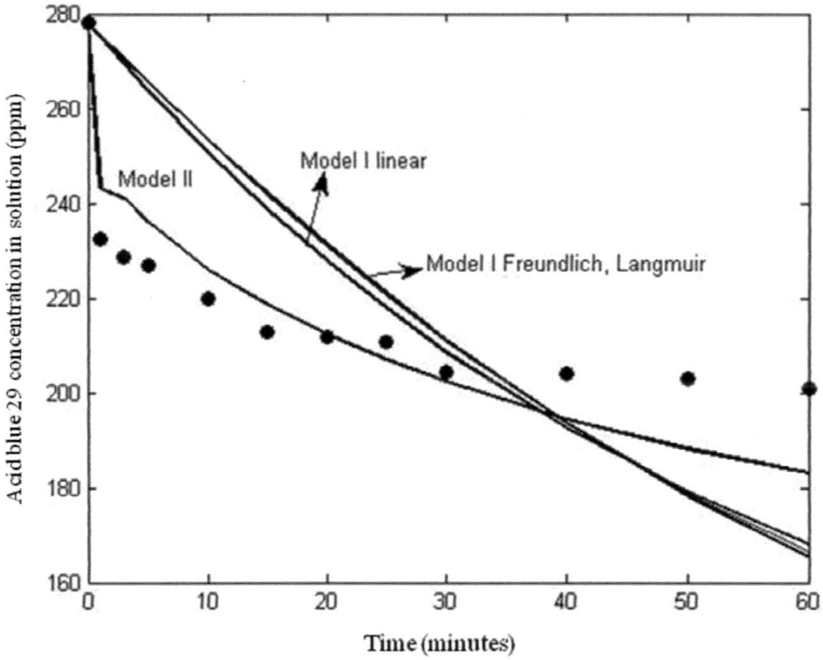

Figure 2. Decreasing the acid blue 29 concentration to the time for the initial concentration of 275 ppm and the ratio of the adsorbent to the solution of 0.001 g/ml.

3.3 The convective coefficient and the effective diffusivity

In observing the suitability of the first kinetic model, the mass transfer coefficient is obtained by solving the set of resulting differential equations. It was found that, even though it hasn't changed significantly, the coefficient decreases with the increasing initial concentration of the dye. For the mass of adsorbent to the volume of solution ratio of 0,001 g/ml, the convective coefficient is $6{,}40.10^{-2}$ cm/minute when the initial dye concentration is 75 ppm and it is $0{,}48.10^{-2}$ cm/minute when the initial dye concentration is 275 ppm. A similar trend also occurs in acid dye adsorption kinetics modeling developed by Aksu et al. [14]. Varying the ratio of the mass of adsorbent to the volume solution showed that, even though it doesn't change significantly, the convective coefficient increases with increasing the ratio. For initial dye concentration of 275 ppm, the convective coefficient is $0{,}48.10^{-2}$ cm/minute when the ratio is 0,001 g/ml and it increases to $0{,}62.10^{-2}$ cm/minute when the ratio is 0,0015 g/ml. Thus this study shows that the value of the convective coefficient is not much influenced by the concentration of the dye in the solution nor the ratio of the mass of adsorbent to the volume of the solution.

Since the second model takes into account the intra-particle diffusion in the adsorbent pores, the outputs of the second model solution are the convective coefficient of the dye in the liquid phase and the effective diffusivity of the dye in the adsorbent particle. From the computation results, itwas found that the convective coefficient is not affected by the initial concentration of the dye nor the ratio of the mass of adsorbent to the volume of the solution. The average value of the convection coefficient is 0.7248 cm/minute. Meanwhile, the effective diffusivity decreased with increasing acid blue 29 initial concentration, from 1.40×10^{-4} cm/minute at initial dye concentration 75 ppm to 1.64×10^{-6} cm/minute at dye initial concentration 275 ppm at mass adsorbent to solution volume ratio 0,001 g/ml. Besides, it was found that the effective diffusivity was not influenced by the changes of the mass adsorbent to the volume of solution ratio.

Decreasing the effective diffusivity at high dyestuff concentrations was also observed in the adsorption of tectilon red acid dyes on granule activated carbon which might be a result of collisions between dye molecules in the pore which caused additional diffusion barriers. Also, at high concentrations, the possibility of agglomeration of dye molecules that have large molecular sizes can also cause micropore-clogging and inhibit diffusion [15].

3.4 Removal percentage.
In this experiment, the removal percentage of the acid blue 29 is defined as the ratio of the difference between the initial dye concentration in the solution and the concentration at a certain time to the initial concentration. In this study, it was found that the higher the initial concentration of dyes, the smaller the percent removal of the acid blue 29. For radio of mass of adsorbent to solution volume of 0.001 g/ml, the percentage of removal reached 86.6% when the initial dye concentration was 75 ppm and decreased to 48.8% when the initial dye concentration was 275 ppm. It was reported that the removal percentage of dyes by adsorption was in the range of 68% – 99% [16].

4. Conclusions
Acid blue 29 dye adsorption onto activated carbon follows the Langmuir correlation and fits with the second model in which the adsorption rate is controlled by convection mass transfer in the liquid phase as well as the intra-particle molecular diffusion that occurs in the pore. It was found that the convective coefficient is independent to the initial concentration of the dye and the ratio of the mass of adsorbent to the volume of solution. On the other hand, the effective diffusivity depends on the initial concentration. The dye removal capacity increase when the initial dye concentration in the solution decrease and reaches 86.6% when the initial dye concentration is 75 ppm.

Acknowledgment
The authors would like to thank Andrew Ronaldi Tandio and Stevani for their contributions to the development and completion of modeling and laboratory works.

References
[1] Robati D, Mirza B, Ghazisaeidi R, Rajabi M, Moradi O, Tyagi I, Agarwal S and Gupta V K 2016 *J. Mol. Liq.***216** 830–35.
[2] Goswani M and Phukan P 2017 *J. Environ. Chem. Eng.***5** 3508-17.
[3] Guo F, Li X, Jiang X, Zhao X, Guo C and Rao Z 2018 *Colloids Surf.***555** 43-54.
[4] Kazeem T S, Lateef S A, Ganiyu S A, Qamaruddin M, Tanimu A, Sulaiman K O, Jillani S M S and Alhooshani K 2018 *J. Cleaner Prod.***205** 303-12.
[5] Noorimotlagh Z, Mirzaee S A, Martinez S S, Alavi S, Ahmadi M and Jaafarzadeh N 2019 *Chem. Eng. Res. Des.***141** 290–301.
[6] Sun J, Liu X, Zhang F, Zhou J, Wu J, Alsaedi A, Hayat T and Li J 2019 *Colloids Surf.* A **563** 22-30.
[7] Abd-Elhamid A I, Kamoun E A, El-Shanshory A A, Soliman H M A and Aly H F 2019 *J. Mol. Liq.***279** 530-539.
[8] Marrakchi F, Bouaziz M and Hameed B H 2017 *Chem. Eng. Res. Des.* **128** 221–30.
[9] Ciesielczyk F, Bartczak P, Zdarta J and Jesionowski T 2017 *J. Environ. Manage.* **204** 123-35.
[10] Marrakchi F, Bouaziz M and Hameed B H 2017 *Colloids. Surf.* A **535** 157-65.
[11] Hadi M, Samargandhi M R and McKay G 2010 *Chem. Eng. J.***160** 408-16.
[12] Auta M and Hameed B H 2011 *Chem. Eng. J.***175** 233-43.
[13] Magdy Y H and Altaher H 2018 *J. Environ. Chem. Eng.* **6** 834-41.
[14] Aksu Z, Tatli A I and Tunc O 2008 *Chem. Eng. J* **142** 23-29.
[15] Walker G M and Weatherley L R 1999 *Water Res.***33**1895-99.
[16] Kausa A, Iqbal M, Javed A, Aftab K, Nazli Z, Bhatti H N and Nouren S 2018 *J. Mol. Liq.* **256** 395-407.

IOP Conf. Series: Materials Science and Engineering **703** (2019) 012044 doi:10.1088/1757-899X/703/1/012044

Utilization of rice straw and used paper for the recycle papermaking

N Suseno [1], T Adiarto[2], M Sifra[1] and V Elvira[1]

[1] Department of Chemical Engineering, University of Surabaya, Raya Kalirungkut, Surabaya, 60293, Indonesia
[2] Department of Chemistry, University of Airlangga, East Java, Surabaya, Indonesia

Email: suseno.natalia@yahoo.com

Abstract. Rice Straw is one of the most important materials that has been used for pulp and paper production due to its abundance and cost-effectiveness. However, the pulping and papermaking characteristics of the mixture of rice straw and used paper have rarely been investigated. In this experiment, delignified rice straw was mixed with used paper in order to make recycle papers with acceptable properties. Soda pulping process was carried out to remove out of lignin. The delignification was designed to measure the effects in terms of sodium hydroxide concentration and temperature, on the cellulose and lignin content of rice straw, and its tensile strength and water absorption. The mass ratio of used paper and rice straw was varied and its physical properties was observed and compared to paper from natural rice straw. The delignification was conducted using NaOH concentrations (4-10% w/w) and temperatures (60-90°C) for 1 hour. The ratio of rice straw to used paper was varied (1:1-1:9). An optimum condition was obtained from rice straw delignification at NaOH concentration of 8% and temperature of 90°C. The results of the optimum condition obtained the highest amount of cellulose content and tensile strength while having the lowest amount of lignin content and water absorption.

1. Introduction

In recent years, the three major problems that would continue to puzzle the development of the paper industry are the shortage of resources, the contamination of the environment and the level of technical equipment. The most dominating factor is the shortage of raw material resources, which is largely due to the contradiction between the structure of the raw materials and the structure of the fibre resources. Thereby, non-wood fibres possess a rich variety of excellent properties in physical and optical aspects, which could be used to improve their products. Production of pulp from non-wood resources has many advantages, such as easy pulping capability, excellent fibres for the special types of paper In some developing countries, about 60-80% of the cellulose fibre comes from non-wood materials, such as bagasse, rice straw, corn straw, bamboo, reed, grass, jute, flax, sisal, and so on [1]. Particularly, in developing country such as Indonesia, the raw materials from non-wood plants are potential to be used in fibre source for papermaking.

Indonesia is the third largest country in the world in producing rice. The total harvested area of rice is 15,697 million acres and the production reaches to 81,073 million tons. The amount of rice straw disposal as by products of paddy milling is 40.537 million tons, about half of total rice production [2].

Rice straw is one of the most important materials that has been used for pulp and paper production due to the high cellulose content. Rice straw contains 33-40% of cellulose, 24-28% of hemicellulose and 2-25% of lignin_[3]. The production of paper is essentially a process of removing the lignin contained in the fibre of raw materials in order to obtain a high cellulose content and low lignin content in pulp or paper [4]. The soda process has been known to be the oldest and simplest pulping process. It is also applicable to leafy and conifer wood and non-wood raw materials, such as agricultural residues [5, 6].

Research of papermaking rice with a variety of processes and additional raw materials have been done by several researchers, among others: Influence of rice straw cooking conditions in the soda-ethanol-water pulping on the mechanical properties of produced paper sheets [7], feasibility of rice straw as a raw material for the production of soda cellulose pulp [5], two-step soda pulping processes of rice straw for the recycle papermaking [8], pulping of wheat straw with caustic potash-ammonia aqueous solutions and its kinetics [9]. The effects of sodium hydroxide content on mechanical and physical properties of rice straw paper [10]. In spite of several studies in rice straw soda pulping, little research has been conducted on the pulping and papermaking characteristics of the mixture of rice straw and used paper or waste paper [11].

In addition, over the last decade, waste minimization, reuse, as well as material recycling have received increased attention due to the governing waste management's regulations and legislations in both national and international scope. As secondary fibre resources, waste paper has an important role to meet the demand of cellulose industries [12]. Waste paper which has been recycled is needed in the process of papermaking, hence, the consumption of waste paper keeps increasing by year. This fact has a positive impact of preserving virgin natural fibre to a great extent. The recycling process of the conventional office paper reduces the abundance of waste paper. Moreover, after being cleaned and reformed, used paper could be converted back into pulp. In this experiment, in order to obtain recycle papers with desirable properties, rice straw was delignified with different NaOH concentrations at different temperatures during soda pulping prior to mixing with office waste papers. The delignification was designed to measure the effects in terms of sodium hydroxide concentration and temperature on the cellulose and lignin content of rice straw, and its tensile strength and water absorption to establish the optimum operating pulping conditions. The ratio of used paper mixture to rice straw was varied in order to observe the physical properties (tensile strength and water absorption) of recycle paper and compared with delignified natural rice straw paper.

2. Experimental

2.1 Materials

The materials include rice straw from Mojokerto, East Java, Indonesia, 95% NaOH (PT Chiwi Kimia), 95% ethanol, benzena p.a, 72% sulfuric acid, and oxylic acid p.a

2.2 Soda pulping process

The wastes of rice straw were cut, milled, and screened prior to pulping. In the soda pulping process of rice straw, various NaOH concentration (4, 6, 8, 10) %, delignification temperature (60, 70, 80, 90)^{o}C were varied. Rice straw was delignified with NaOH solution with a ratio 1:17 (w/v) in a laboratory flask for 1 hour. The resulting pulp was neutralized by washing with water. Delignified rice straw was added with water and blended with motor stirrer in 300 rpm. The mixtures were poured into a filter and pressed by water aided by a gravity force to remove out water. The resulting cake was a sheet of paper which was then dried in the oven for further characterisations, such as the evaluation of chemical composition (cellulose and lignin contents) and physical properties (tensile strength and water absorption).

2.3 Papermaking process

Recycle paper was made by mixing delignified rice straw and used paper with a various mass ratio of 1:1, 1:3, 1:5, 1:7 dan 1:9. Waste paper was cut into small pieces, added with water, and then blended. Delignified rice straw was added with water and then blended before being mixed with blended waste papers. The mixtures were poured into a filter and pressed by water aided by a gravity force to remove out water. The resulting cake was a sheet of recycle paper which was then dried in the oven for further physical properties characterisations (tensile strength and water absorption).

2.4 Paper characterizations
The evaluation of the chemical composition of pulp and paper was characterized chemically in accordance with the applicable Indonesian National Standard (SNI), which are: cellulose (SNI 0444-2009), lignin (SNI 0492-2008) as well as kappa number determination (SNI 0494:2008). Further for physical characterization, such as the measurement of tensile strength and water absorption was conducted. Tensile strength was measured using autograph (SNI ISO 1924-2:2010) for the water absorption measurement with Cobb method (SNI 0499:2008).

2.5 Experimental design
In this experiment the relation of input variables and output variables were regression modelled using a second order polynomial equation.

$$Y_1 = \beta o + \beta_1 . X_1 + \beta_2 . X_2 + \beta_{11} X_1^2 + \beta_{22} X_2^2 + \beta_{12} . X_1 . X_2 \quad (1)$$

where X_1 and X_2 are independent or input variables, i.e. NaOH concentration, delignification temperature. Y_i (i = 1-2) is the response or output variables, i.e. cellulose, lignin content, tensile strength and water absorption. βo is a constant, β_1 and β_2 represent the linear coefficients, whilst β_{11} and β_{22} are the quadratic coefficients, β_{12} is the interaction coefficients. The result of the physical test was analyzed by statistics with 95% significance level. The software Microsoft Excel was performed to analyze the results in the form of analysis of variance (ANOVA).

3. Result and discussion
In this soda delignification process, NaOH concentration, temperature was varied in order to obtain the optimum condition. In Table 1, eight experiments conducted, together with the delignification process conditions and their corresponding with chemical contents (cellulose and lignin) and physical properties (tensile strength and water absorption).

Table 1. Delignification conditions and experimental results of pulping process including cellulose and lignin content, tensile strength and water absorption

C_{NaOH} (%)	T (°C)	Cellulose (%)	Lignin (%)	Tensile Strength (N/mm^2)	Water Absorption (g/cm^2)
4	90	74.22	5.48	2.14	195.67
6	90	76.84	4.78	2.70	195.03
8	90	79.49	3.96	4.36	193.58
10	90	74.58	4.18	4.18	191.03
8	60	71.66	10.78	1.08	253.96
8	70	73.18	10.16	1.13	220.62
8	80	74.78	9.67	1.18	190.98
8	90	79.38	4.58	3.92	185.67

Table 1 shows that for variations of temperature 60°C-90°C in a concentration of NaOH constant at 8% of the increasing temperature will increase the cellulose content followed by the decrease in the lignin content in the paper resulted. The reason is the increasing of delignification temperature will increase the rate of delignification in NaOH solution [9]. Thus, lignin will be more easily dissolved in

NaOH. The highest cellulose of 79.38% and the lowest lignin of 4.58% were obtained at delignification temperature of 90°C.

Tensile strength was generally influenced by the fibre length and molecular bonds. The longer the cellulose fibres and the stronger of molecular bonds would be, the higher the tensile strengths of the paper would be. The removal of the lignin in the cell will increase the molecular bonding between cellulose fibres, thus it would increase the tensile strength. The higher concentration of NaOH used in delignification, the more lignin could be dissolved and decreased the surface roughness. It indicates that the paper has improved fibre structure with fine arrangement to each other, therefore it could be able to improve fibre properties as well as strengthen bonds between the fibres [10]. However, the possibility of the cellulose fibres degradation increases as NaOH concentration, the tensile strength decreases at concentration of NaOH above 8%. The same trend was observed by the variations in delignification temperatures. The increase in temperature would increase the delignification process so it would increase the tensile strength, on the contrary, the water absorption would decrease. The lowest water absorption reaches at 90°C.

The result of experimental data (Table 1) was fitted with a second-order polynomial model by performing multiple linear regressions (equation 1). The results of the experiment have been summarized in Table 2 and p-value presented in Table 3.

Table 2. Linear and non-linear coefficient resulted from the experimental design of dependent variables (pulp composition and physical properties) according to equation (1)

Dependent Variables (Yi)	βo	β_2	β_1^2	β_2^2	$\beta_1.\beta_2$	R^2
Cellulose (%)	1.24	$-1.4x10^{-2}$	-50.72	$6.23x10^{-5}$	0.08	0.93
Lignin (%)	-0.46	$1.7x10^{-2}$	2.94	$1.23x10^{-4}$	$7x10^{-3}$	0.97
Tensile Strength (N/mm^2)	41.81	-1.11	$1.61x10^3$	$6.53x10^{-3}$	2.65	0.87
Water Absorption (g/cm^2)	827.76	-14.50	$2.51x10^3$	$8.49x10^{-2}$	-5.09	0.99

Table 3. p-value (α=5%) resulted from the experimental design of dependent variables (pulp composition and physical properties) according to equation (1)

		p value (α=5%)			
Dependent Variables (Yi)	Intercept	X_2	X_1^2	X_2^2	$X_1. X_2$
Cellulose (%)	0.03	0.74	0.03	0.31	0.03
Lignin (%)	0.10	0.99	0.74	0.03	0.62
TS (N/mm^2)	0.12	$8x10^{-10}$	0.14	0.13	0.12
WA(g/cm^2)	$9.6x10^{-3}$	$9.8x10^{-10}$	0.69	0.03	0.60

Table 2 shows that all the dependent variables were high validity of model fitting with the experimental ($R^2 \geq 0.9$). According to p-value in table 3, it indicated that the independent variables to dependent variables have significantly influenced when $p<0.05$. During the delignification process, the concentration of NaOH in the quadratic function (X_1^2) as well as the interaction between concentration and temperature ($X_1. X_2$) on cellulose content has significant influence ($p < 0.05$). On the other hand, the lignin content would follow the quadratic function of temperature (X_2^2). Furthermore, the tensile strength of the resulting paper follows the linear function of temperature (X_2), while the influence of temperature on water absorption follows the linear function (X_2) and quadratic function (X_2^2).

The effect of the ratio of used paper and rice straw (1:1, 1:3, 1:5, 1:7, 1:9) on tensile strength and water absorption is presented in figure 1 and figure 2. Based on Figure 1, the tensile strength of recycle paper decreases as the percentage of the used paper in the mixture increases. However, the overall tensile resistance of used paper mixture with delignified rice straw was still greater compared

IOP Conf. Series: Materials Science and Engineering **703** (2019) 012044 doi:10.1088/1757-899X/703/1/012044

to natural delignified rice straw. Used paper contains fibril natural fibres, therefore when reformed to the recycle paper, the paper strength will decrease. In the production of paper, the additive needed to make the better performance of paper in term of physical or mechanical properties. During the mechanical pretreatment process, the additive or filler could be removed and replaced by short fibre. The more composition waste paper is used, it will decrease hydrogen bonds resulting to short cellulose fibre source. The composition 1: 1 obtained the highest tensile strength is 8.65 N/mm², while the tensile on natural delignified rice straw paper is 4.36 N/mm². The tensile strength of the recycle paper from the mixture used paper and rice straw was compared to commercial paper. The tensile strength obtained from the commercial standard of paper properties is 10.83 N/mm².

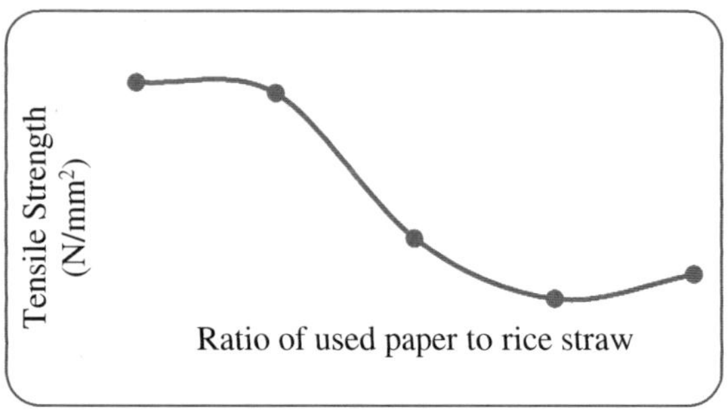

Figure 1. Effect of ratio of used paper to rice straw on tensile strength

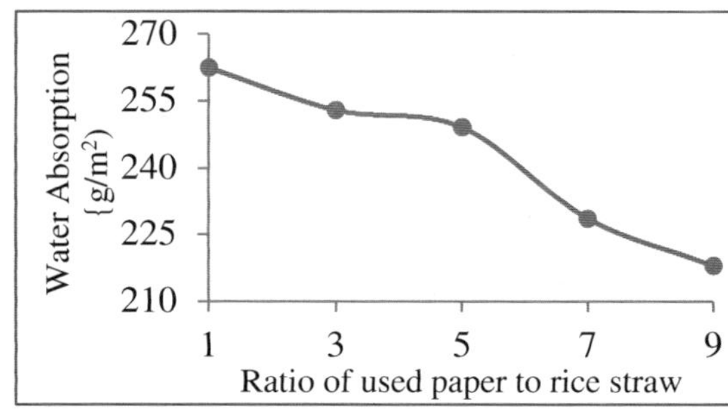

Figure 2. Effect of ratio of used paper to rice straw on water absorption

In Figure 2, the more composition of used paper in the mixture will reduce the amount of water absorption. It may be caused by the fact that there has been already a "sizing agent" additive on used paper. This additive aims to make the paper more hydrophobic, therefore when paper is used for writing, it can prevent excessive spread of ink.

The optimum pulping condition achieves using NaOH concentration of 8% at 90°C based on the relatively the highest tensile strength and the lowest water absorption measured, i.e 8.65 N/mm² and 185.67 gr/m². This Kappa number determined at this condition was 4.5, which the kappa number requirement white paper stock was less than 5. A lower kappa number means that less lignin remains in the pulp, implying that more effectively than a pulp with a higher kappa number [13, 14]. The corresponding measure tensile strength and water absorption commercial recycle paper were 10.60 N/mm^2 and 45 gr/mm^2 respectively. These recycle paper is suitable for writing/printing papers based on SNI 14 0-937-2005. In order to achieve higher grade standard, several additives were still needed to improve physical or mechanical properties in recycle paper.

4. Conclusion

This optimum condition was obtained from rice straw delignification at NaOH concentration of 8%, delignification temperature of 90°C. The highest % cellulose and the lowest % lignin, tensile strength, water absorption were achieved at this optimum condition, such as 79.49%, 3.96%, 4.36 N/mm² and 185.67 gr/m² respectively. For the variable mixture of rice straw and used paper, the highest tensile strength was 8.65 N/mm² for a ratio of 1:1. The lowest water absorption was 218.03 gr/m² for a ratio of 1:9. This investigation showed the potential of using the agricultural waste of rice straw and the used paper for the recycle papermaking. The recycle paper has achieved the adequate properties required for writing paper, however it would be more suitable if it is used as an art or wrapping paper.

Acknowledgment
The author wishes to thank Ms. Dyah Ayu Ambarsari for the technical assistance during the experiments conducted at Polymer and Membrane Laboratory of Surabaya University, East Java, Indonesia.

References
[1] Liu Z, Wang H and Hui L 2018 Pulping and Papermaking of Non-Wood Fibres *Intech Open*
[2] Badan Pusat Statistik *Produksi, Luas Panen dan Produktivitas Padi di Indonesia, 2013– 2017* http://www.pertanian.go.id
[3] Mukherjee A, Banerjee S and Halder G 2018 Parametric optimization of delignification of rice straw through central composite design approach towards application in grafting *J. of Adv. Res.* **14** 11-23
[4] Sjostrom E 1998 *Kimia Kayu, Dasar-Dasar dan Penggunaan* **2** (Yogyakarta: Gajah Mada University Press)
[5] Rodríguez A, Sánchez R, Requejo A and Ferrer A 2010 Feasibility of rice straw as a raw material for the production of soda cellulose pulp *J. of Cleaner Prod.* **18(10)** 1084-91
[6] Harun S and Geok S K 2016 Effect of sodium hydroxide pretreatment on rice straw composition *Indian J. of Sci. and Technol.* **9(21)**
[7] Ardeh S N and Fatehi P 2005 Influence of rice straw cooking conditions in the soda-ethanol-water pulping on the mechanical properties of produced paper sheets (Thailand: Asian Institute of Technology, *Pulp Pap. Technol.*) vol 92 pp 65-69
[8] Sapei L, Purwanto E and Suseno N 2014 Two-step soda pulping process of rice straw for the recycle papermaking *Proc. Bali Int. Semin. Sci. Technol. (Bisstech) II* (Surabaya: UPN Veteran University)
[9] Huang G, Zhang C, Zhongsheng and C.H.E.N 2006 Pulping of wheat straw with caustic potash ammonia aqueous solutions and its kinetics *Chin. J. of Chem. Eng.* **14(6)** 729-33
[10] Hassan A, Salleh A M and Jafferi N 2016 The efects of sodium hydroxide coontent on mechanical and physical properties of rice straw paper *ARPN J. of Eng. and Applied Sciences.* **11** 12
[11] Hurter R W and Byrd M V 2001 Process for Producing a Pulp Suitable for Papermaking from Non-Wood Fibrous Materials *US 6,302,997.* **B1**
[12] Kara S S and Onut S 2010 A Two-Stage Stochastic and Robust Programming Approach to Strategic Planning of a Reverse Supply Network:The Case of Paper Recycling *Expert Syst. Appl.* **37** 6129-37
[13] Pirmahboub H, Rafsanjani A T, Charani P R and Morvaridi R 2015 Eco-Friendly Pulping of Rice Steam by Twin-Screw Extruder: Optimization of Cooking Time and NaOH Concentration *Cellul. Chem. Technol.* **49(5-6)** 485-95
[14] Minu K, Jiby K K, Kishore, V.V.N 2012 Isolation and Purification of Lignin and Silica from The Black Liquor Generated during The Production of Bioethanol from Rice Straw *Biomass Bioenergy* **39** 210-7

IOP Conf. Series: Materials Science and Engineering **703** (2019) 012045 doi:10.1088/1757-899X/703/1/012045

Current Perspectives and Mini Review on Zeolitic Imidazolate Framework-8 (ZIF-8) Membranes on Organic Substrates

P D Sutrisna[1], E Savitri[1], N F Himma[2], N Prasetya[3] and I G Wenten[4]

[1]Department of Chemical Engineering, University of Surabaya, Jl Raya Kalirungkut, Surabaya, 60293, Indonesia
[2]Department of Chemical Engineering, UniversitasBrawijaya, Jl Mayjen Haryono 167, Malang 65145, Indonesia
[3]Barrer Centre, Department of Chemical Engineering, Imperial College London. Exhibition Road, London SW7 2AZ, United Kingdom
[4]Department of Chemical Engineering, Institut Teknologi Bandung, Jl Ganesha 10, Bandung 40132, Indonesia

E-mail: pudod@staff.ubaya.ac.id

Abstract. Zeolitic Imidazolate Framework (ZIF) membranes have been considered as promising membrane for gas separation processes due to their robustness and good separation capabilities. The separation of hydrogen (H_2) from larger gases is one of their potential applications. Among the different types of ZIF, ZIF-8 is considered as potential candidates for H_2 separation owing to its small aperture size and excellent chemical stability. In recent years, ZIF-8 membranes have been fabricated on both inorganic and polymeric substrates. The fabrication of the ZIF-8 layer on polymeric substrates is more challenging than its inorganic counterpart because of the incompatibility issue between organic substrates and ZIF-8. However, the flexible nature of polymeric substrates offers better mechanical stability at high pressure than inorganic substrates. The purpose of this mini-review is to present the state of the art and challenges of research on ZIF-8 membrane synthesis on different polymeric substrates. These challenges include the formation of cracks and defects due to compatibility issues and membrane stability at different operating conditions. Current research results showed excellent gas separation properties of ZIF-8 membranes on polymeric substrates that can surpass the trade-offline of gas permeability and selectivity.

1. Introduction

Carbon Capture and Storage (CCS) is indispensable to enable the ongoing utilization of fossil fuels while reducing the emissions of CO_2 to the atmosphere. CCS is a three-step process including CO_2 capture from other gases in the gas emission source; CO_2 transport to a storage location, and the permanent storage of CO_2. The considerable cost of capture, more than 60 % of the total cost for CCS, is slowing down the implementation of commercial CCS. It is reported that the cost for the capture process already consumes energy up to 40 % of the overall energy cost [1]. The key to reduce the CO_2 capture cost is to increase the separation efficiency of capture material and technique. Therefore, the discovery of new materials with excellent capture of CO_2 develops into the greatest challenge in this field. There are three options available to capture CO_2, i.e. (i) post-combustion, (ii) pre-combustion, and (iii) air separation followed by oxy-fuel combustion captures [2]. Post-combustion capture deals mainly with the separation of CO_2 from flue gas. Pre-combustion capture consists of the production of carbon monoxide (CO) – hydrogen (H_2) mixture by combusting fuel with air, and then the CO is supplied into a catalytic reactor to be converted into H_2 and CO_2. Oxy-fuel combustion involves pure oxygen to burn fuels that can produce pure CO_2 [1]. Until recently, chemical absorption and adsorption are considered as mature technologies to separate gas mixture that contains carbon dioxide [3-4]. However, chemical absorption consumes large amount of chemical solvent especially when treating huge amount of gases [2]. In addition, the simplicity of adsorption in operation and wide option for adsorbent material make adsorption technology developed as a good option for gas separation technique. In spite of these advantages, there are still much research needed to find the best adsorbent material to capture CO_2 from gas mixture. Besides chemical absorption and adsorption, membrane technology achieves a growing interest in gas separation processes by many researchers. This is because membrane can offer very high surface areas for gas separation in a small footprint as well as easy to scale up. However, membrane available today for gas separation processes still cannot achieve the desired separation performance, thus its realization in industrial operation is still limited.

Polymeric membrane gas separation process gains much interest in the current trend on gas separation research because it is believed that this technique has not been explored into its optimum capacity. The application of pure polymeric membrane in the gas separation field is limited by the trade-off phenomenon between membrane permeability and selectivity. A logarithmic correlation between membrane permeability and selectivity for different gas mixtures from the data available in the literature has been developed by Robeson (2008) [5]. The correlation indicates that there is an upper bound line exists in which recent polymeric membrane cannot surpass. It means that until now one main challenge for researchers in the field of gas separation membrane is to find the best membrane that can provide permeability and selectivity that exceed Robeson's upper bound line [6]. In current years, Metal-Organic Frameworks (MOFs) has received a growing interest from many researchers as potential fillers in mixed matrix membranes (MMMs) as well as pure MOFs membranes. MOFs contain metal cations connected by an organic linker. MOFs have coordination network and bonding, which make them less stiff and brittle than zeolite. Thus, it is considered that MOFs are able to show flexibility in terms of gate opening and linker dynamics [7]. There are more than hundreds of MOFs structure that have been synthesized and it is a big challenge to choose suitable MOFs that will be used as membranes.

Zeolitic Imidazolate Framework (ZIF) materials are investigated more widely by many researchers as gas separation membranes or filler for mixed matrix membranes than any other MOFs because of their framework flexibility and high chemical and thermal stability [8]. The bonding in ZIF is a coordination bonding that is weaker than covalent bonding in the zeolite. This means that the bonding between the metal node and organic linker is not kinetically stable. This instability means that substitution by other ligands such as water might happen. However, for some ZIF such as ZIF-8, ZIF-11 and ZIF-69 the stability of their structures can be maintained even in boiling solvents like benzene and water [9]. Hydrophobic ZIF-8 with sodalite (SOD) topology connecting Zn^{2+} metal and organic imidazolate linker is reported has 3.4 Å aperture diameters and it can retain their structure at the temperature up to 400°C. ZIF-8 has properties that make it considered as the most hydrolytically stable porous coordination polymer [10]. However, in the recent report by Zhang, et al. the effective aperture size of ZIF-8 particles is reported as 4.0 to 4.2 Å [11]. Because of this flexibility, ZIFs are considered able to increase the permeability of the polymeric membrane in separating gas mixtures. ZIF-8 emerges as one type of ZIF that are studied and employed widely as membranes in gas separation processes. Due to the simple fabrication process, use of an environmental-friendly solvent, high gas permeance, and its open structure, have attracted many researchers to explore this material. The number of publications related to ZIF-8 used as filler in MMMs and as pure membranes show an increasing trend as shown in Figure 1.There are also a number of reviews on MOFs membranes [12-13], but a review on ZIFs membrane on the polymeric substrate is still required to present the current trend on this membrane.

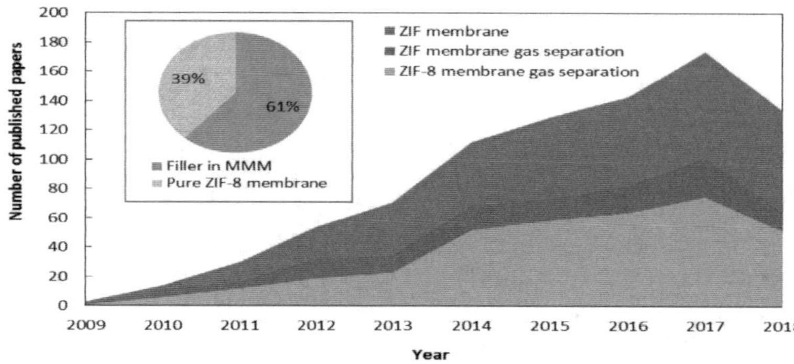

Figure 1.The number of published papers related to the search term of "ZIF membrane", "ZIF membrane gas separation", and "ZIF-8 membrane gas separation" indexed by Web of Science (23 July 2018). The insert is the distribution of ZIF-8 use for gas separation membrane.

2. The Synthesis of ZIF-8 membranes

The fabrication of excellent ZIF membranes, such as ZIF-8 membranes, on the surface of porous substrates for the gas separation process is relatively difficult. The challenges in the ZIF-8 membrane fabrication include the incompatibility issue between ZIF-8 and substrate materials and the flexibility of ZIF-8 materials. In addition, the issue on the reproducibility of the synthesis procedures of ZIF-8 membranes remains a challenging topic recently [12]. The techniques employed to fabricate ZIF-8 membranes can be classified into two techniques, i.e. *in situ* growth technique and secondary growth technique. In addition, inorganic and organic porous supports have also been utilized as the platform to grow ZIF-8. Figure 2A-B present the schematic diagram of *in situ* and secondary growth techniques to fabricate ZIF-8 membranes.

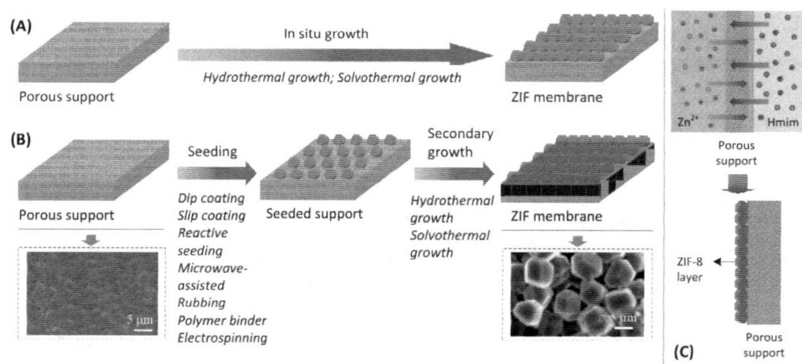

Figure 2. Schematic diagram of ZIF-8 membrane preparation by: (A) in situ growth method; (B) secondary growth method, and (C) counter diffusion method.

2.1 The In situ growth of ZIF-8 membrane on the porous support

Compared to the secondary growth, *in situ* growing technique does not require seed particles. During the synthesis, the porous support is soaked or immersed in the ZIF-8 precursor solution either at room temperature or at elevated temperature [12-13]. In terms of support, unmodified and modified supports have been employed for in situ grow ZIF-8 particles. In addition, both flat sheet and hollow fibre supports have been used to grow ZIF-8 particles.

In general, the growing of ZIF-8 by using this technique involves high temperature and long fabrication time due to the incompatibility issue. To reduce temperature and time, several researchers employed microwave heating and sodium formate to enhance the heterogeneous nucleation of ZIF-8.

Bux et al. (2009) have been successfully fabricated defect-free ZIF-8 membranes using this technique [14]. The ZIF-8 membranes resulted in a good H_2 separation performance. Unmodified porous α-alumina supports were also used as support for the growth of ZIF-8 crystals in the presence of sodium formate [15]. A similar technique was employed into porous polymeric polysulfone support to produce a ZIF-8 membrane with a micrometre and nanometre sized crystals on the support surface [16]. On another line of research, the combination of in situ growth and layer-by-layer techniques was applied on the polysulfone membrane to produce ZIF-8 membranes with low synthesis temperature [17]. In recent years, the growing of ZIF-8 on hollow fibre porous supports has also been started to produce membranes with larger surface area compared to flat sheet membranes. ZIF-8 membrane from concentrated synthesis gel was successfully prepared on the surface of ceramic hollow fibre supports [18]. *In situ* growth on modified support was carried out to create a more efficient ZIF-8 growing process. ZIF-8 film was grown on the inner surface of the APTES modified ceramic hollow fibre using a modified seeding process [19].

2.2 The secondary growth method of ZIF-8 on porous support

Besides its advantages such as the simple fabrication technique, the *in situ* growing technique of ZIF-8 membranes faces some challenges. Such challenges include the difficulty to control the growth of particles on the surface of the substrate as the nucleation and crystal growth occur at the same time [12]. To enhance the performance of the ZIF-8 membrane, several studies have been successfully employed a secondary growth method to produce defect-free ZIF-8 membranes. The secondary growth technique discriminates the nucleation and growth steps, thus offers a better control on the microstructure of the membrane. Rubbing of seed particles on the surface of porous substrates is a relatively common technique used as a precursor during the secondary growing of ZIF-8 crystals. Ge et al. (2012) fabricated ZIF-8 thin film on the surface of polyethersulfone by firstly rubbed the support surface with ZIF-8 seed particles followed by secondary growing technique. Other techniques for seeding are dip-coating and slip coating. One study employed dip-coating to seed YSZ tubular support [20]. Slip coating to seed alumina support was also carried out to synthesize ZIF-8 membrane. Produced membrane showed very good olefin/paraffin separation performance [20-21]. In addition to dip-coating and slip coating, another technique, i.e. reactive seeding, was also used as a precursor before the secondary growth of ZIF-8 membranes [22]. In this technique, microwave-assisted seeding was carried out to enhance the attachment of the ZIF-8 seeding layer to α-alumina support with a rapid nucleation process.

Despite its attractiveness in producing defect-free ZIF-8 membranes, the membranes produced using secondary growth technique experience delamination due to the weak interaction between the ZIF-8 layer and porous support [23]. Hence, support surface modification started to be employed to improve the adhesion of ZIF-8 particles on the surface of porous support [24-26]. A study to improve the attachment of ZIF-8 particles on the surface of porous supports utilized vapour phase modification to add amine functional groups and reduce the pore size of support. This technique could produce the ZIF-8 membrane with a relatively continuous structure [26]. Another technique to improve the attachment of nano-ZIF-8 particles is by using polyethyleneimine (PEI) as a coupling agent. This coupling agent forms hydrogen bonds with ZIF seed crystals and free hydroxyl groups of ceramics (α-Al_2O_3) porous substrate. PEI may also form Zn – N coordination bonds with Zn cations on the surface of the ceramic. The crystal seeds were attached to the porous substrate using dip-coating method [27].

2.3 Novel methods to synthesize ZIF-8 membranes on porous support

In addition to *in situ* and secondary growth methods, several studies offer new and innovative techniques to grow the ZIF-8 layer on the surface of the porous support. Such techniques include counter diffusion technique [28]. The experimental set up of the contra diffusion technique to grow ZIF-8 on nylon support is presented in Figure 2C. ZIF-8 layer on the zinc nitrate with cubic structure was more uniform than fibre-like structure of ZIF-8 crystals grown on the 2-methyl imidazolate surface. ZIF-8 membranes with relatively good gas separation performances could be synthesized in one cycle of crystallization process thus reduce the synthesis time. Another observation from this experiment suggested the different crystal structure formed by changing the position of zinc nitrate and 2-methyl imidazole in the contra diffusion apparatus. Gas separation performance of such membranes could produce gas selectivity higher than Knudsen selectivity, which indicated the formation of defect-free ZIF-8 membranes. Each technique presented above has its own benefits considering the ease of fabrication, reproducibility, and the uniformity of the produced membrane. In general, it can be mentioned that *in situ* growth offers a simpler technique compared to other methods. However, in terms of reproducibility and the uniformity of the membranes, by employing the *in situ* technique will be very hard to produce dense and uniform membrane.

3. Future perspectives and conclusions

Table 1 presents literature data of ZIF-8 membranes that have been synthesized recently. The data compiled the gas separation performance of ZIF-8 membranes fabricated on organic and inorganic supports to compare the performances of these two membranes. Gas permeance and selectivity data for ZIF-8 membrane in Table 1 indicate the roles of ZIF-8 layer thickness and size of ZIF-8 crystals on the gas separation performance of membranes. A thin layer of ZIF-8 on the surface of the substrate ensures high gas permeance through the membrane. Up to now, the ZIF-8 membranes that have been synthesized could achieve H_2 permeance of more than 60,000 Gas Permeation Unit (GPU), which is very high compared to pristine polymeric membranes. On the other hand, the gas selectivity of the membrane should be maintained as high as possible before the membrane can be applied in real applications. In terms of organic support, there is still a possibility to improve the performance of ZIF-8 membranes to produce high gas permeance with good separation capability. In addition, a study on the mechanical strength of ZIF-8 membranes on polymeric supports is still relatively new and indispensable before the membrane can be applied in the industrial scale. Other challenges in the fabrication of ZIF-8 membranes include developing a single-step synthesis method that can be produced in a large scale and tackling the reproducibility issue in the synthesis of the membrane [12]. Thus, the development of pure MOFs membrane, especially the ZIF-based membrane, offers promising performance to surpass the gas permeability-selectivity trade-off. However, some challenges in membrane fabrication, such as method reproducibility and relatively long preparation time, still need to be addressed [34]. Several techniques can be considered to tackle the problems, such as surface modification and post-treatment of the membranes.

Table 1. Selected ZIF-8 membranes synthesized at various substrates and synthesis conditions.

Support	Modification approach	Thickness (μm)	H_2 permeance*)	Selectivity H_2/CO_2	H_2/N_2	Ref.
PD/SWCNT	Contra diffusion	~0.55	6.31	43	NA	[29]
PVDF hollow fibre	TiO_2 functionalised	~1	201	7	7.8	[30]
PVDF hollow fibre	Ammoniation	~40	24.43	12.18	14.31	[24]
PAN hollow fibre	Dehydrogenation	< 1	3.05	6.85	N/A	[25]
Nylon hollow fibre	Contra-diffusion	~2.5	11.3	N/A	4.6	[31]
Nylon hollow fibre	Contra-diffusion	~16	126.2	N/A	3.7	[28]
Porous polysulfone	Micro-nanosized growth	~50	2.0	N/A	12.4	[16]
BPPO hollow fibre	Chemical modification	0.2	20.5	12.8	9.7	[26]
Poly(amide–imide)	Interfacial microfluidic	~0.9	1.23	NA	NA	[32]
BPPO flat sheet	Contra diffusion	~2	7.5	5.1	8.3	[23]
Porous AAO	PDA-modified	< 0.2	287	14	18	[26]
Titania (flat sheet)	Microwave-assisted	~30	0.6	4.61	11.5	[14]
YSZ hollow fibre	Secondary seed growth	~2	15.4	3.85	11	[20]
SS-nets (flat sheet)	PDA-coated surface	~30	266	8.8	15.4	[33]

*) Permeance unit is in ($x10^{-7}$ $mol.m^{-2}.s^{-1}.Pa^{-1}$)

Acknowledgment

The authors gratefully acknowledge the financial support from The Ministry of Research and Higher Education of Indonesia under The National Competitive Fundamental Research Grant under contract No. 004/SP2H/LT/MULTI/L7/2019.

References

[1] D'Alessandro D M, Smit B, and Long J R 2010 *Angew. Chem. Int. Ed.* **49** 6058-82.
[2] Brunetti A, Scura F, Barbieri G, and Drioli E 2010 *J. Memb. Sci.* **359** 115-25.
[3] Maneeintr K, Photien K, and Charinpanitkul T 2018*Chem. Eng. Trans.* **63** 229-34.
[4] Harkin T, Hoadley A, and Hooper B 2011 *Chem. Eng. Trans.* **25** 13-8.
[5] Robeson L M 2008 *J. Memb. Sci.* **320** 390-400.
[6] Dong G, Li H, and Chen V 2013 *J. Mater. Chem. A* **1** 4610-30.
[7] Caro J 2011 *Curr. Opin. Chem. Eng.***1** 77-83.
[8] Shah M, McCarthy M C, Sachdeva S, Lee A K, and Jeong H K 2011 *Ind. Eng. Chem. Res.* **51** 2179-99.
[9] Park K S, Ni Z, Côté A P, Choi J Y, Huang R, Uribe-Romo F J, and Yaghi O M 2006 *Proc. Nat. Acad. Sci.* **103** 10186-91.
[10] Tanh Jeazet H B, Staudt C, and Janiak C 2012 *Dalton Trans.* **41** 14003-27.
[11] Zhang C, Lively R P, Zhang K, Johnson J R, Karvan O, and Koros W J 2012 *J. Phys. Chem. Lett.* **3** 2130-34.
[12] Venna S R and Carreon M A 2015 *Chem. Eng. Sci.* **124** 3-19.
[13] Melgar V M A, Kim J, and Othman M R 2015 *J. Ind. Eng. Chem.* **28** 1-15.
[14] Bux H, Liang F, Li Y, Cravillon J, Wiebcke M, and Caro J 2009 *J. Am. Chem. Soc.* **131** 16000-1.
[15] Shah M, Kwon H T, Tran V, Sachdeva S, and Jeong H K 2013*Micropor. Mesopor. Mater.* **165** 63-9.
[16] Cacho-Bailo F, Seoane B, Téllez C, and Coronas J 2014 *J. Memb. Sci.* **464** 119-26.
[17] Nagaraju D, Bhagat D G, Banerjee R, and Kharul U K 2013 *J. Mater. Chem. A* **1** 8828-35.
[18] Pan Y, Li T, Lestari G, and Lai Z 2012 *J. Memb. Sci.* **390–391** 93-8.
[19] Huang K, Dong Z, Li Q, and Jin W 2013 *Chem. Commun.* **49** 10326-28.
[20] Pan Y, Wang B, and Lai Z 2012 *J. Memb. Sci.* **421–422** 292-8.
[21] Pan Y and Lai Z 2011 *Chem. Commun.* **47** 10275-7.
[22] Kwon H T and Jeong H K 2013*Chem. Commun.* **49** 3854-6.
[23] Shamsaei E, Lin X, Low Z X, Abbasi Z, Hu Y, Liu J Z, and Wang H 2016 *ACS Appl. Mater. Interf.* **8** 6236-44.
[24] Li W, Meng Q, Zhang C, and Zhang G 2015 *Chem. – A Eur. J.* **21** 7224-30.
[25] Li W,Yang Z, Zhang G, Fan Z, Meng Q, Shen C, and Gao C 2014 *J. Mater. Chem. A* **2** 2110-8.
[26] Shamsaei E, Low Z X, Lin X, Mayahi A, Liu H, Zhang X, and Wang H 2015 *Chem. Commun.* **51** 11474-77.
[27] Bux H, Feldhoff A, Cravillon J, Wiebcke M, Li Y S, and Caro J 2011 *Chem. Mater.* **23** 2262-9.
[28] Yao J, Dong D, Li D, He L, Xu G, and Wang H 2011*Chem. Commun.* **47** 2559-61.
[29] Zhang S, Wang Z, Ren H, Zhang F, and Jin J 2017 *J. Mater. Chem. A* **5** 1962-6.
[30] Hou J, Sutrisna P D, Zhang Y, and Chen V 2016 *Angew. Chem. Int. Ed.* **128** 4015-9.
[31] He M, Yao J, Li L, Wang K, Chen F, and Wang H 2013 *ChemPlusChem* **78** 1222-5.
[32] Ge L, Zhou W, Du A, and Zhu Z 2012 *The J. Phys. Chem. C* **116** 13264-70.
[33] Huang A, Liu Q, Wang N, and Caro J 2014 *J. Mater. Chem. A* **2** 8246-51.
[34] Julian H, Sutrisna P D, Hakim A N, Harsono A O, Hugo Y A, and Wenten I G 2019 *Polym.-Plast. Tech. Mater.* **58** 678-89.

The Influence of water and catalyst leach process toward propane oxidation on MoVTeNb catalyst

R K Widi

Department of Chemical Engineering, University of Surabaya, Jl. Raya Kalirungkut, Surabaya, Indonesia, 60293

E-mail: restu@staff.ubaya.ac.id

Abstract. The effect of the water stream to the propane oxidation on diluted MoVTeNb catalyst has been investigated. The present work has elucidated that careful operation of high throughput instrumentation can be used in various beneficial ways to speed up the discovery process of improved catalysts in other forms than enabling efficient trial-and-error testing of compositional variations of a given catalyst system. The result shows that the addition of massive amounts of water to the feed should have a negative influence on the kinetics, as water will compete with all other polar molecules in the system for adsorption sites. This work also investigated the effect of catalyst leach process toward propane oxidation. From the result, it can be described that catalyst leach process tends to reduce the phase of the catalyst that responds to the total oxidation of propane. This work also proposed the reaction network and gave the comparison between the propane oxidation reaction kinetic using leached and un-leached catalyst. The result showed that the activation energy of the acrylic acid formation on the leached catalyst was slightly higher than that of on un-leached catalyst. On the other hand, the activation energy of the carbon dioxide formation on the leached catalyst was much higher than that of on un-leached catalyst. It can be described that the leaching process to the catalyst can reduce the phase of the catalyst responsible for the total oxidation of propane.
Keywords: *MoVTeNb catalyst, propane, oxidation, water, leach process*

1. Introduction

The oxidation process using a catalyst has developed into a very important technology in the modern chemical industry. One of the important processes that utilize this technology is the functionalization of aromatics and alkanes, due to the advantageous economy and the easy processing. Acrylic acid is one of the most important catalytic oxidation products, in which its consumption in 2020 is expected to reach 8,000-kilotons. Acrylic acid is widely used in the industry as intermediate for polyacrylates and methacrylate esters production. These products can be used for various purposes, such as super adsorbents. Acrylic acid is presently factory-made from propylene in two steps via acrolein in a gas phase using special catalysts which the yield up to 80% [1]. Despite the promising yield of the product, the price of propylene has seen attention being drawn to develop a method for producing acrylic acid alternatively by direct oxidation of propane in the presence of a catalyst. One of the promising catalysts that are widely studied is a mixed metal oxide, such as MoVTeNb catalyst [2-7].

The problem that is still faced in the utilization of this catalyst is the control of the complexity of the reaction network consisting of consecutive reactions and parallel reactions [3, 8-10]. The phase-in MoVTeNb which is responsible for the production of acrylic acid from direct oxidation of propane is referred to as "M1" phase [2, 4, 11-13]. Mo and V are very suitable for the oxidation of alkanes via alkenes to carboxylic acids [14, 15]. Some results of previous studies have reported the proposed reaction mechanism [3, 9, 16], as well as the structure of the catalyst during the reaction [4, 17-21]. The presence of water in this oxidation reaction provides a reaction advantage as has been reported by previous researchers [3, 22-25].

In this paper, we will discuss the effect of the presence of water on catalytic reactions and the effect of the leaching process on diluted MoVTeNb oxide catalyst. Water is believed to have an indirect role in the process of propane oxidation reactions to acrylic acid. The leaching process is believed to effect on the phase composition of M1 and M2 in the catalyst which has a very important role in the oxidation reaction mainly related to the reaction network of propane to acrylic acid.

2. Experimental

2.1. Materials

The materials used for catalyst synthesis were Telluric acid (Aldrich), Ammonium metavanadate (Riedel-de Haën), Niobium ammonium oxalate (Aldrich), Ammonium heptamolybdatetetrahydrate (Merck), Aerosil300 (Degussa). The gases used for reactants were oxygen and propane (MOX).

2.2. Instrumentation

Nanoflow catalytic reactor is one of the equipment that is used for this high throughput experimentation study.

2.3. Procedure

Preparation of MoVTeNb oxide catalysts and kinetic studies was carried out according to the procedure described previously, and some kinetics data were also reported previously [4, 5, 7]. The catalytic reaction was done by presenting a small quantity of catalyst (0.1 g) mixed in SiO_2 to a total volume of 0.5 ml. The high total flow rates (15 mlN/min, GHSV = 1800 h^{-1}). The reaction rate is calculated by Equations presented below:

$$\text{Reaction rate (mol/h. gcat)} = \frac{\%vol\ x\ total\ flow\ x\ d\ x\ 60}{Mr\ x\ catalyst\ weight} \quad (1)$$

Where Mr is molecular weight and d is density.

To study the effects of water dependence, the concentration of water varies and other concentrations are fixed. To determine activation energy, the oxidation reaction was studied at four different temperatures, namely, 653, 663, 673 and 683 K. After taking the natural logarithm of the equation, it can be seen that a plot of ln k versus 1/T should be a straight line whose slope is proportional to the activation energy.

$$k = Ae^{-\frac{E_A}{RT}} \quad (2)$$

Where k is the rate constant, T is the absolute temperature (in Kelvin), A is the pre-exponential factor, Ea is the activation energy for the reaction, and R is the universal gas constant.

3. Result and Discussion

The characteristic of the MoVTeNb oxide catalyst has been reported in our previous work [4]. The catalyst consists of two main phases namely M1 and M2 phases. M1 phase is more related to the structure of Mo_5O_{14} and is responsible for the oxidation process of propane to acrylic acid, while the M2 phase is more related to the orthorhombic variant of HTB structure which is responsible for the total oxidation process of propane to COx. Based on catalyst characterization, it shows that the M1 phase is more dominant than the M2 phase.

3.1. Effect of water

There are several opinions about how water affects the performance of the catalyst and the distribution of products, such as by facilitating desorption of oxygenating from the catalyst or modifying the catalyst structure. The present study has substantiated that the modification of the catalyst surface required for partial oxidation by water concentration in the feed is a fact for the systems studied here. The beneficial effect of water in helping to wash off the partial oxidation products and to protect them from over-oxidation by blocking sites next to the adsorption site of the freshly formed products and so inhibiting the access of activated oxygen from neighboring sites have also been identified in the present study.

To interpret this point, kinetic investigations were lead with water as the varied feed component. The rate of propane disappearance and acrylic acid formation were calculated and plotted against the water concentration (see Figure 1 and 2).

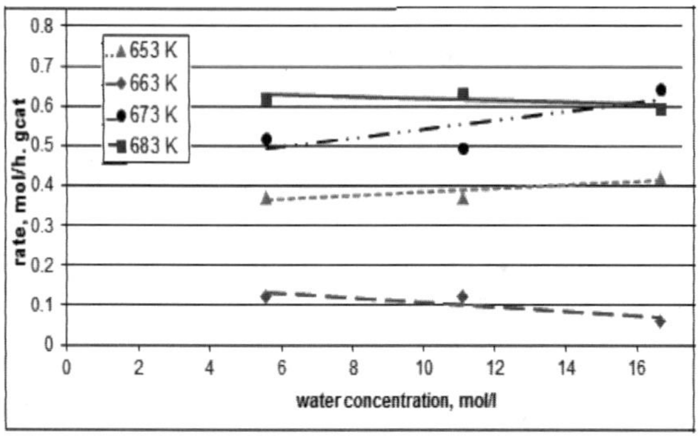

Figure 1. Rate of propane disappearance with respect to water concentration at 653, 663, 673, and 683K

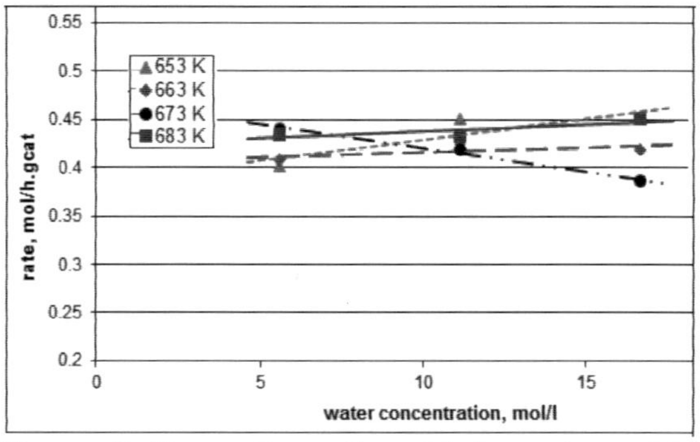

Figure 2. Rate of acrylic acid formation with respect to water concentration at 653, 663, 673, and 683 K

The reaction stoichiometry for propane to acrylic acid is$C_3H_8 + 2\ O_2 \rightarrow C_3H_2COOH + 2\ H_2O$. From the result above, the rate of propane disappearance (r_{prop_dis}) and acrylic acid formation (r_{AA_form}) from the selective oxidation of propane over this catalyst with respect to water concentration can be determined by the equation (see table 1).

In all cases, a small value of the reaction order with respect to water was obtained. This is unexpectedly taking into account that water is a reaction product in all processes considered here and that it will control the processes of C-H activation leading to (intermediate) OH groups. The addition of massive amounts of water to the feed should have a negative influence on the kinetics, as water will compete with all other polar molecules in the system for adsorption sites. It has to be assumed that the integrity of the catalysts requires its surface to be covered to a substantial extent with water. On such a surface the additional water created during the reaction of propane will not affect greatly the coverage and hence will not lead to a significant reaction order detectable by formal kinetics. In contrast to current opinion, the low values of reaction order do not imply that water cannot interfere with the reaction kinetics.

For acrylic acid a strong dependence of the reaction order on temperature is observed, showing at low temperatures a slightly beneficial and for high temperatures a slightly inhibiting effect of water. As the C-H chemistry should not be relevant, it may be concluded that at low temperatures the beneficial washing off effect is at work whereas at high temperatures the site competition effect prevents the optimization of the surface coverage with other species than water indicating that water desorption must be a strongly activated process.

3.2. Effect of leach process

For further development of the catalyst system, two conclusions can be drawn. It is very important to design the dilution of the active phase in such a way as to maximize the hydrothermal stability and to optimize the heat management of the entire catalyst. Concerning the active phase, it became evident that a single phase is essential as then the chances to bring about a single type of active sites, capable to perform minimum consecutive total oxidation the selective oxidation of propane. To this end, leaching and nanostructuring of the surface are suitable concepts. To clarify this point, a reaction mechanism has been proposed (Scheme 1, [7]) and then it will be compared the amount of energy activation of the propane oxidation reaction using unleached and leached catalysts.

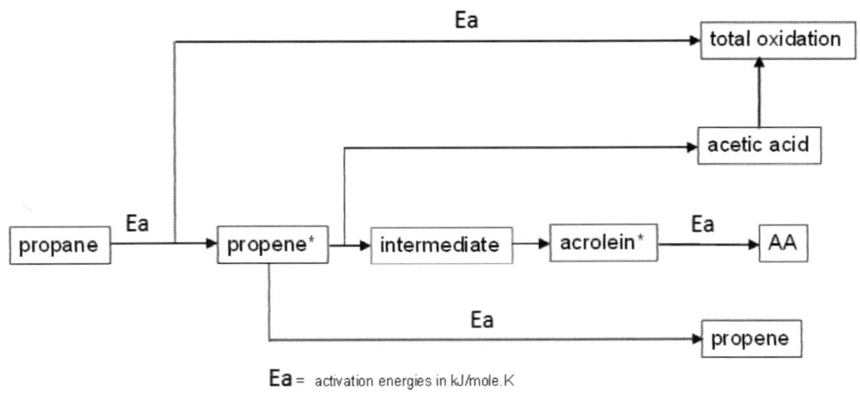

Scheme 1: Schematic representation of the reaction network

Overall, this propane oxidation reaction consists of three parallel processes paths. These processes are one main path that produces the desired product, acrylic acid. The other pathway is the total oxidation of propane to carbon dioxide, and another pathway is the process of desorption of propene. One thing that is quite good in this reaction process is that the total oxidation pathway of propane requires a fairly large activation process, so it is technically quite difficult to occur. The proposed reaction mechanism is consistent with the presence of 2 types of active catalyst sites referred to as M1 for the partial oxidation pathway to produce acrylic acid and M2 for the total oxidation pathway.

To further understand the effect of washing or leaching the catalyst on the presence of phases responsible for the formation of acrylic acid, activation energy calculations for each process are presented in table 1. This washing is intended to reduce the M2 phase which leads to a total oxidation reaction. It is expected that the reduction of the M2 phase in total in the catalyst material will be dominated by the M1 phase which pushes towards selective oxidation so that the desired product will be produced more.

Tabel 1. Comparison of activation energy in each process of selective propane oxidation using leached and unleached MoVTeNb [5,7]

Processes	Rate of reaction (r)		Activation Energy (kJ mol^{-1})	
	Unleached catalyst	Leached catalyst	Unleached catalyst	Leached catalyst
Propane disappearance	$k[C_3H_8]^{1.00} [O_2]^{0.24}$	$k[C_3H_8]^{1.00} [O_2]^{0.21}$	21,8	27,3
Propene formation	$k[C_3H_8]^{1.00} [O_2]^{0.00}$	$k[C_3H_8]^{1.00} [O_2]^{0.00}$	62,7	40,9
COx formation	$k[C_3H_8]^{1.00} [O_2]^{0.21}$	$k[C_3H_8]^{1.00} [O_2]^{0.31}$	123,2	177,3
Acetic acid formation	$k[C_3H_8]^{0.22} [O_2]^{0.24}$	$k[C_3H_8]^{0.83} [O_2]^{0.42}$	46,9	91,6
Acrylic acid formation	$k[C_3H_8]^{0.50} [O_2]^{0.11}$	$k[C_3H_8]^{0.49} [O_2]^{0.09}$	32,9	30,3

From table 1, it can be described that the process of removal (conversion) of propane requires slightly higher energy in the reaction using leached catalysts. It can be logically explained that the washing process causes several active catalytic sites both M1 and M2 phases to be washed away. The reduced active site will complicate the process of propane removal/conversion. However, in the process of using leached catalysts, it appears that the formation of propene is easier. Propene is an intermediate product which will immediately turn into the next oxidation product, acetic acid, and acrylic acid. The formation of acetic acid which is a by-product requires higher energy, while the formation of acrylic acid as a product of the main reaction pathway requires lower energy. This proves that the washing process is quite successful in eliminating the M2 phase which is responsible for the total combustion of propane. This is supported by the calculation of the activation energy of COx formation which shows a very high increase. This suggests that COx formation is difficult. On the other hand, converting propane towards propene and continuing towards acrylic acid is even easier. This confirms that the washing process succeeded in reducing the M2 phase which is responsible for the total oxidation reaction of propane to COx. In addition the washing process successfully maintains the M1 phase which is responsible for the selective oxidation reaction of propane to acrylic acid.

4. Conclusion

In general, this study has provided a phenomenological framework for understanding the action of MoVTeNb in propane oxidation catalysis. It was found that there are two types of reaction sites present, one of which requires the presence of excess water to function. The extent to which these sites will be present depends on the details of the bulk material nanostructuring and can be modified by synthetic and post-synthetic steps. One of these steps is the washing process. Parallel reaction pathways for partial oxidation need to be treated more carefully so that acrylic acid can be produced optimally.

Acknowledgment

The authors wish to acknowledge the following: 1. Ministry of Research, Technology and Higher Education of the Republic of Indonesia, Research Grant (Hibah Penelitian Terapan Unggulan Perguruan Tinggi contract number 004/SP2H/LT/MONO/L7/2019) for partial funding. 2. COMBICAT University of Malaya for some data collection.

References

[1] Cavani F 2010 Catalytic selective oxidation: The forefront in the challenge for a moresustainable chemical industry *Catal. Today* **157** 8 – 15.

[2] Naumannd'Alnoncourt R, Csepei, L I, Hävecker M, Girgsdies F, Schuster M, Schlögl R, and Trunschke A 2014 The reaction network in propane oxidation over phase-pure MoVTeNb *J. Catal.***311** 369 – 385.

[3] Heine C, Havecker M, Trunschke A, Schlogl R, and Eichelbaum M 2015 The impact of steam on the electronic structure of the selective propane oxidation catalyst MoVTeNb oxide (orthorhombic M1 phase) *Phys. Chem. Chem. Phys.***17** 8983 – 8993.

[4] Wagner J, Timpe O, Hamid F, Trunschke A, Wild U, Su D, Widi R K, Abd Hamid S B, and Schlögl R 2006 Surface texturing of Mo–V–Te–Nb–Ox selective oxidation catalysts *Top.Catal.***38** 51 – 58.

[5] Widi R K 2012 Kinetic investigation of carbon dioxide, acetic acid, acrylic acid formation on diluted and leached MoVTeNb catalyst *Indones. J Chem.***12** 131 – 134.

[6] Widi R K, Abdul Hamid S B, and Schlogl R 2008 Effect of diluent and reaction parameter on selective oxidation of propane over MoVTeNb catalyst using nanoflow catalytic reactor *J. Nat. Gas Chem.***17** 130 – 134.

[7] Widi R K, Abdul Hamid S B, and Schlogl R 2009 Kinetic investigation of propane oxidation on diluted Mo1–V0.3–Te0.23–Nb0.125–O x mixed-oxide catalysts *React. Kinet. Catal. L.***98** 273 – 286.

[8] Pyrz W, Blom D, Shiju N, Guliants V, Vogt T, and Buttrey D 2009 The effect of Nb or Ta substitution into the M1 phase of the MoV(Nb,Ta)TeO selective oxidation catalyst Catal. Today **142** 320 – 328.

[9] Heine C, Hävecker M, Stotz E, Rosowski F, Knop-Gericke A, Trunschke A, Eichelbaum M, and Schlögl R 2014 Ambient-Pressure Soft X-ray Absorption Spectroscopy of a Catalyst Surface in Action: Closing the Pressure Gap in the Selective n-Butane Oxidation over Vanadyl Pyrophosphate *J Phys. Chem.*C **118** 20405–20412.

[10] Eichelbaum M, StoˈBer R, Karpov A, Dobner C-K, RosowskiF, and Trunschke A, 2012 The microwave cavity perturbation technique for contact-free andin situelectrical conductivity measurements in catalysis and materials science *Phys. Chem. Chem. Phys.***14** 1302 – 1312.

[11] Schlogl R 2011 Active Sites for Propane Oxidation: Some Generic Considerations *Top.Catal.***54** 627.

[12] Naumannd'Alnoncourt R, Kolen'ko Y V, Schlogl R, and Trunschke A 2012 A New Way of Probing Reaction Networks: Analyzing Multidimensional Parameter Space *Comb. Chem. High Throughput Screen.***15** 161 – 169.

[13] Amakawa K, Kolen'ko Y V, Villa A, Schuster M E, Csepei L-I, Weinberg G, Wrabetz S,Naumannd'Alnoncourt R, Girgsdies F, Prati L, Schlögl R, and Trunschke A 2013 Multifunctionality of Crystalline MoV(TeNb) M1 Oxide Catalysts in Selective Oxidation of Propane and Benzyl Alcohol *ACS Catal.***3** 1103 – 1113.

[14] Kardash T, Plyasova L, Bondareva V, Andrushkevich T, Dovlitova L, Ischenko A, Nizovskii A I, and Kalinkin A V 2010 M5O14-like V–Mo–Nb oxide catalysts: Structure and catalytic performance *Appl.Catal.A.***375** 26 – 36.

[15] Freund H-J, Meijer G, Scheffler M, Schlögl R, and Wolf M 2011 CO Oxidation as a Prototypical Reaction for Heterogeneous Processes *Angew. Chem. Int. Ed.* **50** 10064 – 10094.

[16] Helander M, Greiner M, Wang Z, and Lu Z 2010 Pitfalls in measuring work function using photoelectron spectroscopy *Appl. Surf. Sci.***256** 2602-2605.

[17] Meyer J P, Zilberberg K, Riedl T, and Kahn A 2011 Electronic structure of Vanadium pentoxide: An efficient hole injector for organic electronic materials *J. Appl. Phys.***110** 033710.

[18] Greiner M, Chai L, Helander M, Tang W M, and Lu Z H 2012 Transition Metal Oxide Work Functions: The Influence of Cation Oxidation State and Oxygen Vacancies *Adv.Funct. Mater.***22** 4557 – 4568.

[19] Kolen'ko Y, Zhang W, Naumannd'Alnoncourt R, Girgsdies F, Hansen T, Wolfram T, Schlögl R, and Trunschke A 2011 Synthesis of MoVTeNb Oxide Catalysts with Tunable Particle Dimensions *Chem.Cat.Chem.***3** 1597 – 1606.

[20] Hävecker M, Wrabetz S, Kröhnert J, Csepei, L-I, Naumannd'Alnoncourt R, Kolen'ko Y, Girgsdies F, Schlögl R, and Trunschke A 2012 Surface chemistry of phase-pure M1 MoVTeNb oxide during operation in selective oxidation of propane to acrylic acid *J. Catal.***285** 48 – 60.

[21] Sanfiz A C, Hansen T, Teschner D, Schnörch P, Girgsdies F, Trunschke A, Schlögl R, Hong L M, Abd Hamid S B 2010 Dynamics of the MoVTeNb Oxide M1 Phase in Propane Oxidation *J. Phys. Chem. C.***114** 1912-1921.

[22] Manganas D, Roemelt M, Hävecker M, Trunschke A, Knop-Gericke A, Schlögl R, and Neese F 2013 First principles calculations of the structure and V L-edge X-ray absorption spectra of V2O5 using local pair natural orbital coupled cluster theory and spin–orbit coupled configuration interaction approaches *Phys. Chem. Chem. Phys.***15** 7260-7276.

[23] Manganas D, Roemelt M, Weyhermüller T, Blume R, Hävecker M, Knop-Gericke A, DeBeer S,Schlögl R, and Neese F 2014 L-edge X-ray absorption study of mononuclear vanadium complexes and spectral predictions using a restricted open shell configuration interaction ansatz*Phys. Chem. Chem. Phys.***16** 264-276.

[24] Ulbricht R, Hendry E, Shan J, Heinz T, and Bonn M 2011 Carrier dynamics in semiconductors studied with time-resolved terahertz spectroscopy *Rev. Mod. Phys.***83** 543 – 586.

[25] Nguyen T, Aouine M, and Millet J M 2012 Optimizing the efficiency of MoVTeNbO catalysts for ethane oxidative dehydrogenation to ethylene *Catal. Commun.***21** 22 – 26.

Requirements analysis for the disaster logistics inventory information system to improve the effectiveness and efficiency of handling emergency response periods

N U Handayani, D P Sari, Y Widharto and G Basyir

Industrial Engineering Department, Faculty of Engineering, Diponegoro University

E-mail: naniekh@ft.undip.ac.id

Abstract. The exact specifications of system requirements are very important factors in the system development process. Any deficiency at this stage will affect the quality of the output of the information system developed. Therefore, system analysts must focus on how requirements are defined and modeled. Building a new information system is a complex process that consists of many steps that must be done before the final product is prepared for the customer. It is important to meet the needs and expectations of customers to maintain project sustainability in the future. There are several approaches to develop new information systems in which different strategies, methodology, modeling techniques or life cycles can be used.This study aims to analyze the need for the development of a website-based disaster logistics inventory information system. This research produces an analysis of information system requirements both from functional and non functional aspects. Based on functional analysis obtained the minimum specifications that must be owned by information systems, while the non-functional analysis obtained the minimum data needs to run a website-based disaster logistics inventory information system.

1. Introduction

Mount Merapi is one the most active mountain in the world for its relative short periodic and high intensity of eruption around 3 - 7 years [1]. The biggest eruption of Mount Merapi occurred in 2010 with 3 times bigger eruption power than the previous one with the launch of a glowing cloud with the radius of 14.5 km. According to the *Badan Nasional Penanggulangan Bencana (BNPB)* - National Emergency Response - Mount Merapi eruption occurred lasted for 14 days since October 26, 2010. Based on the data of *BNPB* on the eruption of Mount Merapi in 2010, it recorded 277 deaths and missing, 186 injured and 159.977 people were evacuated.

According to *Pusat Vulkanologi dan Mitigasi Bencana Geologi (PVMBG),* the type of Mount Merapi eruption was cyclical and it would reoccur in a certain period of time, so that Mount Merapi has a periodic potential disaster. Therefore, the local government must be ready in handling the eruption [2, 3, 4, 5, 6]. The local emergency response (BPBD) Sleman area, Special Region of Yogyakarta is a government institution that has the duty to deal with the disasters occur within the province of The Special Region of Yogyakarta.

Management of relief items for disaster victims from the start of procurement or shelter assistance, inventory, and distribution is part of Humanitarian Logistics (HL) [7]. There are four things that must be considered in handling HL, namely a) the uncertainty of distribution routes (road conditions), security, changes in facility capacity, and uncertainty in demand for logistics needs; b) communication and coordination difficulties due to disruption of communication, involvement of many parties consisting of government, civilians and Non-Government Organizations (NGOs), and there is no access to real-time and accurate request information; c) efficient and timely distribution is difficult; d) limited resources so that they cannot overcome the scale of the disaster [8].

One obstacle in the disaster logistics process in Indonesia is the limited access to information on the needs and supplies of relief goods. Delay in access to information causes the handling of disasters to be ineffective and inefficient. For example, related parties are often late to know the scale of the disaster, the number and type of assistance both goods and services needed, and the distribution of locations of refugees [9]. For this reason, the Regional Disaster Management Agency (BPBD) in charge of handling disasters in the regions needs to have an integrated inventory information system (IIS) between the center of the aid post in BPBD Sleman Regency and all refugee camps. The IIS needs to be equipped with an inventory database per item/type of relief item, so that the relevant parties can find out the type and amount of relief items available in the main warehouse. In addition, IIS can be used as a medium of communication between the BPBD, donors, and those responsible for refugee barracks (users). Donors can find out the type of relief items needed so they can provide assistance according to the needs of victims, while users can provide data on the needs of relief items in each barrack. Thus, the distribution of disaster relief items can be carried out according to their needs, namely in terms of time, number, and type.

2. Research Methods

2.1. Humanitarian Logistics (HL)

Various types of disasters need to be managed by different solutions. Logistics is the most important element in humanitarian assistance efforts. How to manage the logistics of humanitarian assistance will determine whether the disaster management operation is a success or failure? However, logistics is often the most expensive activity of any disaster aid. Based on the study, it is estimated that the logistics costs for disaster management are around 80% of total costs in disaster relief [10].

Logistics management for disaster management is known as humanitarian logistics or often referred to as humanitarian aid logistics. Humanitarian logistics is an activity of planning, implementing, and controlling the flow of humanitarian aid in an efficient, cost-effective and humanitarian aid supply supported by information regarding the distance and location of the main warehouse and refugee barracks in order to reduce the suffering of disaster victims [11]. Optimizing the logistics performance of humanitarian assistance requires that all relationships between parties or actors involved in disaster management be managed through an integrated approach efficiently and effectively in coordinating inter-organizational performance, eliminating redundancy, and maximizing the efficiency of the entire emergency supply chain.

Disaster management is often described as a process consisting of several stages, namely: mitigation; preparation; response; and reconstruction [12]. The four stages are called the disaster management cycle. The involvement of disaster logistics management at the preparation, response and reconstruction stages as presented in figure 1.

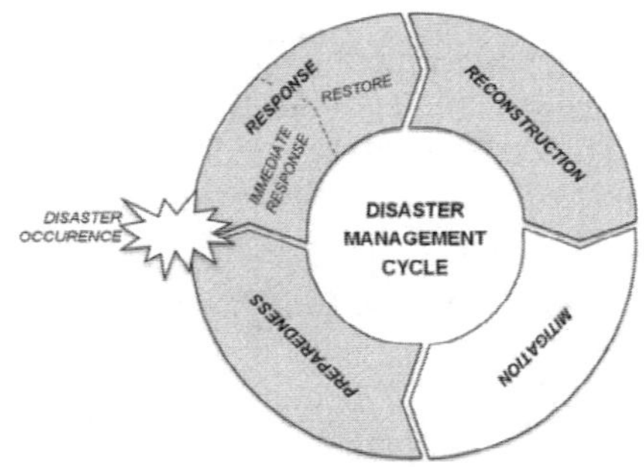

Figure 1. Disaster management cycle [12].

2.2. Stakeholders in Humanitarian Logistics

Humanitarian logistics includes a number of activities and involves many parties, starting from preparation, planning, procurement, transportation & distribution activities, storage, tracking, and customs clearance. Stakeholders in humanitarian aid logistics activities are [13]: a.) Domestic and foreign donors, government, companies, citizens and NGOs; b.) National NGOs, *PMI*, and *BNPB/ BPBD*; c.) Transportation service providers: land, air, sea, river and railroad; d.) Warehousing service provider; e.) Transportation management company; f.) Customs and Excise; g.) Relief recipient.

2.3. Inventory Information System

The system is a series consisting of two or more components that are interconnected and interact with each other to achieve goals. Systems are usually divided into smaller sub-systems that support larger systems [14]. Meanwhile, information is data that is presented in a certain form to help make decisions [15]. So, the information system is a combination of the components that collect, process, store, and provide the output of every information needed in business processes, Information systems in the form of applications consisting of software, databases and related manual processes [16].

The inventory information system provides detailed data about items of goods available, suppliers, quality of goods, the number of items that must be ordered and the number of items that have expired [17]. Information systems for disaster relief goods inventory are needed to facilitate access to information between stakeholders so that handling disaster logistics becomes more effective and efficient [9].

3. Requirements Analysis

3.1. User Characteristics

Characteristics of users are parties who will carry out activities on this information system. These entities are super-admin, admin refugee posts, and donors.

- Superadmin is the head or employee of the natural disaster management agency whose task is to manage and supervise the logistical assistance contained in the posts when natural disasters occur. Superadmin can manage user data, user assistance posts, users of refugee posts, goods, refugee posts, aid posts and transaction reports contained in the system.
- Admin of the refugee post is an employee/volunteer who is in a refugee post whose duty is to manage the demand for goods/logistics materials and validate the items that have been received.
- Donors are parties involved in providing voluntary assistance to the community, whether individuals or groups of people without aiming to benefit from their activities.

3.2. Device Requirements

The need for minimum specifications on the Merapi relief inventory information system consists of the needs of minimal specifications for hardware, software and brain ware.
Hardware used as a means of support in the form of:

- Computers with minimum specifications of Intel Pentium IV processors, 512 MB (Random Access Memory) and 80 GB Hard disks.
- Monitor, keyboard, and mouse.
- Modems or similar devices to connect computers to internet networks.

The software needed to run the information system created is:

- Microsoft Windows XP as the minimum operating system on a computer or other similar operating system.
- Web browser to run information systems.
- XAMPP is software that includes the Apache program as a web server, MySQL as a database server and PHP as a web programming language.

3.3. Functional Needs

Functional requirements are needs related to the business processes of the system made. From the observations made, the functional requirements in the system are made, here is a list of system functional requirements.

- The system can add, change, delete and display user data in refugee posts carried out by Superadmin
- The system can add, change, delete and display user data to donors carried out by Superadmin.
- The system can add, change, delete and display relief items data carried out by Superadmin.
- The system can add, change, delete and display data on refugee posts conducted by Superadmin.
- The system can display reports on the sending of relief items carried out by Superadmin, and the admin of the refugee post
- The system can display data reports on receipt of relief items carried out by Superadmin and admin of refugee posts
- The system can add, change, delete and display the data on aid items carried out by Superadmin, admin of refugee posts and donors
- The system can add, change, delete and display the data requested by the refugee post
- The system can add, change, delete and display data requests from refugee posts conducted by Superadmin
- The system can add, change, delete and display delivery data carried out by the refugee post admin
- The system can add, change, delete and display the purchase of relief goods through the government funding budget carried out by Superadmin
- The system can add, change, delete and display imported goods received from donors carried out by Superadmin
- The system can display reports of items received and items sent during a certain period of time carried out by Superadmin.
- The system can add, change, delete and display items sent by donors.
- The system can add, change, delete and display data items that have been received by Superadmin by donors

3.4. User Menu Structure

The menu structure is the initial design for software applications and website systems that focus on interactions with users. In the user menu structure, there are features that can be accessed by each type of user.

The structure of the login menu is a page for users to enter into information systems. Users can enter according to the user's access level. Access levels in this system include superadmins, admin and donors.

The superadmin menu structure that can be accessed after logging in to a superadmin user. Login is done by using an email and password from superadmin, then superadmin will enter the dashboard that has a menu list on the side. Menus that can be accessed by superadmin include:

a. Add, change and delete admin data for refugee posts and donors.
b. Add, change and delete system user group data.
c. Add, change and delete item category data.
d. Add, change and delete funding source data.
e. Add, change and delete warehouse data.
f. Add, change and delete post data.
g. Add, change and delete item data.
h. Add, change and delete inventory data.
i. Add, change and delete incoming order data.
j. Add, change and delete donation data.
k. Add, change and delete item request data.
l. View reports of incoming orders and goods requests.
m. View and change BPBD office data.

n. View and change account data such as names, emails, and passwords.
o. Exit the system

The admin menu structure of the post that can be accessed after logging in with the email and password from the admin, then the admin will go to the dashboard which has a menu list on the side that matches the access rights. Menus that can be accessed by superadmin include:
a. See admin data for refugee posts and donors.
b. View data group system users.
c. Add, change and delete item category data.
d. View funding source data.
e. View warehouse data.
f. Add, change and delete post data.
g. Add and change unit item data.
h. Add, change and delete inventory data.
i. Add and delete item request data.
j. View and change BPBD office data
k. View and change account data such as names, emails, and passwords.
l. Exit the system

Donor menu structure that can be accessed after logging in using e-mail and password from the donor, then the donor will enter the dashboard which has a menu list on the side. Menus that can be accessed by superadmin include:
a. Look at the item category data.
b. View warehouse data.
c. See post data.
d. Look at item data.
e. View inventory data.
f. Add and delete donation data.
g. View and change account data such as names, emails and passwords.
h. Exit the system

4. Conclusion
This research produces an analysis of information system requirements both from functional and non functional aspects. Based on functional analysis obtained the minimum specifications that must be owned by information systems, while the non-functional analysis obtained the minimum data needs to run a website-based disaster logistics inventory information system.

5. Acknowledgments
This research was financially supported by Applied Research Grants from the Directorate of Research and Community Service (DRPM). Ministry of Research, Technology, and Higher Education, Republic of Indonesia.

References
[1] Zamroni M I 2011 *Jurnal Penanggulangan Bencana* **2**
[2] Handayani N U Rinawati D I Wiguna Y K 2015 *Proc. of Joint Int. Conf. on Electric Vehicular Tech. and Industrial, Mechanical, Electrical and Chemical Engineering* (ICEVT & IMECE) 2015 – IEEE 5C2-5-1 - 5C2-5-4
[3] Rinawati D I, Sari D P, Handayani N U and Siwi B R 2018 *MATEC Web of Conf.* **154** 01050 1-6
[4] Altay N and Green W G 2006 *Euro. J. Oper. Res.* **175** 475– 493
[5] Balcik B, Beamon B, Krejci C, Muramatsu K and Ramirez M 2010 *Int. J. Prod. Econ.* **126** 22–34
[6] Zhou Q, Huang W and Zhang Y 2012 *Saf. Sci.* **49** 243–252
[7] Kaynak R and Tuger A T 2014 *Procedia – Soc. Behavior. Sci.* **109** 432-437.
[8] Beamon B M and Balcik B 2008 *Int. J. Public Sect. Manage.* **21** 4-25

[9] Rossum J V and Krukkert R 2010 *Jurnal Teknik Industri* **12** 25-32
[10] Van Wassenhove L N 2006 *J. Oper. Res. Soc.* **57** 475–489
[11] Thomas A S and Kopczak L R 2005 *From logistics to supply chain management: The path forward in the humanitarian sector* (San Francisco: Fritz Institute)
[12] Cozzolini A 2012 *Humanitarian Logistics: Cross-Sector Cooperation in Disaster Relief Management* (New York: Springer)
[13] Kovacs G and Spens K 2007 *Int. J. Phys. Distrib. Logist. Manage.* **37** 99-114.
[14] Romney M B and Steinbart P J 2015 *Accounting Information System* (15th ed.) (England: Pearson Educational Limited)
[15] Gellinas U J, Dull R B and Wheeler P R 2012 *Accounting Information System* (9th ed.) (USA: South-Western Cengage Learning)
[16] Satzinger J W, Jackson R B and Burd S D 2012 *Systems Analysis and Design in a Changing World* (6th ed.) (Boston: Joe Sabatino)
[17] Oetomo B S 2002 *Perencanaan dan Pembangunan Sistem Informasi* (Yogyakarta: Andi)

IOP Conf. Series: Materials Science and Engineering **703** (2019) 012048 doi:10.1088/1757-899X/703/1/012048

Anchored instruction ITS: a novel approach to make learning programming interesting and effective

B Hartanto[1] and J Reye[2]

[1]Department of Informatics Engineering, University of Surabaya, Raya Kalirungkut, Surabaya, 60293, Indonesia

[2]Science and Engineering Faculty, Queensland University of Technology, Australia

E-mail: budi@staff.ubaya.ac.id

Abstract. Learning programming is known to be troublesome to the students in the computer science department. The problem is originated in the students' inability to understand the logic in developing the program. Many teachers advise the students to practice writing the program to improve the students' skill. However, many students do not enthusiastically follow this advice due to the unattractive activity in writing the program as well as the lack of help when they have problems in writing the program. In this research, we proposed a system that increases the attractiveness of the program development activity and is able to provide help when the students have problems in it. These goals were performed by integrating an anchored instruction in an Intelligent Tutoring System (ITS). The attractiveness of the anchored instruction and the effectiveness of the Intelligent Tutoring System were analysed using a pre- and post-test, and a survey. From the tests and the survey, we found that the anchored instruction can increase the students' interest in practicing programming; the Intelligent Tutoring System could help the students in writing the program and improve their knowledge in programming (24.67% improvement). The research showed that anchored instruction Intelligent Tutoring System can be used as an alternative approach to make learning programming interesting and effective.

1. Introduction

Teaching and learning programming are known to be hard [1, 2, 3]. In order to have a good knowledge in programming, students should practice writing computer programs. Unfortunately, many students find that writing the program for the given exercises are not interesting and too difficult. Students often have problems in understanding the logic of the program to make. As a result, many students do not spend enough time and effort to practice programming.

Some works have been performed to make learning programming becomes interesting. Some of the works incorporated games in the learning and some others involved the usage of modern technology such as mobile devices. The incorporation of games in learning programming can be in the form of creating games [4] or playing games to understand computer programming topics [5].

In this research, we investigated another approach of making learning programming becomes interesting using anchored instruction. Anchored instruction has been used to motivate students and make learning become more interesting by emphasizing generative learning anchored in meaningful contexts [6,7]. In anchored instruction, the learning activities are designed around a story or situation that includes a problem. The students are expected to use their knowledge and skills to solve a certain problem posed in the story.

Anchored instruction has successfully been used to increase the students' motivation in learning mathematics, science, and statistics, but it has never been tried in the programming domain. It is interesting to know if anchored instruction can also be used to increase students' interest in learning programming.

The usage of anchored instruction in learning programming can be optimized if the system can also help the students to understand the logic of the program to be developed. Providing an individual tutor that can help students in their specific problem can make learning becomes more effective [8]. However, it will be resourced impractical to provide as many tutors as the number of students. One possible solution is to create a system that can provide specific help to the students' problems as if a human tutor will do. This system is called an Intelligent Tutoring System.

An Intelligent Tutoring System (ITS) is a computer program that can tailor its assistance and teaching to the student current knowledge [9]. Some ITSs have been built in the programming domain, for example: PHP ITS [10], J-LATTE [11], and JITS [12]. Unfortunately, there was no evidence that these ITSs were designed to make learning programming interesting.

An ITS that can assist the student individually and make the learning process becomes interesting is a good candidate to change the student perception that learning programming is hard and dull. This paper proposed a novel approach to make learning programming becoming interesting and effective by incorporating anchored instruction in an Intelligent Tutoring System. The pedagogy that we used in our Intelligent Tutoring System required the students to write programs to learn programming.

2. Previous work on anchored instruction and Intelligent Tutoring System (ITS)

Anchored instruction is an instructional framework of which the instruction is anchored or situated in the context of meaningful problem-solving context to overcome the inert knowledge problem [6]. It is the context that makes learning becomes interesting because it reflects the true nature of problems in the real world.

In anchored instruction, the learning activities are designed around a story or situation that includes a problem or issue. Some works that utilized the concept of anchored instruction can be seen in Prado and Gravoso [6], and also in Kumar [7]. The usage of anchored instruction in this research increased the students' interest to the subjects, decreased their tensions in learning, and improved their skill in problem solving

Intelligent Tutoring System works like a human tutor. The system can predict the student's current knowledge and use the information to provide the most suitable instruction or exercise. The ITS can also interpret the student responses and offer hints to help the student to understand the material or to complete the given exercise. Therefore, the ITS can be considered as a private computer-based tutor.

Some ITSs, such as JITS, J-LATTE, or PHP-ITS have been built in the programming domain [10-12]. JITS were created to help students in their first programming course to learn programming in Java. JITS was designed to recognize any small Java program and offer intelligent feedback when there is no authored solution available [12]. As a result, JITS can only produce a syntax error correction or assistance system. These features may not give additional assistance to the students because these features are already provided in many compilers.

Other ITSs, such as J-LATTE and PHP ITS provide assistance to the students when they have problems in the syntax and logical error. Assisting students to understand the logic of the program is important because many students have problems in understanding the logic of the program or how to fix a logical error in the program development. J-LATTE and PHP ITS start the tutoring process by giving some programming questions to the students. After selecting a certain question, the students can write the program on the developed application. Some buttons are provided to let the students choose some actions, such as submitting the program or requesting help.

PHP-ITS was created to help students learn PHP. PHP-ITS used predicates as a set of constraints to check if the students' programs had achieved the expected goal or not. The goal itself was divided into several sub-goals. These sub-goals were used to give specific feedback to the students. Although JITS, J-LATTE, and PHP ITS were created to help the students in learning programming, they were not designed to make the learning process become interesting. The learning process in the ITSs were similar to the one in the classroom, which some students considered uninteresting. Many exercises were presented without context and were not designed to inspire the students of their possible use to solve real world problems.

3. Anchored instruction ITS

Having a potential to make learning become interesting, anchored instruction can be used as the pedagogy in the developed Intelligent Tutoring System, named CSharpTutor. Some aspects that make learning becomes effective were also included in the design of CSharpTutor. The anchored learning in CSharpTutor was implemented using a role playing.

In the role playing, the student has a role as a programmer who works in a company. When the student starts CSharpTutor in Visual Studio, a greeting from a virtual Head of Development is displayed. The Head of Development will ask the student to show his capability in writing programs for some given cases to be accepted or to get a better position in the company.

The program for the case given by the Head of Development can be written by the student in Visual Studio directly. Unlike any other existing ITS, CSharpTutor was developed and integrated with Visual Studio, the development environment that is used by the targeted students. This integration enables CSharpTutor to read the student code in Visual Studio directly and gives advice about how to create the program for the given case.

As an ITS, CSharpTutor tries to understand the student program and provides feedback upon the student request to the logical error in the program. Understanding the logic of the student program was performed using first order logic. The solution of the case was represented into goal and sub goals in the first order logic representation. The student program was parsed and translated to predicate logics. The predicate logics were checked if they can satisfy the goal or sub goals. This information was used to give feedback to the student or to advance the student to the next case.

The following is the example of the goal that must be satisfied by the student program to convert the temperature value from Celsius to Fahrenheit:

```
goal← var(CelsID, celsius),
      val(CelsID, Cels),
      var(FahrID, fahr),
      val(FahrID, valFahr),
      eq(ValFahr, 9.0/5.0 * Cels + 32),
      print(FahrID)
```

After completing a program for a case given by the Head of the Development, the student can submit his work to the Head of the Development, and the Head of the Development will assign another case to be solved by the student. The submission can be performed through CSharpTutor. After several correct program submissions, the Head of the Development will promote the student to a higher position in the company. With this promotion, the student will be given a new case that needs more advance programming skill.

There are three levels that exist in CSharpTutor. The topics assigned for each level are:

- Level 1: variables, assignment, and input output statements
- Level 2: simple condition
- Level 3: nested condition, repetition, and one-dimensional array

We also named the level with a certain position in IT company, such as Junior Programmer, Senior Programmer, and Analyst Programmer. The story about being a programmer and to achieve a higher position in the company were chosen because many students enrol in the computer science department with an expectation to be able to work in a programming-related working area when they graduate one day. Aligning the role playing with the student expectation can increase the student interest and motivation in learning programming.

CSharpTutor contains 16 cases that can be completed by the student. However, the student may not be asked to complete all cases in CSharpTutor. Based on the student model that was developed when the student submitted his/her programs, CSharpTutor can suggest the most suitable tasks for the student current knowledge. For example, if the student makes unnecessary mistake in writing the program for a case in certain level or ask too much help from CSharpTutor then CSharpTutor will use this information to update the student model. The student model is also updated based on the student submitted program. Using this student model, CSharpTutor will select the next case that should be given to the student.

The topics are represented as variables in Bayesian Network along with other variables created from the cases' sub-goals. The Bayesian Network is used to estimate the probability of the student knowledge of a certain topic. Using Bayesian Network, CSharpTutor can update the belief for the variables in the network based on the new evidence. For example, if the evidence shows that the student can write the program for a certain case correctly, we can use Bayesian Network to increase our belief that the student knows the topics that are related to that case.

All cases in CSharpTutor are connected to a certain context. Having an exercise in a context encourages and motivates students to complete the exercise because they can see the "real" application of the program that they made [13]. It is necessary to change the real cases to simpler version to make them on par to the student knowledge. CSharpTutor does not have any out-of-context tasks that commonly be seen in many programming books such as "swap the values in two variables" or "find the average of *n* numbers", etc.

The assistance in CSharpTutor is designed to be adaptive and multilevel. In this case, CSharpTutor checks the condition of the student program using code analyser [14] to determine the sub goals that cannot be accomplished by the students and display the appropriate assistance to the student. The assistance is displayed in multilevel. When the student has the problem for the first time, a general level assistance will be displayed. A more detail level of assistance will be given if the student keeps having problems with the same issue. CSharpTutor also provides a minimal feedback by just responding "yes/no" when the student asks if his/her program is correct or not. This feature is provided to encourage the student to find the solution him/herself before he/she asks the detail help from CSharpTutor. All of this helps can be obtained by pressing a certain button in CSharpTutor.

4. Methods used to measure the effectiveness and the interestingness of anchored instruction ITS

In evaluating the effectiveness and the interestingness of anchored instruction ITS, 36 students in Programming class voluntarily tried CSharpTutor. Before the students started CSharpTutor, they had to answer 20 multiple choice programming questions that were used a pre-test. These questions were presented again in random order as a post-test. In addition to these data, CSharpTutor also recorded the submitted students' programs, the information about which tasks that were completed, the date and time when the students wrote the program, the length of time needed to complete a program (omitting the idle time when the students did nothing in the development environment), and the number of help requested by the students. All of these data were used to measure the effectiveness of CSharpTutor. On the other hand, the interestingness of the anchored instruction ITS was measured through a questionnaire. This questionnaire had to be filled by the students after they completed the post-test.

5. Result and discussion

5.1. The effectiveness of CSharpTutor in helping student learn programming

The effectiveness of CSharpTutor in helping the students learn programming was evaluated by analysing the students pre- and post-tests. The students' average score in pre-test was 64.17 and the students' average score in post-test increased to 80 (24.67% improvement). A one-tailed paired t-test with a 95% confidence interval was used to check if the average score in post-test was significantly better than the average score in pre-test (see Table 1). The null hypothesis for this test was: the students' scores in post-test were not higher than the students' scores in pre-test. The p-value of one tailed paired t-test could be found by dividing the p-value of two-tailed paired t-test by 2, that was 0.00007. This p-value showed a strong evidence that we could reject the null hypothesis. This means that the students' average score in post-test was significantly higher than the students' average score in pre-test.

This finding shows that anchored instruction ITS can be used in an ill-defined domain such as programming and produces as effective result as the one in the well-defined domain [6,7]. The integration of anchored instruction to an ITS is necessary to check the correctness of the student program and use it to advance the story in the anchored instruction. The ITS is also needed to provide assistance to the student in writing the program.

Table 1. Paired t-test of the students' post-test and pre-test.

	Mean	Std. Dev.	Std. Error Mean	95% confidence interval of the difference		t	df	Sig. (2-tailed)
				Lower	Upper			
Pair 1. Post Test – Pre Test	15.83%	18.80%	3.13%	9.47%	22.20%	5.05	35	0.000015

We also measured the contribution of CSharpTutor in supporting students to learn programming. Some statements in a questionnaire that we used to measure this aspect were:

- CSharpTutor assistance is helpful
- The number of tasks given in CSharpTutor is ideal
- Overall, CSharpTutor helps the students in learning programming

Most students responded positively to the statements above. 67% of the students agreed or strongly agreed that CSharpTutor assistance was helpful. The number of tasks in CSharpTutor was considered ideal or almost ideal by 77% students. Because of the adaptive tasks used in CSharpTutor, the students who can do well in the pre-test and can submit correct codes got fewer tasks to be completed than the students who did the opposite. This positive feedback showed that the students were satisfied with the usage of adaptive number of tasks given in CSharpTutor. When the students were asked about the overall performance of CSharpTutor in helping them learn programming, 92% of the students agreed or strongly agreed that CSharpTutor could help them learn programming.

5.2. The interestingness of anchored instruction ITS

We also used questionnaires to measure the interestingness of anchored instruction ITS. We used a five level Likert scale that ranges from strongly disagree, agree, neutral, agree, and strongly agree. The following statements were used in the questionnaire:

- The role playing in CSharpTutor is good
- The approach of giving a higher level goal in writing the program is good
- The approach to learn programming by writing program is good
- The approach of giving the students a chance to develop answers themselves by providing multilevel help is good

From the questionnaire, 75% of the students agreed or strongly agreed that the role-playing approach was good. 25% of the students gave a neutral opinion, and none of them disagreed or strongly disagreed. The students also gave positive responds when they were asked about the approach of giving higher level goal to be achieved when they wrote a program. Writing a correct program became an intermediate goal to achieve higher level goal that was being promoted to a better position in the company. 89% of the students agreed or strongly agreed with this approach. Again, none of the students disagreed or strongly disagreed.

All students (100%) also agreed or strongly agreed that the best way to learn programming was by writing programs. An interesting approach such as the one used in CSharpTutor could encourage the students to practice programming. Most of the students (97%) also agreed or strongly agreed that the students should develop the answers themselves before they got help from CSharpTutor. This was performed in CSharpTutor through a multilevel and a "yes/no" help.

6. Conclusion and further work

The findings from this research indicated that anchored instruction ITS can effectively help student learn programming. In addition, learning programming in anchored instruction ITS makes the process of learning itself become more interesting. The usage of the ITS becomes the key point in making the success of the system. The ITS guides the students to understand the logic of the program to be written.

Other educational related factors also give contributions in the success story of helping the student understands programming. These factors are minimizing the cognitive overload by integrating CSharpTutor to the development environment, the usage of case based exercises that provide an idea of the real world application of the program, the adaptive feedback to help the students to understand the logic of the program, and the multilevel feedback to give the students a chance to think the solution themselves.

For further work, the story on the anchored learning can be changed to another story that is still in the students' interest, for example to the one that includes modern technology or science fiction. The ITS can be extended to another development environment or to another mode in Visual Studio such as to Windows Forms or to Windows Presentation Foundation (WPF) application. The topics in the ITS can also be added to a more advanced topic such as method and class. A more extensive evaluation can be conducted by increasing the number of participants and the usage of cohort to compare the result.

References

[1] Hartanto B 2015 Enhancing the student engagement in an introductory programming: a holistic approach in improving the student grade in the informatics department of the University of Surabaya *Int. Conf. on Soft Computing, Intelligence Systems, and Information Technology* ed R.Intan, C Chi-Hung, H N Palit, L W Santoso (Bali, Indonesia: Springer) pp 493–504

[2] Teague D M and Roe P 2009 Learning to program: from pear-shaped to pairs *Proc. of the 1st Int. Conf. on Computer Supported Education* vol 2 (Lisboa: Portugal) pp 151–8

[3] Koulouri T, Lauria S and Macredie R D 2014 Teaching introductory programming: a quantitative evaluation of different approaches *ACM Trans. Comput. Educ.* **14** pp 26:1–28

[4] Kurkovsky S 2013 Mobile game development: improving student engagement and motivation in introductory computing courses *Comput. Sci. Educ.* **23** pp 138–157

[5] Villani A N, Hermiz H T and Cutler R 2013 Using a serious game approach to teach operator precedence to introductory programming students *17th Int. Conf. on Information Visualisation* pp 523–6

[6] Prado M M and Gravoso R S. 2011 Improving high school students' statistical reasoning skills: a case of applying anchored instruction *Asia-Pac. Educ. Res.* **20** pp 61–72

[7] Kumar D D 2010 Approaches to interactive video anchors in problem-based science learning *J. Sci. Educ. Technol.* **19** pp 13–9

[8] Chen L, Eugenio B D, Fossati D, Ohlsson S and Cosejo D 2011 Exploring effective dialogue act sequences in one-on-one computer science tutoring dialogues *Proc. of the 6th Workshop on Innovative Use of NLP for Building Educational Applications* (Portland: Oregon) pp 65–75

[9] Conati C 2009 Intelligent tutoring systems: new challenges and directions*Proc. of the 20th Int. Joint Conf. on Artificial Intelligence* (Pasadena) pp 2–7

[10] Weragama D and Reye J 2014 Analysing student programs in the PHP intelligent tutoring system *Int. J. Artif. Intell. Educ.* **24** pp 162–88

[11] Holland J, Mitrovic A and Martin B 2009 J-LATTE: A constraint-based tutor for Java *Proc. of the 17th Int. Conf. on Computers in Education* ed S C Kong *et al* (Hongkong) pp 142–6

[12] Sykes E 2007 Developmental process model for the Java intelligent tutoring system. *J. Interact. Learn. Res.* **18** pp 399–410

[13] O'Grady M J 2012 Practical problem-based learning in computing education *ACM Trans. Comput. Educ.* **12** pp 10:1–16

[14] Hartanto B and Reye J 2013 Incorporating anchored learning in a C# intelligent tutoring system *Doctoral Student Consortia Proc. of the 21st Int. Conf. on Computers in Education* ed W Chen, *et al* (Indonesia) pp 5–8

IOP Conf. Series: Materials Science and Engineering **703** (2019) 012049 doi:10.1088/1757-899X/703/1/012049

The evaluation of academic website using eye tracker and UEQ: a case study in a website of xyz

A H Kusumo and M Hartono

Department of Industrial Engineering, University of Surabaya, Raya Kalirungkut, Surabaya, 60293, Indonesia

E-mail: argohadi@staff.ubaya.ac.id

Abstract. The current study evaluated the academic website using eye tracking and User Experience Questionnaire (UEQ). A total of 46 participants had been recruited for this study but then the data were reduced to 33 participants due to random response errors of an item in UEQ followed by reducing in eye tracker's data. Eye tracker had a gaze plot (scan path) data and task completion time as time to first view area of interest (AOI) where these data were used to evaluate important gaze plot and whether the provision of information in website had been clearly done. The gaze plot gave insight where important link, information and picture should be put in the website. Furthermore, Kruskal Wallis and Mann Whitney U test were also applied to see the difference between each task which were used to improve sub menu's name. At the same time, UEQ data showed that only novelty should be improved to be more creative, inventive, leading edge and innovative. Thus, this recommendation provides more input in designing website for academic use.

1. Introduction

Web is playing a significant role in diverse application domains such as business, education, industry and entertainment. As a result, there are increasing concerns about the ways in which websites are developed and the quality of information delivered [1]. The growth of the web is profoundly changing the way people interact with information and with people. This has led to an expansion of opportunities for the web on different vectors, including the massive production of contents [2]. This must be taking into account in the new website development of academic website. As a case study, a website of XYZ has been evaluated. This evaluation involved the students of IE who used the website and access the important information such as lecturer data (email and area of interest), learning outcomes, downloaded forms, requirement for final project etc. Therefore, user experience measurement was required.

The first requirement for an exemplary user experience is to meet the exact needs of the customer, without fuss or bother. Next comes simplicity and elegance that produce products that are a joy to own, a joy to use [3]. Without the user experience aspect of the user, the user will leave the website that is actually important [4]. User experience measurement can be done by eye tracker data which have been collected that demonstrate that eye movements are intimately related to the moment-to-moment cognitive processing activities of readers [5]. When users reach a Web page, they can scan the page and obtain a comprehension of it in a few seconds. Cognition refers to the ability of the human mind to acquire and manage information [6] and comprises different mental processes such as attention, memory, perception, problem solving and learning [7]. By understanding sighted users' visual understanding of Web page complexity in relation with the time of task completion, important information should reveal with reference to the cognitive effort required for interaction with that page. To use eye tracking, participants' visual attention was measured and analyzed based on the gaze plot (scan path) and heat maps eye gaze analysis [8]. Other than using UEQ, user experience can be measured using user experience questionnaire (UEQ). UEQ has 6 scales, consisting of Attractiveness, Perspicuity, Efficiency, Dependability, Stimulation, Novelty [9]. Eye tracking's method is called objective measurement, while UEQ is referred to as subjective measurement [10]

The current study focused on evaluating XYZ new website as an example of academic website by analyzing eye tracking and UEQ. Gaze plot (scan path) of eye tracker's data will be used to analyze navigational behavior and mean difference in time to completion task as time to first view can be utilized to develop sub menu's name. At the same time, mean value of each scale in UEQ will be used to develop interface of the website. In this study, conformity of cognitive aspect between eye tracking and UEQ was also analyzed.

2. Materials and methods

This research used two methods, that is, eye tracking as an objective measurement and user experience questionnaire as a subjective measurement. Experimental procedure for eye tracker is explained below:

Eye movements were recorded by using Gaze point eye tracker. This eye tracker generates raw gaze-point location data at the camera field rate of 60 Hz. The position of GP3 was below the screen, centered and as close to the bottom edge of the screen as possible. This was approximately arm's length (65 cm) from the user. The device was approximately 40 cm below eye level and pointing a tan upwards angle towards the eyes [11]. Obtained data were analyzed using the Gaze point Analysis system. It provided powerful method for collecting and analyzing eye-gaze data. Visualization could be performed through gaze replay, gaze plots, heat maps, and area of interest (AOI). Gaze replays enabled to watch the test session video again with the user's eye motion overlaid over a recording of the changing computer screen image. In these, a blue dot could be seen or user's eye, moving around a page. Gaze plots are another very valuable analysis tool in eye tracking technology. These compile the eye gaze of one user on one page. These do not combine more than one page or user in the representation. Light blue dots on the page show a number of important data items, including: 1) where the user's fixations were, 2) numbers in the dots depicting the order in which the user looked at the items, and 3) the size of the dot denoting how long the user looked at the item. Larger dots mean longer looks. There is another analysis tool in eye tracking technology that deserves a mention here, called Areas of Interest, AOIs, or Look Zones. This feature is meant to help with doing quantitative analysis [12].

This study recruited 46 participants based on Nielsen theory which stated that the minimum participant of eye tracking usability study is 6 Qualitative eye tracking (watching gaze replays) and 20 for Quantitative user testing [12].

This study used XYZ website as a research object. Participants were asked to fulfill 5 tasks. Task 1: participants were asked to explore the website for 30 seconds. This task aimed to enable participants to assess the appearance of the website and to find out the elements on the website that attract the attention of participants through gaze plot (scan path). Task 2 was to find lecturer data, task 3 to find learning outcome, task 4 to find courses assignment's form, and task 5 to find Final Project requirement. Another 4 tasks will give a prior knowledge which area can be found faster. The matrix used was time to first view or fixation which measured the length of time taking for the respondent to reach AOI for the first time. Hence, we can evaluate whether the important information can be seen faster. Vertegaal explained [13], fixation information can be used to measure the attention that individuals have paid to stimuli.

The UEQ contains 6 scales with 26 items [9] as explained below. The questionnaire and the analysis can be downloaded online [14]. Attractiveness means overall impression of the product, perspicuity means how easy to get familiar or learn the product. Efficiency means solving the tasks without unnecessary effort, while dependability means feeling in control of the interaction. Stimulation means how exciting and motivating it is to use the product, while novelty means how innovative and creative the product is.

Mean values between -0,8 and 0,8 represent a more or less neutral evaluation of the corresponding scale. Values > 0,8 represent a positive evaluation and values < -0,8 represent a negative evaluation. The range of the scales is between -3 (horribly bad) and +3 (extremely good) [9].
To detect such more or less random or not serious answers, a simple heuristic was used. The idea to detect random or not serious answers is to check how much the best and worst evaluation of an item in a scale differs. If there is a big difference (>3), this is seen as an indicator for a problematic data pattern. Such situations can also result from random response errors or a misunderstanding of an item. Thus, it makes no sense to consider a response as problematic if this occurs just for a single scale. Answers will be removed from the data set that shows a value of 3 or higher in the Critical [9] which is then followed by eliminating the data from eye tracker.

To compare cognitive aspects between eye tracker and UEQ, the average first time to view of 4 tasks was analyzed, then it was compared with mean value of each cognitive scale in UEQ (we suggested perspicuity, efficiency, dependability as a cognitive scale). Nielsen [15] explained, If the web page survives 10 second judgment, users will look around a bit. However, they are still highly likely to leave during the subsequent 20 seconds of their visit. Only after people have stayed on a page for about 30 seconds does the curve become relatively flat. People continue to leave every second, but at a much slower rate than during the first 30 seconds. Therefore, we used 30 second as a standard time. When average time to first view was more than 30 seconds, but the UEQ result had high mean value, we called it discrepancy for cognitive aspect.

The differences in the completion time for tasks 2,3,4,5 (H1) were also analyzed. Since data were not normally distributed, we used Kruskal Wallis test which is a nonparametric alternative to a one-way ANOVA [16]. Further analysis using Mann Whitney U Test was used to see the difference between each task [17]. Statistical significance was analyzed through the use of SPSS 18.

3. Results

As was stated in the methods, a total of 46 participants were recruited for this study but the data were reduced to 33 participants due to random response errors or a misunderstanding of an item in UEQ [9] which then followed by eliminating the data from eye tracker.

The first set of analyses examined the gaze plot from eye tracker's gaze replay in response to task 1. We presented gaze plot as a navigational behavior started from path 1 as the most first viewed area in website, then continued to path 2, path 3, path 4, and path 5, respectively. As shown in Fig. 1, The most first viewed area in website sequentially were 1) home page which then the participants scrolled down through the picture which containing the news link, 2) profile as main menu, 3) each of main menu which was explored by participants.

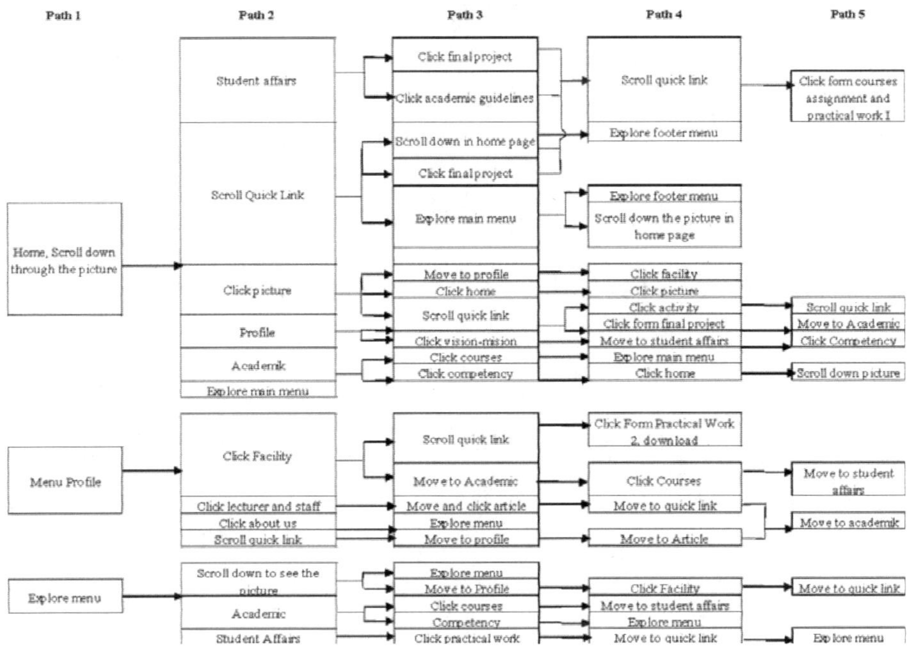

Figure 1. Gaze plot analysis of eye tracker (navigational behavior)

While scrolling on the homepage, the gaze moved to the quick link. This was mostly done not only on path 2 but also on the next path (3,4, and 5) since this quick link is located on all pages. While participants explored the quick link, they would click on the assignment and practical courses I form, the final project form, and practical work 2 from. It can be seen that quick links have an important role here. Gaze plot to profile menu could be seen in path 1, 2 and 3. In this menu, sub menu facility and lecture and staff were the most clicked and viewed. The Academic and Student Affairs were the next main menu frequently seen. After looking at the academic main menu, participants would click sub main menu, that is, courses and competency, whilst in student affairs main menu, participants clicked final project, guidelines practical work, practical work I. They also explored the menu one by one in the first path. The picture containing the news link in the home page was also clicked by some participants.

In response to UEQ as shown in table 1, perspicuity, efficiency dependability got high mean value since it was above the neutral value 0,8. At the same time, stimulation had a high value while novelty had a value that was close to a neutral value where improvement will be needed.

Table 1. UEQ result

UEQ Scales	Mean	Variance	UEQ Scales	Mean	Variance
Attractiveness	1,737	0,34	Dependability	1,833	0,35
Perspicuity	1,432	0,87	Stimulation	1,742	0,30
Efficiency	1,833	0,32	Novelty	0,811	0,67

The obtained completion time of 4 tasks from 33 participants can be seen in table 2. Completion time was obtained by putting area of interest (AOI) in the destination website page for every task that called time to first view. It can be seen that tasks 2 and 5 consumed less time than tasks 3 and 4 since task 2 and 5 have familiar sub menu's name according to participants. Sub menu for task 3 was considered unfamiliar since participants were unaware that learning outcome was a part of the competency as well as practical work 1 form was also part of the document.

Table 2. Eye tracker result in response to task 2 to 4

Task	Time to 1st View (s)	Task	Time to 1st View (s)
Task 2	16,826	Task 4	31,051
Task 3	42,897	Task 5	17,372

To see the differences between each task, Kruskal Wallis were applied. These tests revealed that there were significant differences between time for completing each task since asymp. Sig 0,00004 (asymp. Sig <0,05). To understand which time of the task was different, Mann Whitney U test was applied. The result of the test can be seen in table 3. It highlighted that time in completing task 2 and 5 had no significant difference as well as task 3 and 4.

Table 3. Mann whitney U test result for each task

Task	Mean Rank	Asymp. Sig. (2-tailed)	Conclusion
2 vs 3	24.27 vs 42,73	0,00009	Significant difference
2 vs 4	27.35 vs 39.65	0,009	Significant difference
2 vs 5	33,73 vs 33,27	0,923	No significant difference
3 vs 4	36,79 vs 30,21	0,164	No significant difference
3 vs 5	42,21 vs 24,79	0,0002	Significant difference
4 vs 5	39,97 vs 27,03	0,006	Significant difference

Moreover, verification of cognitive aspects in eye tracker and UEQ has been done. As stated in the methods, 30 seconds was determined as standard time. It was interesting to note that the lowest discrepancy was 30,30% between eye tracker's result and scale of perspicuity in UEQ. The detailed results were presented in table 4. Our experiments were in line with the meaning of perspicuity itself which showed how easy the participants get familiar or learn the website. The findings above gave recommendations of website evaluation that can be seen in table 5.

Table 4. Discrepancy between eye tracker and cognitive aspects of UEQ results

Cognitive scale in UEQ	Respondent Discrepancy	Percentage (%)
Perspicuity	10	30,30
Efficiency	18	54,54
Dependability	18	54,54

Table 5. Recommendation

Element	Recommendation
Gaze plot	• As discussed above, some participants had difficulty finding the learning outcomes (LO) and forms needed. Therefore, since LO is also important for the users, we suggest to put LO link in home page as a picture. As analyzed in path above, participants frequently scrolled down the home page through the picture. Therefore, link as a picture such as important news, accreditation, student's life, and upcoming events should be added to attract more user see this page. • Downloaded form should be put in sub menu according to the needs. Final project form should be put in the sub menu final project while practical work form should be put in sub menu practical work. We also suggest to add more sub menu in practical work which are Practical work 1, 2, and courses assignment form. • Quick link placement is very useful. We suggest adding more important links such as competency, curriculum, courses and eliminate less important link such as archives and categories. Practical work 1, 2, and courses assignment form will remain here.
Novelty	• Website needs to be designed in a more creative, inventive, leading edge and innovative way. • Each page of the website should be designed more colorfully and not monotonous and more pictures should be added. • Font should be changed from robot to more pleasant and web safe font such as arial,

AvenirNextLTPro-Regular, verdana, etc.
- Paragraph arrangement should be improved. At this time, paragraph was looked too monotonous hence it was difficult to read.

4. Discussions

The aim of this study is to evaluate academic website using eye tracking and UEQ. As analyzed in the result, the data can support this evaluation. Eye tracker as an objective measurement was used to analyze gaze plot that showed navigational behavior of the participants. This is important to see how they explored the website. A subjective measurement using UEQ showed that appropriate recommendations were needed for near neutral value of scale.

Our study failed to prove that there is a 100% conformity of cognitive aspect between eye tracking and UEQ. This can be related to standard time setting and the type of assignment given. In relation to standard time as explained by Nielsen [15], users often leave web pages in 10–20 seconds and continue to leave every second, but at a much slower rate than during the first 30 seconds. If we determine standard time was 10 - 20 seconds, level of discrepancy will be higher, more than 30,30%. Therefore, further experimental investigations are needed to estimate this standard time. In addition, further studies are also needed to explore what kind of task used for eye tracking studies which has conformity with UEQ cognitive. To avoid elimination of participants data, analysis need to be done immediately after the participants complete the UEQ.

5. Conclusions

The website is currently very important for various fields, especially education. A lot of information can be obtained by students especially from the website so that an appropriate website evaluation and development is needed. This study successfully evaluated websites using two methods, namely eye tracking as an objective measurement and UEQ as a subjective measurement. This paper has highlighted eye tracker data, that is, gaze plot and time to first view. Gaze plot could give insight where to put important information, picture or link in the website. Kruskal Wallis and Mann Whitney U test was applied to understand significant difference of time to complete each task which was used to give recommendation of changing and adding sub menu. At the same time, UEQ data showed that only novelty which had a value near neutral, hence website needs improvement to be more creative, inventive, leading edge and innovative.

In this study, cognitive level of discrepancy between eye tracker's task and perspicuity in UEQ was 30,30% due to 30 seconds time standard. This has not been able to prove 100% of the conformity of the cognitive aspect between eye tracker and UEQ. Despite the fact that there are limitations of measurement of cognitive aspects, the further studies need to develop standard time and type of task for eye tracking studies to measure cognitive aspects.

References

[1] Lopes R and Carrico L 2008 *Proc. Int. Cross Disciplinary Conf. on Web Accessibility (Beijing)* (New York: ACM) pp 5-14
[2] Iqbal M and Warraich N F 2012 *Pak. J. Libr. Inf. Sci.* **13** 1-11
[3] Norman D and Nielsen J 2006 The definition of user experience (UX) Online: https://www.nngroup.com/articles/definition-user-experience/
[4] Izabal S V, Aknuranda I and Az-Zahra H M 2018 *Jurnal Pengembangan Teknologi Informasi dan Ilmu Komputer* **2** 3224-32
[5] Rayner K 1998 Eye movements in reading and information processing: 20 years of research. *Psychological Bulletin* **124** 372–422
[6] Tsianos N, Germanakos P, Lekkas Z, Mourlas C and Samaras G 2009 Eye-tracking users' behavior in relation to cognitive style within an e-learning environment *Proc. IEEE Int. Conf. on Adv. Learning Tech. (Riga)* (Piscataway: IEEE) pp 329–33
[7] Solso R L, MacLin M K and MacLin O H 2004 Cognitive Psychology vol 7 (Boston: Allyn & Bacon)
[8] Nisiforou E A, Michailidou E and Laghos A 2014 *Proc. Int. UAHCI* (Crete) (Basel: Springer) pp 46–57
[9] Schrepp M 2019 User Experience Questionnaire Handbook Version 7 Online: https://www.ueq-online.org/Material/Handbook.pdf

[10] Mahardhika W, Wibirama S and Ferdiana R 2016 Analisis parallax scrolling pada web storytelling dan online shop menggunakan eye tracking dan kuesioner user experience *Electronic Theses and Dissertations Gajah Mada University*
[11] Hennessey C, Tam J and Robillard J 2014 Quick start guide Gaze point research inc Canada
[12] Pernice K and Nielsen J 2009 How to Conduct Eyetracking Studies Nielsen Norman Group Online: http://www.nngroup.com/reports/how-to-conduct-eyetracking-studies/
[13] Vertegaal R and Ding Y 2002 *Proc. Conf. on Computer Supported Cooperative Work* (New Orleans) (New York: ACM) pp 41–8.
[14] Schrepp M 2012 User experience questionnaire Online: https://www.ueq-online.org/
[15] Nielsen J 2011 How Long Do Users Stay on Web Pages? Online: https://www.nngroup.com/articles/how-long-do-users-stay-on-web-pages/
[16] Ghoodjani A 2016 Why should I use a Kruskal Wallis Test? (Tehran: Avicenna Research Institute)
[17] Hidayat A 2017 Tutorial Uji Mann Whitney U Test dengan SPSS Online: https://www.statistikian.com/2014/04/mann-whitney-u-test-dengan-spss.html/amp

IOP Conf. Series: Materials Science and Engineering **703** (2019) 012050 doi:10.1088/1757-899X/703/1/012050

Computer vision system in measurement of the volume and mass of egg using the disc method

M Widiasri, L P Santoso and J Siswantoro

Department of Informatics Engineering, Faculty of Engineering, University of Surabaya
Jl. Kali Rungkut, Surabaya, 60293, Indonesia

E-mail: monica@staff.ubaya.ac.id

Abstract. One important process in egg production is the sorting process to determine the grade of the egg. The volume and mass of the egg are the factors needed in this sorting process. But the conventional way to measure egg volume and mass can damage egg and takes a long time. This paper proposes a Computer Vision System (CVS) to measure egg volume and mass quickly, accurately, precisely without damaging the egg. The disc method is a method used to calculate the volume of an object with a circular cross-section such as an egg. CVS was designed to calculate the volume using the disc method and calculate the egg mass using density and regression models, based on images captured in real-time or images that have been captured previously. The validation process is carried out using 50 egg samples by comparing results from the proposed method with manual measurements. To compare the time needed by CVS and manual measurement, an average test was used. To test the output accuracy of the volume and mass of egg, relative absolute error, ANOVA test, and correlation test were used. Whereas to test the output precision, the coefficient of variation was used. Based on the results of the testing conducted, CVS in the proposed method successfully measuring the volume and mass of egg quickly, accurately, precisely, without damaging the egg.

1. Introduction

Egg volume and mass are important factors in the egg production process. One process that uses volume and mass of the eggs as its criteria is the sorting process. Because of the importance of knowing the volume and mass of eggs, the calculation method used must be precise and efficient. One method commonly used to measure egg volume is the water displacement method, where the volume of spilled water is the same as the volume of eggs put into it. But this method is considered to be less accurate, can damage the eggs, and takes a long time [1]. The using of the computer vision system (CVS) is a better alternative to the existing manual method. Computer Vision System (CVS) can measure accurate, precise, quick, and non-destructive egg volume [2]. To measure the volume of eggs, the method proposed used is a two-dimensional CVS, where the camera will capture the cross-sectional image of the egg. Assuming that eggs are an axisymmetric object, the egg volume can be calculated by rotating the cross-section of an egg around its long axis [3]. We can then use various methods of calculating the volume of solids of revolution, namely the cone method [4], methods based on the Pappus theorem [5], Simpson's rules [6], and the disc method [7][8]. The disc method assumes that the volume of a solid of revolution is the sum of the volume of all the thin cylinders that made up the object. The disc method used by Bridge et al. [7] to calculate the volume of Florida Scrub-Jay bird eggs. Koc [8] also uses the disc method to calculate the volume of watermelons. The result from a study by Bridge [7] shows that the volume calculation is accurate. However, the CSV system used cannot automatically align the image obtained parallel to the main axis. This will cost extra time for the user to align the egg manually. The system proposed by Koc [8] was less accurate because as the size of the watermelon increased, the image processing method overestimated the volume. This was because of the change in distance between the camera and the watermelon surface. Although the distance between the camera and the surface where the object lays was constant, the distance between the watermelon surface and the camera reduced with increasing watermelon size.

Mass has a very close relationship with volume. The information about egg's mass was not only used in the grading process but was also used to determine egg-yolk to egg ratio, eggshell thickness, and egg hatchability [9]. For egg mass calculations, if the egg volume is known, the egg mass can be obtained from multiplying the volume and density of eggs [5]. In addition, estimation of egg mass can also be done using a linear regression model [9]. This system uses diameter data obtained from digital image processing to estimate the mass of "White Leghorn Line" chicken eggs.

A similar system that has been developed, there is no system that can calculate the volume and mass of objects simultaneously. This study aims to propose a 2D CVS for predicting the volume and mass of egg using the disc method from the image of the egg. The proposed CSV system can automatically align the image obtained parallel to the main axis. The rest of the paper is organized as follow, section 2 explains the proposed computer vision system. The experimental results and discussion are provided in Section 3. The conclusion is explained in Section 4.

2. Proposed computer vision system

Computer Vision System proposed consists of hardware part and software apart. Hardware in the system is mainly used to take a picture of the egg. Software in the system is used to process the image captured, and perform mathematical calculations to find the volume and mass of the eggs.

2.1. Hardware part

The hardware consists of a camera, LED lighting, black background, and a computer to operate the software, as can be seen in Figure 1. The camera used in this system is a portable webcam "Logitech C170" with a resolution of 5 MP and connected to a computer via a USB cable. This webcam is also equipped with a clamp, making it easier to place. The webcam is placed 30 cm above the egg, with lighting coming from a 5 watts LED lights placed 40 cm above the webcam. The system uses dark-colored cotton fabric. This cover cloth serves to block light from outside the system during the image capture process.

2.2. Software part

The processing phase consists of image capture, initial processing, segmentation, opening-closing, rotation, extracting the size of the egg object, volume calculation using the disc method, and mass calculation.

1. Image Capture. The image capture process is carried out using a portable webcam placed right above the image. The captured image has an RGB color space and is then stored on the hard drive in jpg format with a size of 640 × 480 pixels.

2. Initial processing. Initial processing begins with the color transformation of the image from the RGB color space to the gray scale. The filtering process is done using a Gaussian filter with a 7x7 kernel size, aims to clear the image of existing noise.

3. Segmentation. The segmentation process separates an object from the background. This process begins with the search for the Threshold (T) value automatically. Pixels in gray scale images that have an intensity greater than T will be recognized as objects (white) with binary value 1, otherwise, it will be recognized as a background (black) with binary value 0.

IOP Conf. Series: Materials Science and Engineering **703** (2019) 012050 doi:10.1088/1757-899X/703/1/012050

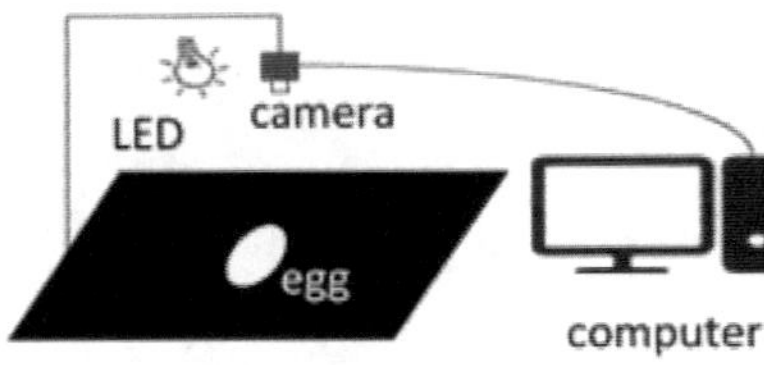

Figure 1. Hardware used in CVS

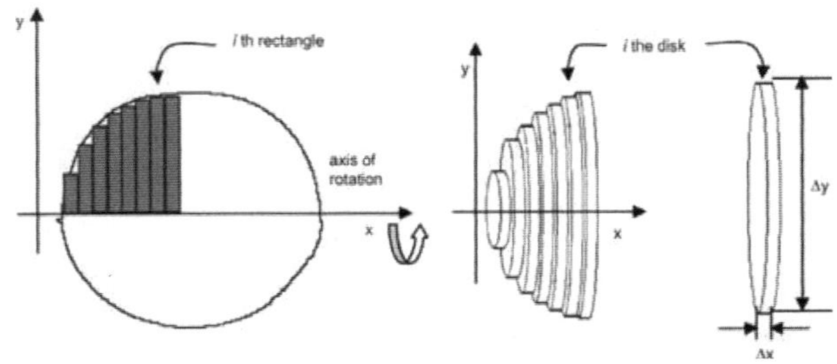

Figure 2a. Cross-section of the object divided into n rectangles; b: cylindrical rotation results on the x-axis

4. Opening-closing. An opening-closing process is performed to remove white spots on the background and black dots on the foreground.
5. Rotation. An image that was not aligned properly will be rotated until its axis parallel to the x or y-axis.
6. Feature extraction. This process aims to find the major axis and the minor axis of the egg. First, we need to locate the leftmost pixel, the rightmost pixel, the topmost pixel, and the lowest pixel of the egg object in the image. The location of the leftmost pixel can be found by scanning the image from the leftmost column (x=0) to the right until the first white pixel found. This will be x_l. The position of the rightmost pixel (x_r), the topmost pixel (y_t), and the lowest pixel (y_l) can be found by similar fashion. The length of the x-axis (Δx) is the difference between x_r and x_l. The length of the y-axis (Δy) is the difference between y_t and y_l. The major axis is the longest between Δx and Δy.
7. Volume measurement. After the length of the major axis and the minor axis of the egg are known, the calculation of volume using the disc method. Volume measurement can be obtained by dividing the cross-section of the object above the major axis into n rectangles with width Δx of 1 pixel and length ½ Δyi with i = 1, 2, 3, ..., n as shown in Figure 2a. Then the rectangles are rotated around the x-axis to produce a cylinder with a height of Δx and a cylinder radius of ½ Δyi as shown in Figure 2b.
The volume of egg (V) can be calculated using Eq. 1, with a k scale factor that can be calculated from the ratio of the object length in cm to the object length in pixels (from the image captured by the camera). In this experiment, the scale factor value k obtained from the average scale factor of five randomly selected samples.

$$V = \frac{\pi}{4} k^3 \sum_{i=1}^{n} (\Delta y_i)^2 \tag{1}$$

8. Mass measurement. Mass measurement is done using two methods, namely, using density and linear regression models. Density is a measurement of mass per unit volume of matter. The mass of eggs (m) was determined using volume (V) and density (ρ) by Eq. 2,

$$m = V * \rho \tag{2}$$

Linear regression is a statistical model that serves to predict the value of non-dependent variables (y) based on independent variables (x) (Franklin and Haribaran, 1994). Eq. 3 represents the common equation of linear regression.

$$y = a + bx \tag{3}$$

From 20 randomly selected egg samples, the density is found to be 1.07 gr/ml, and the regression model is

$$y = 7.3148 + 0.448\, x \tag{4}$$

IOP Conf. Series: Materials Science and Engineering **703** (2019) 012050 doi:10.1088/1757-899X/703/1/012050

3. Result and discussion

The validation process of the proposed method is carried out using 50 egg samples from 9 different angles (0°, 30°, 45°, 60°, 90°, 120°, 135°, 150°, and 180°). The result of the proposed method will be compared by manual measurement using water displacement for volume and using digital scales for mass.

3.1. Average Test

The average measurement from 9 different angles, the average of measurement from the experiment, and the time needed by the proposed method will be compared with manual measurements, as shown in Table 1.

Table 1. The average measurement result

Angle	Volume(ml)	Mass (gr)-density	Mass (gr)-regression	Time (ms)
0°	57.79	61.85	61.91	311.59
30°	57.44	61.48	61.58	315.03
45°	57.11	61.13	61.27	312.52
60°	56.75	60.74	60.93	312.09
90°	56.49	60.46	60.69	316.14
120°	56.50	60.48	60.70	318.5
135°	56.54	60.52	60.74	312.22
150°	56.79	60.79	60.97	313.68
180°	56.99	61.00	61.16	319.52
Average measurement of CVS	56.93	60.94	61.11	314.59
Manual measurement	57.87	61.62	61.62	47962.8

As can be seen in Table 1, the results obtained from the proposed method are very close to the results obtained from manual measurement, with differences less than 1.62% in volume and less than 1.1% in mass on average. The average time to measure the volume and mass of the egg using CVS takes 314.59 ms. This is much faster than the time needed manual measurement of 47962.8 ms. This proves that using CVS to measure egg volume and mass can be done more quickly without damaging the egg.

3.2. Accuration Test

To test the accuracy of the result of the volume and mass of egg measurement using CVS, the relative absolute error, ANNOVA test, and correlation test were used.

3.2.1. Relative absolute error test. The absolute error used in this experiment is the magnitude of the difference between the manual measurement value and the measurement value by CVS. The percent error is the relative error expressed in term of per 100, by the following equation:

$$Relative\ absolute\ error = \frac{|manual\ measurement - measurement\ by\ CVS|}{manual\ measurement} * 100\% \quad (5)$$

The relative absolute error test results for this experiment are shown in Figure 3. An experimental result is said to be accurate if it yields a relatively small absolute error value. Based on the experimental results obtained the absolute relative error value is <5%. This proves that the proposed method is accurate.

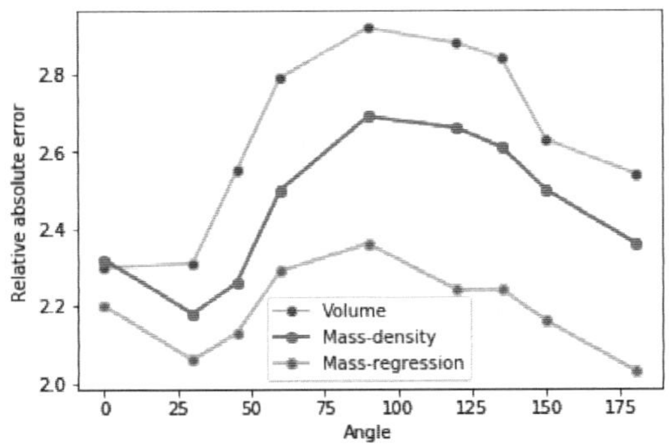

Figure 3. The relative absolute error result

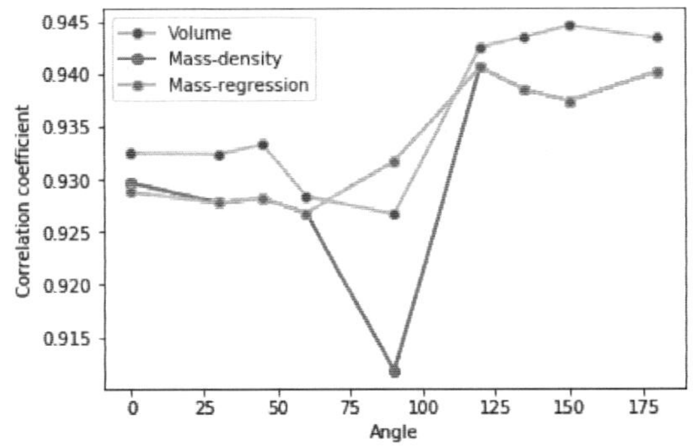

Figure 4. The correlation coefficient result

3.2.2. Correlation test. Correlation is a method of statistical analysis used to measure relationships between variables using a value called the correlation coefficient (R). Two data are said to have a relationship if they have R^2 values that are close to 1. The result of the correlation test in this experiment shown as in Figure 4. All the result of this experiment is close to 1. Thus, this proves that data from proposed method calculations have a relationship with manual data measurement.

3.2.3. Anova test. Anova is a statistical method used to compare the mean of several populations (usually more than 2 populations)[10]. The hypothesis used in ANOVA in this study is:
H0: $\mu_1 = \mu_2 = \ldots = \mu_{10}$
H1: at least two of the means are not equal
Where μ_1 to μ_9 are average measurements from 9 different angles and μ_{10} is manual measurement as shown in Table 1. To test the hypothesis, use the calculation of the f value using Eq. 6, where MK_k is mean square between populations and MK_d is mean square values within population:

$$f = \frac{M_K}{MK_d} \quad (6)$$

H0 is rejected significantly α if the value $f > f\alpha$, the value of $f\alpha$ can be seen in the distribution table F. The results of Anova test in this experiment are shown in Table 2.

Table 2. The Anova test result

	Volume	Mass-density	Mass-regresion
f	0.357	0.294	0.272
Significant	0.955	0.976	0.982

The significant value of the Anova test for volume is 0.955, for the mass data density method is 0.976 and for linear regression method is 0.982, so it can be concluded that H0 failed to be rejected. It proves that the mean square values of measured and calculated results are equal or not different. Thus the proposed method is accurate.

3.3. Precision Test
The coefficient of variation is useful to see the distribution of data from the calculated average if the percentage of data distribution is small (<1%) it can be stated that the data is precise. The coefficient of variation (CV) is the comparison between standard deviation (s) and the average value $\bar{x}$ expressed in percentage. The coefficient of variation is expressed by the formula,

$$CV = \frac{s}{\bar{x}} * 100\% \quad (7)$$

The overall value of the CV of the proposed method can be seen in Figure 5. Based on the above equation, the CV value for all data measurement is <1%, so it can be stated that the volume and mass data resulted from this method are precise.

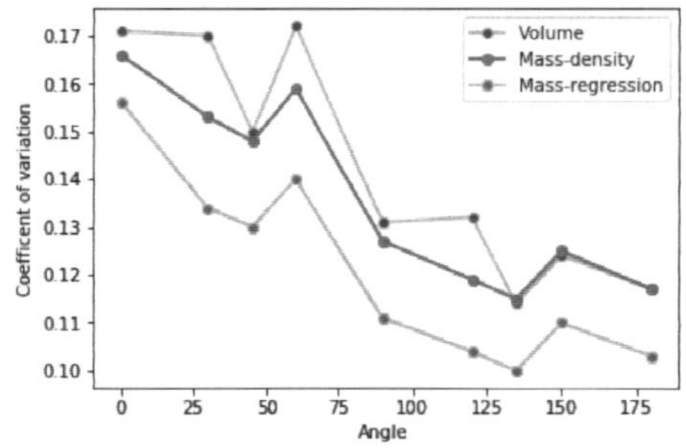

Figure 5. The coefficient of variation result

4. Conclusion

This study proposes a 2D CVS for predicting the volume and mass of an egg using the disc method from the image of an egg. The proposed method has been tested using 50 egg samples from 9 different angles (0°, 30°, 45°, 60°, 90°, 120°, 135°, 150°, 180°). The difference in the orientation of the angle of the egg used turned out to yield similar results. Based on the results of the experiments, CVS in the proposed method successfully measuring the volume and mass of eggs quickly, accurately, precisely, and without damaging the egg. The next study will be to develop CVS to calculate the volume and mass of several eggs simultaneously.

References

[1] Rush S A, Maddox T, Woodrey M S, Fisk A T and Cooper R J 2009 A precise water displacement method for estimating egg volume *J. Field Ornith.* **8** pp 193-7

[2] Siswantoro J, Hilman M and Widiasri M 2017 Computer vision system for egg volume prediction using backpropagation neural network *IOP Conf. Ser.: Mater. Sci. Eng.* **273** 012002

[3] Marsden J and Weinstein A 1985 Calculus II *Springer* pp 98-104

[4] Wang T Y and Nguang S K 2007 Low-cost sensor for volume and surface area computation of axis-symmetric agricultural products *J. Food Eng.* **79** pp 870-7

[5] Soltani M, Omid M and Alimardani R 2014 Egg volume prediction using machine vision technique based on pappus theorem and artificial neural network *J. Food Sci. Technol.* pp 1-7

[6] Siswantoro J, Prabuwono A S and Abdulah A 2012 Kerangka kerja penentuan volume telur menggunakan computer vision dan aturan simpson *SNASTIA 2012* pp 1-5

[7] Bridge E S, Boughton R K, Aldredge R A, Harrison T J E, Bowman R and Schoech S J 2007 Measuring egg size using digital photography: testing Hoyt"s method using florida scrub-jay eggs *Field Ornitho* **78** pp 109–16

[8] Koc A B 2007 Determination of watermelon volume using ellipsoid approximation dan image processing *Postharvest Biol. Technol.* **45** pp 366-71

[9] Asadi V, Raoufal M H and Nassiri S M 2012 Fresh egg mass estimation using machine vision technique *Int. Agrophys.* **26** pp 229-34

[10] Walpole R E, Myres R H, Myres S L, Ye K 2011 *Probability & Statistics for Engineers & Scientists 9th Ed.* pp 507-508

AUTHOR INDEX